C. R. ALDER CHURCH

THE

THRESHOLD OF SCIENCE

Elibron Classics
www.elibron.com

Elibron Classics series.

ISBN 1-4212-3339-8 (paperback)
ISBN 1-4212-3338-X (hardcover)

This Elibron Classics Replica Edition is an unabridged facsimile of the edition published in 1891 by Charles Griffin and Co., London.

THE THRESHOLD OF SCIENCE.

THE THRESHOLD OF SCIENCE:

A VARIETY OF

Simple and Amusing Experiments

ILLUSTRATING

SOME OF THE CHIEF PHYSICAL AND CHEMICAL PROPERTIES OF SURROUNDING OBJECTS, AND THE EFFECTS UPON THEM OF LIGHT AND HEAT.

BY

C. R. ALDER WRIGHT,

D.SC. (LOND.), B.SC. (VICT.), F.R.S.,

LECTURER ON CHEMISTRY AND PHYSICS IN ST MARY'S HOSPITAL MEDICAL SCHOOL, LONDON.

With Numerous Illustrations.

LONDON:

CHARLES GRIFFIN AND COMPANY,

EXETER STREET, STRAND.

1891.

CONTENTS.

INTRODUCTION.

It was originally intended that this little book should be simply a revised edition of *The Magic of Science*, by the late James Wylde, some time Lecturer at the London Polytechnic Institution; but as nearly thirty years have elapsed since the first appearance of that work, it has been found necessary to recast entirely the whole subject-matter, and to produce a completely different book.

This course has been rendered requisite, not only on account of the progress of Science generally during the last three decades, but also and more especially because of the altered position in which the study of Physical Science now stands in reference to ordinary education. Thirty years ago, systematic instruction in this subject was hardly to be met with in England outside the walls of a limited number of universities, colleges, and medical schools. As a matter of every-day education requisite for all classes, and not merely for a comparatively small number of students preparing for special walks in life, Physical Science had little or no existence; and when scientific experiments were shown in public, and popular lectures on Science delivered, the matter was treated more from the point of view of the conjurer and showman or caterer for amusement, than from that of the teacher of a branch of knowledge as valuable as a means of developing the faculties as any other kind of mental training known. At the present day, public opinion in reference to the educational value of Physical Science has so far changed, that a higher kind of instruction is now largely available for the masses in the numerous Science Classes in existence all over the kingdom, and as an item in the general curriculum of Board Schools for even junior scholars, than was generally obtainable thirty years ago in most Public Schools and Colleges even of the first class; although it must be admitted that the system of instruction in useful knowledge by the employment of experiment and inductive reasoning thereon as a means of mental training, has not yet superseded the older methods (which regarded the knowledge of the dead languages as the *sine quâ non* of mental development) to as great an extent in this country as in some others, where the value of sound instruction in science as an element of ordinary education was generally recognised a quarter of a century ago quite as fully as is now the case in Great Britain. Still, in a great number of our Schools and Colleges the purely scholastic mode of training has been of late years largely supplemented by the institution of a "modern side," where science is recognised as competing successfully with ancient literature as a necessary element in education and as a means of mental training; and signs are not wanting that this modification of our general educational system is likely ere long to be greatly extended. Thus, for example, the British Association for the Advancement of Science recently appointed a Committee for the purpose of inquiring into and reporting upon the present methods of teaching Chemistry and Physical Science; and in the report of this Committee (autumn of 1889), it is stated that "it cannot be too strongly insisted that elementary Physical Science should be taught from the first as a branch of mental education, and not mainly as useful knowledge. It is a subject which, when taught with this object in view, is capable of developing mental qualities that are not aroused, and indeed are frequently deadened, by the exclusive study of languages, history, and mathematics. In order that the study of Physical Science may effect this mental education, it is necessary that it should be employed to illustrate the scientific method of investigating nature, by means of observation, experiment, and reasoning with the aid of hypothesis; the learners should be put in the attitude of discoverers, and should themselves be made to perform many of the experiments. The lessons ought to have reference to subjects which can be readily understood by children, and illustrations should be selected from objects and operations familiar to them in every-day life."

In the following pages the object aimed at is to provide a kind of "Playbook," which, in addition to affording the means of amusement, shall also to some extent tend in the direction of that course of mental education advocated by the British Association Committee; so that whilst the juvenile philosopher finds pastime and entertainment in constructing simple apparatus and preparing elementary experiments, he may at the same time be led to observe correctly what happens, to draw inferences, and make deductions therefrom; and, by comparing together the results obtained in different cases, may gradually acquire some degree of elementary training in the proper exercise of those mental qualities which systematic scientific education chiefly calls forth; and in addition may pick up a variety of pieces of information as to the "why and wherefore" of things connected with the phenomena of nature generally, and more particularly with matters of everyday occurrence, especially the various arts and manufactures, and the innumerable scientific applications nowadays met with on all sides of everyday life.

Not being written with a view to aid in "preparing for examinations" nor in accordance with any particular syllabus of requirements, the subject-matter is not arranged in any specially methodical order, but rather in such a way as to connect together somewhat analogous phenomena, quite irrespective of whether they would or would not be juxtaposed in a more systematic text-book; to aid in elucidating the general principles underlying the observed manifestations and effects, by comparing together the results of various experiments, numerous cross references are inserted, by the use of which many phenomena alluded to in different sections are shown to be related to one another or to have some common cause. Accordingly, although the book is not written specially with the expectation of its being used as a class-book for elementary instruction, it is nevertheless believed to be capable of being to some extent usefully employed in that direction, *i.e.*, as a sort of primer that might be referred to with the object of developing an inclination towards scientific study and logical habit of thought. It has not been attempted to deal with the whole range of physical science and elementary chemistry, but only with the general outward properties of natural objects and the things surrounding us, and the ways in which these become modified by the action of one thing upon another, either alone or when aided by the application of heat; and as a sort of further study, with some of the general phenomena connected with Heat and Light. Should the occasion arise, other branches of physical science, such as Acoustics, Magnetism, Electricity, and the Physical Forces of Nature generally, will be similarly treated in a companion volume.

It is quite a mistake to suppose that costly appliances and an elaborately fitted laboratory are indispensable to carry out scientific experiments, more especially those of an elementary character. Of course, for the illustration of the more recondite points, such as exact measurements and quantitative physical and chemical determinations, such accessories are generally requisite; although, even in such matters as these, wonderfully accurate valuations can often be made by careful and experienced hands with eminently simple appliances. For certain experiments where rare, and consequently costly, materials are indispensable (*e.g.*, experiments requiring the use of platinum vessels, palladium foil, &c.), it is obviously impracticable to obtain the desired result without incurring expense; but for by far the larger majority of the experiments described only simple and inexpensive apparatus are required, such as test-tubes and flasks, glass tubing and perforated india-rubber corks, stands of various shapes and such like, obtainable for at most a few shillings, or even pence, from any of the numerous dealers in scientific apparatus.

For certain chemical experiments where evil-smelling gases like sulphuretted hydrogen, or corrosive and injurious vapours like bromine and hydrofluoric acid, are necessarily evolved, when a properly fitted "fume chamber" or "stink cupboard" is not available, a room with door and windows wide open, so as to produce a draught and thorough ventilation, is required; or preferably a table *outside* the window in the open air, so that whilst a supply of gas can be led to the table by a short length of flexible tubing in order to obtain heat when required, the evolved vapours, &c., may not become a source of inconvenience or damage inside the house. Wherever liquids of a corrosive nature are employed, such as sulphuric acid (either pure or diluted with water), the same kind of precaution is desirable to avoid damage to furniture by the "accidents" that will inevitably occur with beginners by breakages of various sorts, and upsets; for similar reasons an old suit of clothes is to be recommended for wear whilst experimenting, more especially with chemicals. One of the advantages, and not the least of them, to be gained by making physical and chemical observations and experiments is, that a considerable degree of manual dexterity in arranging and handling apparatus, sometimes of a somewhat fragile character, is gradually cultivated and acquired; the tyro, at first awkward at manipulating unfamiliar objects, soon becomes more expert, especially if he endeavours to form the habit of doing everything as carefully and neatly as possible; the importance in after life of the early cultivation of such a habit, and of the acquisition of manual dexterity and the free "use of the hands and fingers" without awkwardness, can hardly be overestimated, these being of very great advantage in all cases, and simply invaluable in a large variety of professions and special callings.

C. R. ALDER WRIGHT.

CHEMICAL LABORATORY,
ST MARY'S HOSPITAL MEDICAL SCHOOL,
LONDON, W.

SCIENTIFIC AMUSEMENTS.

§ 1. States of Matter.

CHAPTER I.

GENERAL DISTINCTIONS BETWEEN SOLIDS, LIQUIDS, AND GASES; AND NATURE OF PHYSICAL AND CHEMICAL ACTIONS INVOLVING CHANGE OF STATE, OR PASSAGE FROM ONE CONDITION TO ANOTHER.

A very little observation of the inanimate objects of nature and art surrounding us is sufficient to show that one notable point in which they differ from one another is in the varying **textures** and **degrees of compactness** they possess. The air, through which the hand may be moved with the greatest facility, but which, nevertheless, is capable when put in motion by a fan of readily moving about objects not too heavy, represents the lightest and rarest class of bodies, spoken of as **gases**; water, which can only be kept from flowing in all directions by means of a containing vessel, is an example of a much less rare, but intensely soft class of substances, spoken of as **liquids**, of which olive oil and spirits of wine (alcohol) afford two other examples; whilst bricks and stones, chairs and tables, more dense still and resistant to the touch, belong to the class of **solids**.

Between the most mobile liquids and the hardest and most rigid solids may be ranged a number of substances of intermediate texture; when the liquid characteristics predominate as a whole, as in the case of treacle, soft pitch, and such like substances, the bodies are said to be **viscid** liquids; when, on the other hand, they have sufficient coherence to retain their form without the aid of a containing vessel, but are easily altered in shape by pressure, they are called **plastic** solids. Some solids require but little force to alter their shape; well-tempered stiff clay, wax, and lead are examples of bodies of this kind, which are said to be **soft**; others require much more powerful pressure to mould them into different form, but can be readily shaped when the force employed is sufficiently great. Copper, iron, and most of the ordinary metals are examples of this kind of solid, and owe much of their usefulness in ordinary life to the property of being sufficiently pliable to be capable of being wrought into shape by appropriate tools, whilst sufficiently hard and tough to bear rough handling and resist wear and tear. Such bodies are said to be **malleable**, *i.e.*, they can be flattened out and extended by blows from a hammer, or by powerful squeezing machines and rolls; when they cohere or stick together sufficiently to enable them to be drawn into wire, they are said to be **ductile**; the stronger the wire, the greater is said to be the **tenacity** of the body, *i.e.*, the greater is its power of sticking together under the influence of forces tending to pull it asunder. Some solids are almost entirely destitute of malleability and ductility, although possessing considerable hardness and rigidity. When only moderate degrees of pressure are applied they resist completely; but when subjected to greater force they splinter and break up. Such substances are said to be **brittle** when easily broken in this way, *e.g.*, most kinds of glass. When the effect of pressure is to alter the shape as long as the pressure is applied but no longer, the body springing back to its original shape when released, they are termed **elastic.** As regards the quality of elasticity, both liquids and gases possess this property usually to a greater extent than solids; thus, if water be strongly compressed by a powerful pump, it shrinks in volume as a piece of india-rubber similarly treated would do, though not to the same relative extent; on releasing the pressure the water instantly springs back to its original bulk. Exactly the same thing occurs with air and all other gases, this class of substances being, however, far more readily compressible and expansible than water and other liquids.

The following table will serve as illustration of the different degrees of texture and compactness exhibited by different classes of ordinarily occurring substances :—

Gases, e.g., the air. Bulk for bulk much lighter than most liquids and solids; excessively easily set in motion, and eminently mobile. Readily compressible and highly elastic.

Mobile liquids, e.g., water and ether. Much heavier bulk for bulk than gases, but considerably lighter than most ordinarily occurring metals, such as lead, silver, copper, &c. Compressible and elastic, but requiring far more force to compress them than gases.

Viscid liquids, e.g., treacle. Requiring a containing vessel, but only flowing sluggishly when poured.

Soft plastic solids, e.g., wax, putty, modelling clay, and lead. Sufficiently stiff to retain form without a containing vessel, but readily moulded into any required shape by moderate pressure.

Tough plastic solids, e.g., copper, brass, wrought iron, silver, gold, &c. Completely stiff under ordinary conditions, but exhibiting plasticity, ductility, and malleability when subjected to a sufficient degree of compelling force. In general, substances of this class become softer and more plastic on heating; the property of wrought iron to be "forged" whilst red hot depends on this circumstance. Many of them exhibit "tenacity" to a very high extent, being, in fact, the most tough and tenacious substances known.

Elastic solids, e.g., steel watch-springs, whalebone, thin filaments of glass. When strained or bent out of shape, solids of this kind immediately return to their original shape on being released.

Brittle solids. When strained, compressed, or otherwise treated in such a fashion as to tend to alter the shape, such substances snap and break asunder easily. Brittleness and elasticity are often exhibited simultaneously by the same substance, *e.g.*, glass.

Hard or *rigid solids, e.g.*, flintstones. Most kinds of building stone and rock, and gems, especially the diamond. Bodies exhibiting great power of resistance to forces tending to alter their shape. Great hardness is often combined with a considerable degree of brittleness, a sudden blow often sufficing to shiver an intensely hard substance.

Plasticity.

One of the very earliest of the arts practised by semi-civilised man appears to have been the moulding of vessels for various purposes out of natural clay, this material possessing the property of being highly plastic whilst moist, but becoming tolerably rigid and coherent when dry. Bricks simply dried in the sun, and similar dried clay building materials, are still used in hot climates where a high degree of permanence in building is not indispensable; but when such bricks or tiles are strongly heated or fired, they become far more lasting, and (saving for a certain degree of brittleness) almost as indestructible as stone.

As already mentioned, the property of plasticity under considerable pressure is exhibited by most of the metals in common use. In order to impress the devices visible on the two sides of a coin, such as a penny piece, a shilling, or a sovereign, the copper, silver, or gold employed must be squeezed very hard indeed between moulds (termed *dies*) by means of machinery; but the principles involved in the art of embossing devices by pressure upon flat pieces of metal in a powerful press, so as to convert them into coins or medals, are identical with those concerned in the stamping of a device upon a piece of softened sealing-wax by means of an ordinary seal (Expt. 2), the main difference being that, in the latter case, the wax is first heated so as to render it plastic enough to take an impression, and that much less pressure is required to produce the device.

Some soft solids, like *putty* and *clay*, are naturally plastic without heating, so that by the pressure of the hands alone they can be moulded into all sorts of shapes; good clay of proper consistency, when thus moulded into form, becomes dried and hardened, or **indurated**, by exposure to air, and especially by subsequent heating by fire; the art of **pottery**, founded on this property, is one dating from the very earliest periods, and one of the most important manufactures that have descended to us from prehistoric times. Little but patience and some degree of manual dexterity is requisite to enable any one in the possession of a lump of properly tempered clay to exercise this art in a rough way, and to fashion such articles as pots and saucers, or even to mould and model more artistic and ornamental objects; although the after processes of firing, glazing, and the like, usually requisite to enable the objects thus fashioned to become permanent, often require a considerable degree of skill and care to carry them out properly.

Expt. 1. To mould Crucibles out of Clay.—Clay pots so prepared as to stand the fire without cracking are largely employed in the arts for the purpose of melting substances of various kinds, such as metals and glass. To prepare such **crucibles** of the best kind, a particular sort of clay known as "fire-clay" is necessary; but small crucibles of sufficient firmness for some experiments can often be made from a good sample of tenacious brick clay free from stones and well "tempered" by continual rolling-out between the hands and kneading together; finally the mass is worked into the shape of a small plant-pot (without a hole in the bottom) or thick cup, and set aside to dry in the air. After some days the drying may be completed before a fire or in the oven, when (if the clay is sufficiently good) the crucibles can be heated red hot in the fire without cracking, and may be used for melting lead, and such like operations requisite in certain kinds of experiments.

Crucibles of various materials and shapes can be cheaply purchased at the instrument dealers, so that except for the amusement of making one's own apparatus it is not absolutely necessary to prepare them oneself. Thin porcelain crucibles (fig. 1) are used for some kinds of laboratory operations; for others, thicker and coarser ones of clay mixed with sand or powdered plumbago (blacklead) to enable them to stand heat better (fig. 2). In order to lift hot crucibles, specially shaped metal tongs (fig. 3) are con-

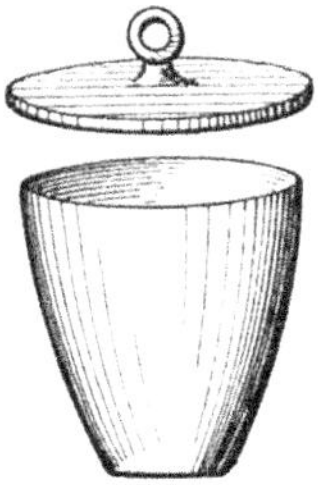

Fig. 1. Porcelain Crucible, and Cover.

Fig. 2. Plumbago Crucible.

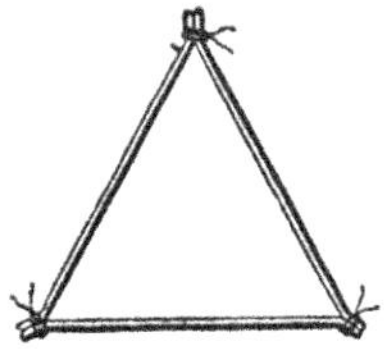

Fig. 5. Wire Triangle.

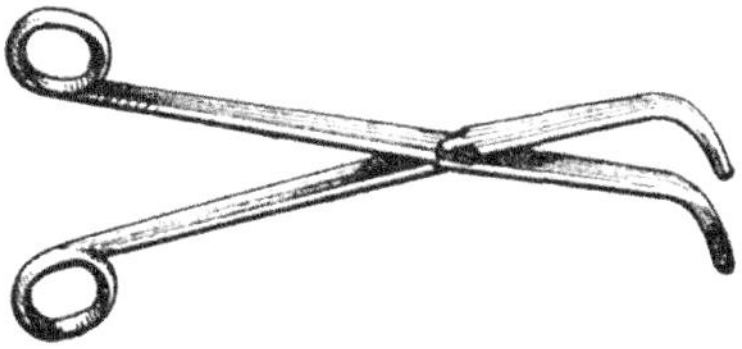

Fig. 3. Crucible Tongs.

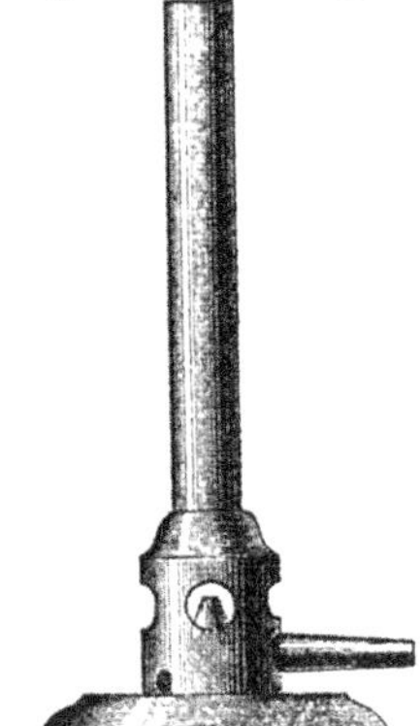

Fig. 4. Bunsen Lamp.

Fig. 8. Spirit Lamp.

Fig. 7. Tripod.

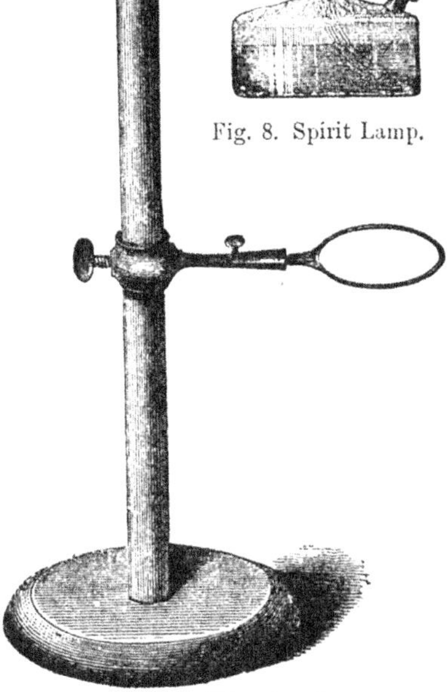

Fig. 6. Ring Stand.

venient; with small ones, unless they are required to be heated extremely hot, a fire is not necessary, as a "Bunsen gas-lamp" (fig. 4) will furnish enough heat for most purposes when placed so that the flame laps round the crucible supported by a triangle of wire (fig. 5), on the ring of a retort stand (fig. 6), or tripod (fig. 7). When gas is not available, a spirit-lamp (fig. 8) may often be employed instead; the cap serves to prevent loss of spirit by evaporation when the lamp is not in use, whilst the stoppered orifice permits of the introduction of fresh spirit as required.

Expt. 2. To take Impressions in Sealing-Wax.—Some kinds of solids possess the property of becoming plastic when warmed or heated, although when allowed to cool again they become hard as at first; sealing-wax affords a good example of this. To take a good clear impression of a well-cut seal or of a coin, &c., appears to be a tolerably simple operation, but is really one requiring some considerable amount of care in execution; the wax should be as little as possible smoked by the flame in melting, and the whole mass of melted and softened wax on the paper, &c., intended to support the impression should be stirred together by a sort of circular motion of the rod of wax, so that all parts of it may be about equally hot and therefore equally soft; *if this is not attended to, a bad impression will probably result*, parts of the wax being too cool and partially hardened again to "give" sufficiently when the seal is applied. On the other hand, if the sealing-wax is too hot and fluid, and especially if the seal is hot also, the wax may stick to the seal so firmly as to render it difficult to remove it after complete cooling without breaking the wax; whilst in taking a cast of a large piece of metal (such as a good-sized medal) the chilling action of the metal is apt to harden the wax so quickly that it has not time to penetrate into all the crevices and interstices of the device, and so a blurred and imperfect impression results.

Expt. 3. To make Plaster Casts.—Certain solids, when powdered and stirred up with a little water, possess the power of forming a semi-liquid mass which shortly "sets" hard, somewhat as a jelly solidifies on cooling. Plaster of Paris and various kinds of cement owe their usefulness to this property. To take an impression of a flat object, such as a coin in plaster of Paris, the simplest method of proceeding is to get the lid of a pill-box a little larger than the coin, and lay it flat on a table, saucer-fashion. A little good plaster is now mixed with water to a thick cream in a cup by means of a spoon, and the box-lid filled by pouring the "slip" into it. The coin is previously prepared by being very lightly oiled or greased on the surface to prevent it sticking to the plaster; and when the slip begins to thicken and solidify is carefully placed on the surface of the thickened slip and gently pressed down. After being left awhile, the plaster will have hardened sufficiently to permit of the coin being cautiously picked off by means of a pin or a pair of tweezers, &c., when an accurate impression of the lower face of the coin will be obtained. In this impression, as with an ordinary seal, the hollows and projections of the original coin will be reversed; so that, for example, the head on a half-crown piece will appear not in "*relief*" as on the coin, but sunk in, or in "*intaglio*."

Another mode of proceeding is to put the coin, &c., on a table or plate, and then pour over it the fluid slip, taking care that no air-bubbles are left sticking to the coin, which would damage the impression. In operating in this way it is often convenient to tie a strip of paper round the coin so as to form a sort of border, sticking up a quarter of an inch or more, which prevents the slip from flowing over the edge of the coin, and causes the cast to be of the same diameter as the coin itself. If the cast is required to be larger than the coin, the latter may conveniently be placed in a pill-box of the required size, lying flat on the bottom of the box and concentric with the sides, and the slip then poured in over the coin so as partly to fill the box.

After the plaster has become solid, by slowly drying it and then well oiling it, or better still, soaking it in melted paraffin wax until completely imbued therewith and allowing to cool, the cast is obtained in such a condition that fresh plaster slip will not permanently stick to it; so that by taking a second cast from the mould thus prepared, in the same way as before, an exact duplicate of the original object is obtained, the hollows and relief being now reversed a second time.

Statuettes and similar solid objects are cast in plaster by preparing hollow moulds of metal or other material in two or more pieces hinged together, so that when the slip is poured into the hollow of the closed-up mould, it solidifies therein and is finally removed as a solid block by taking the outer mould to pieces (*vide* Expt. 5).

By mixing with the plaster a little Prussian blue, yellow-ochre, or other coloured pigment in proper proportion, tinted casts may be prepared of any required shade according to taste. A polished appearance may be given by applying strong soap-water to the surface and allowing it to dry, and then gently rubbing with a soft cloth. The appearance and almost the hardness of marble may be given by applying solution of alum in the same way, or by mixing alum-water with the plaster of Paris in first preparing the "slip" for the casting.

Clay tobacco-pipes are prepared after a somewhat similar fashion. A long thin piece of clay is rolled out and a wire carefully threaded through it; this is placed in that part of a hinged mould that is to form the stem; some more clay is placed in the adjoining part of the mould forming the bowl, and a well-oiled solid metal plug forced in so as to produce the hollow of the bowl, and join it on to the stem; the plug and wire being then withdrawn, the mould is opened and the soft clay shaped into a pipe removed and set by to dry and harden; after which it is baked to complete the operation.

Expt. 4. To mould a Human Hand.—In order to take a cast of a friend's hand or foot, a piece of flat board should be provided with the upper surface well oiled; the hand to be modelled is also oiled, and then held so as to rest lightly on the board, palm downwards. Plaster slip is then poured over the hand, and when a thin coating has been formed, two or three pieces of stout tape are placed over this coating with the ends projecting outwards some inches. More slip is then poured over the coating so as to imbed the tapes firmly in the resulting mass, and the process continued until the mould is sufficiently thick. When the plaster has thoroughly set, the whole mass is gently lifted up by means of the tapes, and detached from the hand, care being taken not to break or crack the mass whilst liberating the fingers and thumb. In this way a mould of the hand is obtained from which an exact copy of the hand itself may be prepared by thoroughly drying the mould and well oiling it (*see* Expt. 3), and then turning it upside down and filling the cavity with fresh slip, to which (if required) a little vermilion or rouge or carmine-lake may be added so as to give a light red or pink tint; whilst this slip is setting a piece of tape is thrust into it, so that when completely solidified the mould and cast may be separated from one another by gently pulling the tapes.

Caution.—Do not attempt to take a cast of the face in this way; in the first place there is very great danger of stifling the subject by plugging up the mouth and nostrils so as to render it impossible to breathe; and in the second place, unless the forehead and eyebrows, and in the case of a gentleman the whiskers and moustache, be cut very short or shaved, it will very probably be impossible to detach the cast on account of the hairs being imbedded therein, in which case the plaster would have to be cut away by means of a chisel.

Expt. 5. To make a Wax Lemon.—Cut a lemon in two halves

with a sharp knife and lay one half flat on the table and pour over it plaster slip just about to set, until it is coated about an inch thick; or fill a well-greased cup with slip and then press into it the half lemon until the flat cut surface is level with the slip and the mouth of the cup. In this way you will ultimately obtain a plaster mould of one half of the lemon. Now do exactly the same with the other half, and when the two moulds are thoroughly set, apply them together and fasten them by tying or by a piece of wire bent into a clip (fig. 9), so that the hollow parts of each half mould coincide. Scoop out a small hole in the plaster in order to pour in melted wax and let out air, but before thus filling the mould, press a little soft plaster (slip just solidifying) into the interstices where the two half moulds meet, so as to fill up the crevice and prevent the wax running out thereat. The softened surface-layer of a piece of common yellow soap soaked in water will also serve conveniently as a "caulking" to make the mould tight at this part. The wax should not be too hot, just enough to enable it to flow readily into the mould and not to set before it has filled it up. When the wax is cold the plaster moulds are carefully separated and a wax model of the lemon obtained; this requires a little paring and trimming at the line where the two halves meet and the part where the wax was poured in, after which it may, if required, be tinted yellow to render it a perfect facsimile of the original lemon. The plaster moulds can be used over and over again for making additional wax models, but care must be taken to damp the plaster so as to prevent the wax sticking to it. Fruit of all kinds and many other solid objects may be moulded in precisely the same way.

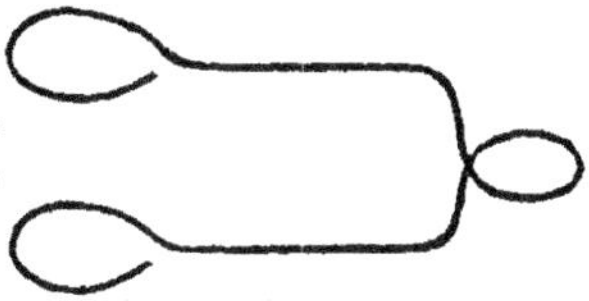

Fig. 9. Wire Clip.

Expt. 6. To take Impressions in Gutta-Percha.—Gutta-percha is a very convenient material for taking casts, more especially those intended to be employed in turn as moulds so as to produce replicas of the original object. A sufficiently large piece is well softened in very hot water, and well rolled and kneaded in the hands, to which it will not stick if they are previously thoroughly wet. The coin, &c., to be copied is placed on a flat table with wetted surface, and the plastic gutta-percha pressed down on to it thoroughly, and finally covered with a plate or flat wet board with a weight on, and allowed to stand. After some time the gutta-percha will have cooled and set, when the coin can readily be removed from it, the mould being somewhat elastic; but if moved too soon the impression will be damaged and the mould mis-shaped. Another way of operating is to form the plastic gutta-percha into a ball and lay it on a plate, flattening it down to a disc; the coin is then placed on this and well pressed into it, and finally covered with a flat board and weight as before.

When the gutta-percha mould thus obtained is filled with plaster of Paris slip and the latter allowed to harden, on removal of the plaster cast an exact facsimile of one side of the coin used is obtained. The plaster cast should be trimmed at the edges and dried, and then rubbed with *bronze powder* of coppery, silvery, or golden hue, according to the metal of which the coin is made. A second cast being made of the other side of the coin and similarly treated, the two casts enable both sides of the coin to be viewed together without having to remove the coin and turn it over. If antique coins, seals, &c., can be borrowed from friends for the purpose of thus copying, with a little trouble an interesting collection of seals and impressions may soon be accumulated.

Expt. 7. To make Casts in Fusible Metal.—Rose's "fusible metal" is a mixture of three different kinds of metal, as follows:—

Bismuth, ...	8 parts.
Lead, ...	4 to 5 ,,
Tin, ...	2 to 3 ,,

To prepare it obtain quantities of these three metals in about these proportions. First melt the lead in a small crucible, then add the bismuth, and finally the tin, and stir well together with a piece of tobacco-pipe stem, or a thick knitting-needle. This mixture of metals possesses the remarkable property of melting much more easily than any of the three constituents from which it is prepared, whence its name. If made in exactly the right proportions, it will melt in boiling water. The melted metal may be placed in the lid of a pill-box, and the object to be copied pressed on to it as in taking a gutta-percha or plaster cast; metal objects, such as coins and medals, should be previously well greased or blackleaded to prevent any chance of the fusible metal sticking permanently to them. Or the melted metal may be poured into a plaster mould, so that when set hard it will form a replica of the object from which the mould was taken. Owing to the low temperature at which fusible metal melts, it is possible to use it for obtaining impressions of fine lace, skeleton leaves, and such like delicate objects without burning them. By depositing copper by means of electricity on the moulds thus formed, beautiful metal copies of the patterns thus impressed in the fusible metal may be obtained; or plaster casts may be struck from the metal moulds thus produced.

Expt. 8. To take Casts in Sulphur.—Melt some sulphur in a ladle, taking care that it does not take fire; should it do so cover the ladle with a card so as to exclude air, which will extinguish the flame. When completely melted the heat must be withdrawn, otherwise the sulphur will thicken again when over-heated, and become too thick to pour properly. The melted sulphur may be used to take casts just as the nearly-fluid plaster of Paris "slip" in Expt. 3. Silver coins must not be thus treated, as they will be blackened, and probably blacken the cast too; other metals should be slightly oiled to prevent the sulphur sticking to them; if the mould is of plaster, it should be previously moistened with water.

Cohesion and Adhesion.

The difference between a mass of solid stone and the same substance reduced to powder by repeated blows with a hammer and triturating the fragments in a mortar, or between a bar of iron, and the same after being reduced to small fragments by a file, is that in the one case the component particles are so situated with reference to one another that there is great **cohesion** between them, and consequently the solid mass possesses great power of resistance to forces tending to break it asunder; whilst in the other case the granules, dust, or filings do not cohere together at all. In some cases subjecting a powdered substance to powerful pressure will cause its particles to cohere together to some considerable extent; well polished perfectly flat glass plates laid one on top of another will sometimes cohere (without any cement) so perfectly that it is almost impossible to separate them without breaking the glass. Freshly-cut clean surfaces of soft solids will often cohere perfectly on pressing together; thus two bright surfaces of freshly-cut lead or india-rubber will often stick together with great force in this way. Liquids exhibit cohesion as well as solids; one result of which is the rounded shape assumed by drops of dew in the morning on plants, or by drops of water scattered over a dusty surface.

Adhesion is the term applied to the tendency of dissimilar kinds of matter to stick together, cohesion being precisely the same kind of thing with substances when the constituent particles are of the same kind of matter. When the hand is dipped into water it becomes *wet*, the water adhering to the skin; a well-greased finger, however, may be dipped into water without being wetted on account of the smallness of the adhesion between the greasy film on the skin and the water. The use of cements, paste, glue, solders, and such like uniting materials is due to the circumstances—firstly, that the cement, &c., adheres firmly to the surfaces to be united; and secondly, that as it sets or dries this **adhesion** still continues, whilst the solidification of the cement or solder, or the drying up of the paste or glue, gives sufficient **cohesion** to the film of uniting material to prevent its separating in the way that two pieces of wood would do if freshly glued together, and moved before the glue is dry, or two pieces of metal soldered together but handled before the solder has set hard by cooling. In all cases where two solid objects are thus cemented together, the strength of the junction depends on the degree of cohesion of the cementing material and its adhesion to the surfaces of the solids to be united; if either of them be weak, the joint will bear but little rough usage. Thus glass and wood cannot be effectively glued together; a strip of wood glued to a window or mirror will easily break away, the rupture usually occurring at the surface of the glass, the glue parting therefrom owing to the small adhesion between glass and glue. On the other hand, two flat pieces of wood well glued together and then forcibly separated will frequently not divide along the line of the glue; one piece of wood will often split sooner than this will happen, leaving part still glued to the other piece. It is possible to split a sheet of paper (such as a bank-note) into two by cementing it firmly between two flat surfaces and then forcibly separating them; the cohesion of the paper is less than its adhesion to the film of cement, so that the paper splits before the junctions will yield.

Changes of State produced by Variation of Temperature.

One of the commonest observations in nature is that liquid water, when subjected to considerable cold, as in winter time, changes its physical condition of fluidity to that of solidity, the mobile liquid **freezing** to a mass of ice, which **melts** again to liquid water on warming; whilst on the other hand, when heated in an appropriate vessel, water **evaporates** or disappears from view, becoming transformed into an invisible vapour somewhat resembling the air and readily miscible therewith, but unlike air being capable of reappearing in the liquid form on slightly cooling again, making its appearance as mist, dew, or larger visible drops of water, according to the circumstances under which the cooling takes place.

Experience teaches us that other bodies besides water exhibit the same properties; by the application of heat of sufficient intensity, almost all known solids can be melted (*i.e.*, transformed into liquids), and most known liquids can be vaporised (*i.e.*, transformed into gaseous forms of matter). With some substances such "changes of state" require only comparatively moderate degrees of heat, whilst with others much higher temperatures are requisite, such as are obtainable by condensing the heat of the sun considerably by the aid of a "burning glass," or by the application of fire, or by means of powerful electric currents, this latter source of heat being the most powerful one yet known so far as terrestrial phenomena capable of measurement are concerned.

Changes of State produced by Alterations of Pressure.

Some very curious effects are producible under certain circumstances by the application of considerable amounts of pressure to bodies of various kinds. As already stated, powdered substances can often be thus converted into solid masses of considerable strength when an intense pressure is used to bring the particles sufficiently near to one another to enable them to cohere together; here, however, no change of state (*i.e.*, from solid to liquid or *vice versâ*) occurs. If a block of ice, just at the temperature at which ice melts (*vide* Expt. 28), be subjected to strong pressure, a portion of the ice becomes liquefied to water; but on relieving the pressure this melted ice instantly solidifies again; in this case the application of pressure *assists* the effect of heat in converting a solid into a fluid. Many other substances are affected by pressure in quite a different way; the pressure *prevents* their melting, so that more heat must be applied to them when under pressure before they liquefy than would be necessary under the ordinary atmospheric pressure only. A simple rule is always followed in such cases; some substances, like water, become lighter on freezing, and heavier on melting, bulk for bulk, so that the solid body floats on the liquid, as ice on water; with other substances the solid form is heavier than the liquid and sinks therein. With the first class of bodies the effect of pressure on the solids is to tend to compress them into a smaller volume, and so render them heavier than before, bulk for bulk; as this is exactly what heat tends to do in such a case when it produces melting, it results that *the effect of pressure is to assist heat*, and therefore the greater the pressure the less heat will be requisite to produce fusion. With substances of the second class the case is just the opposite; applying pressure to the solid as before tends to squeeze it smaller; but as this is now the opposite effect to that produced by heat during melting, it results that with bodies of this kind *more heat must be applied to counteract the effect of pressure* in order to melt them.

Still more marked effects are produced when gases are strongly compressed by a powerful pump or by generating them by chemical action in a confined space; the gas becomes converted into a liquid, just as steam condenses to water on cooling; but in the case of many compressed gases becoming liquefied, cooling is not indispensable as it is with steam. Some gases require to be cooled as well as compressed before they will liquefy; with a few a very enormous amount of cooling and most intense pressure are both necessary to produce the effect.

Change of State produced by Causes involving Chemical Action.

The term "change of state" in the narrowest sense is only applicable to such alterations of condition as take place with water when moderately heated so as to produce steam or aqueous vapour, or chilled so as to freeze to ice; or, conversely, when ice melts or steam condenses again to liquid water; in each case one substance only is acted upon, producing only one product when the action is complete.* But in a wider sense "change of state" includes a number of actions where, by the application of heat or electricity, solids give rise to liquid or gaseous products, liquids to gases, or, conversely, gases to liquids or solids; the most marked feature in each case being that the number of products formed is not the same as the number of kinds of materials employed. Such actions are spoken of as **chemical changes**, and are divided into two classes, viz., those where **decomposition** or **breaking up** occurs, in which cases the number of products is greater than the number of different materials used; and those where **combination** or **synthesis** occurs, in which cases the number of products is less than the number of materials.

Thus when water is heated to a much higher temperature than that required to convert it into steam, or when it is subjected to the action of electricity under certain conditions, the water (or steam) disappears, and in its stead there are obtained two entirely new substances, both of which are gases analogous to the atmosphere, but quite different from each other and from steam; one of these gases is termed *hydrogen* and the other *oxygen;* both of them will be frequently met with in our subsequent experiments, being capable of formation in other ways besides this "decomposition" of water into its two constituents. On the other hand,

* Under certain conditions solid ice, liquid water, and a small amount of aqueous vapour may be simultaneously present.

if hydrogen and oxygen prepared in any convenient way are mixed together in certain proportions and then heated, a violent explosive action takes place (Expt. 208), much heat being developed; and, as a result, instead of the two dissimilar gases, hydrogen and oxygen, we have steam produced, which on cooling becomes liquid water. Here "combination" has taken place.

Besides these two fundamental kinds of chemical action, others are also known (in some cases, but not always, leading to change of state) where the number of products is the same as the number of materials; all these changes, however, may be regarded as being really the nett result of changes of decomposition and of combination taking place in succession: thus, for example—

Expt. 9. To cover a Steel Knife with Copper.—Dissolve in water a little of the substance termed *chloride of copper*, and in the liquid place a steel knife blade or other polished steel object. In a short time a red coating of metallic copper will be visible on the surface of the steel; and after some length of time all the copper originally contained in the solution will be thus deposited, so that no coating of copper can be any longer obtained on placing another bright steel blade in the liquid. When this is the case, the liquid will contain a substance termed *chloride of iron*. Here, then, there were two original substances, steel (or iron) and chloride of copper; and there result the same number of products, viz., copper and chloride of iron. The total chemical action may be viewed as the sum of two actions taking place in succession; firstly, an action of decomposition whereby the chloride of copper becomes broken up into two constituents known as *copper* and *chlorine* (just as water gives rise to oxygen and hydrogen); and secondly, an action of combination, whereby the chlorine thus produced instantaneously combines with the iron forming chloride of iron, the copper being left in the free state as metal.

Expt. 10. To produce a Combustible Gas from a Watery Fluid and Zinc.—In the last experiment no visible change of state on the whole occurred, as we started with a solid and a liquid (steel and chloride of copper solution) and ended with a solid and a liquid (copper and chloride of iron solution). In the following case, however, a solid becomes apparently liquefied and a liquid gasefied as the result of chemical change.

On to some small fragments of granulated zinc (Expt. 16) pour a little diluted *hydrochloric acid;* the zinc will rapidly dissolve, and a copious *effervescence* or bubbling will take place, owing to the escape of bubbles of a peculiar kind of gas which will take fire on bringing a light to them. This gas is in fact the *hydrogen* already referred to as a constituent of water. We start with two materials, solid zinc and liquid solution of hydrochloric acid; we end with two products, gaseous hydrogen and a liquid, which is a solution of the substance called *chloride of zinc.* In this experiment, as in the preceding one, the total action may be regarded as the sum of two changes; firstly, the hydrochloric acid, being a compound of chlorine and hydrogen, may be regarded as broken up into these two constituents; and secondly, the chlorine may be viewed as combining with the zinc to form chloride of zinc as fast as it is produced, just as it was regarded as combining with the iron in the last experiment. The result of the two actions jointly may be regarded as a **displacement** of one substance from combination with a second by a third, which combines with the second and forms a new product in so doing, whilst the first substance is set free. Thus in Expt. 9 *the iron displaces the copper* from combination with chlorine, forming a compound of iron and chlorine and setting the copper free; and in Expt. 10 *the zinc displaces the hydrogen* from combination with the chlorine, forming a compound of zinc and chlorine and setting the hydrogen free. Chemical changes of this kind are accordingly spoken of as actions of **displacement** or **substitution.**

A somewhat more complex kind of chemical action is that spoken of as **double decomposition** or **double displacement.**

Expt. 11. To produce a Yellow Solid by Double Displacement from two Colourless Fluids.—Boil up in a *test-tube* (a thin glass tube made for the purpose, sealed up at one end, obtainable at the instrument dealers) as much *chloride of lead* as will lie on a three-penny piece with a tablespoonful of distilled water; let the whole stand for an hour or two, and decant off some of the clear colourless liquid into a wine-glass, and add to it a few drops of a solution of *iodide of potassium* (also colourless). The liquid will become yellow, and on standing a yellow solid substance will gradually subside to the bottom, being said to be "precipitated" from the liquid.

In this experiment the two materials employed are respectively chloride of lead, made up of the two constituents *chlorine* and *lead;* and iodide of potassium, made up of the two constituents *iodine* and *potassium.* The total action may be regarded as made up of two pairs of subordinate ones, viz., the chloride of lead is decomposed into chlorine and lead, and the iodide of potassium is similarly decomposed into iodine and potassium; whilst the lead and the iodine thus formed combine to produce *iodide of lead,* which being solid and not dissolving easily in cold water, makes its appearance in the solid state; simultaneously the chlorine and the potassium set free combine to form *chloride of potassium,* which being easily soluble in water does not precipitate in the solid form, but remains dissolved.

In order to boil a liquid in a test-tube, a *test-tube holder* is convenient, such as that represented in fig. 10, consisting of a brass spring-clip furnished with a handle. In default of such a holder a piece of paper may be twisted round the upper part of the test-tube, as in fig. 11. The tube should never be held pointing towards the face of the operator, or towards any other person, because sometimes steam is generated so rapidly that a quantity of the hot fluid is projected bodily out of the tube, which thus becomes a sort of miniature steam-gun; dangerous scalding might readily be brought about if the boiling fluid struck any one on the face or elsewhere, especially when acids or other corrosive fluids are contained in the test-tube. Fig. 12 represents a method of holding a test-tube by means of a clamp-stand instead of by the operator's hand.

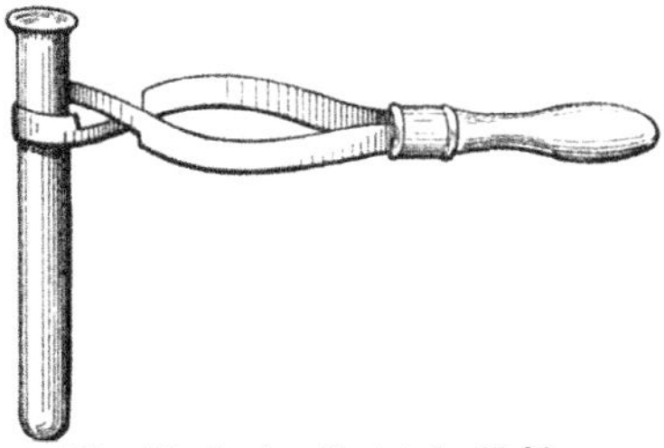

Fig. 10. Spring Test-tube Holder.

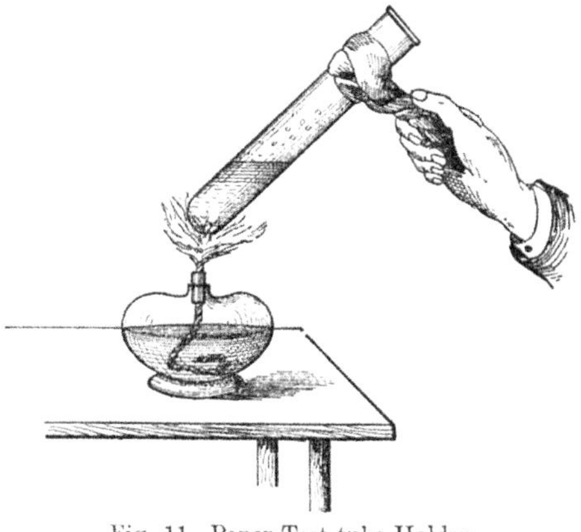

Fig. 11. Paper Test-tube Holder.

Expt. 12. To produce variously Coloured Precipitates from the same Solution.—The following experiments further illustrate the production of solid substances (as precipitates) by double decomposition on mixing two liquids together. Into six test-glasses (fig. 13), or ordinary wine-glasses, respectively pour a tablespoonful of the solutions of the following compounds:—(1) common salt (otherwise known as *chloride of sodium*) *; (2) *potassium chromate*;

* Chemists often use indifferently the terms "chloride of sodium" and "sodium chloride" to indicate common salt; and similarly in other cases,

(3) *potassium iodide;* (4) caustic soda (*sodium hydroxide*); (5) sulphuretted hydrogen (*hydrogen sulphide*); (6) *sodium phosphate.* Now add a few drops of solution of *silver nitrate* to each. A

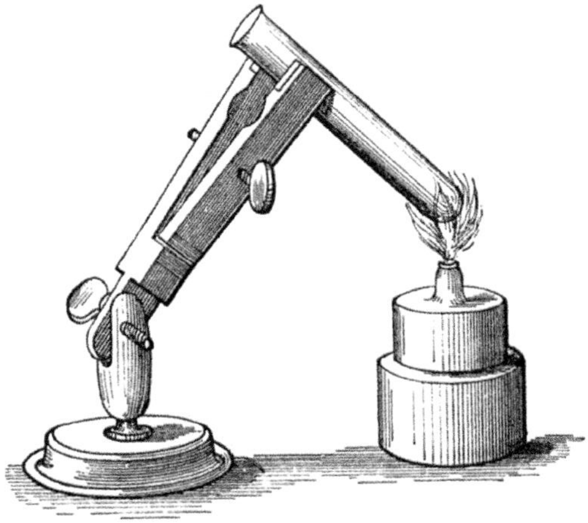

Fig. 12. Clamp.

precipitate will be formed in each glass by double decomposition, the colour being different in each case. In the first glass a *white* precipitate will be formed, sodium chloride and silver nitrate producing *sodium nitrate* which remains dissolved, and *silver chloride* which precipitates. In the second glass a *dark red* precipitate of *silver chromate* will be formed together with *potassium nitrate* in solution. In the third glass *pale yellow silver iodide* is precipitated, whilst *potassium nitrate* is again formed in solution. In the fifth glass *black silver sulphide* is precipitated and *hydrogen nitrate* (nitric acid) produced in solution. In the fourth glass *brown silver hydroxide* is precipitated, and in the sixth *yellow silver phosphate, sodium nitrate* in solution being also formed in each case.

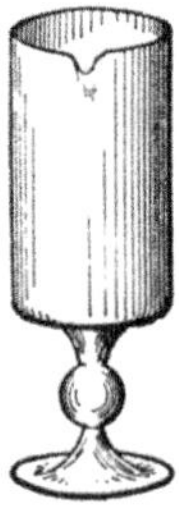

Fig. 13. Test Glass.

Expt. 13. To produce a Gas Smelling like Rotten Eggs by Double Decomposition.—Put into a test-tube a few grains of powdered *iron sulphide* and pour over them a little diluted *hydrochloric acid* (solution of hydrogen chloride). Double decomposition will take place, resulting in the evolution of the gas *hydrogen sulphide* (the solution of which

such as "iodide of potassium" and "potassium iodide," and so on. Of the two forms of expression, perhaps the latter is the preferable one.

in water was used in the last experiment), whilst the complementary product *iron chloride* remains in solution. If the acid is not too much diluted it is not necessary to heat the test-tube in order to cause the action to take place.

Expt. 14. To produce a Black Precipitate by means of a Gas by Double Decomposition.—Hold over the mouth of the test-tube in the last experiment a piece of white blotting-paper that has been dipped into a strong solution of *silver nitrate* (or of *lead nitrate*). The paper will turn black because a precipitate of black *silver sulphide* (or *lead sulphide*) is formed, just as in the fifth glass in Expt. 12, *hydrogen nitrate* being produced as the second product.

Many substances give precipitates of characteristic colours and special properties when treated with appropriate "reagents" in this kind of way, and upon actions of this kind are based a large proportion of the various "tests" used in analytical chemistry. Thus the property of sulphuretted hydrogen gas to blacken paper soaked in nitrate of silver solution, or other analogous compound, is made use of in testing coal-gas to see if it is free from admixture with that particular gas, which is not the case if the coal-gas have been imperfectly purified during manufacture: if the test-paper is blackened the gas is still impure.

In all such cases the total chemical change may be regarded as the final result of a succession of actions of decomposition and combination taking place between the various materials involved in the reaction; and in similar fashion the most complex chemical actions known may be viewed as the final results of analogous series of changes.

Double Combination is the term applied to another kind of chemical action where two different substances act upon one another so as to produce two new products different from each other and from each of the original substances, and where the total action may be regarded as consisting of the breaking up of one of the substances into two constituents, each of which thus combines with part of the second substance. Thus in Expt. 243 the volatile fluid known as "*carbon disulphide*," and consisting of a compound of the two materials *carbon* and *sulphur*, is burnt (like spirits of wine), producing a flame and generating heat; the action of burning consisting in the chemical change taking place between the carbon disulphide and the oxygen of the air, the result of which is to form two new compounds, *carbon dioxide* and *sulphur dioxide*, both of which are gases: the change may be regarded as made up of two stages; first the carbon disulphide breaks up into carbon and sulphur; and secondly, each of these then combines with oxygen, whence the term "double combination."

Double Evolution is the name applied to a less frequently occurring kind of chemical action where there are still two original substances and two different products, and where the action is the exact opposite of double combination; thus, suppose that carbon dioxide and sulphur dioxide could so act on each other that the oxygen contained in each was set free whilst the carbon and sulphur combined together forming carbon disulphide, then this would be an action of the kind in question. In point of fact this particular change does not take place; but certain chemical actions are known of a similar nature, *i.e.*, where two compounds each containing oxygen react on one another in such a fashion that oxygen is evolved from each, the other constituents of each original compound either combining together, or being further changed by subsidiary chemical actions.

On the whole, the essential difference between physical and chemical actions leading to change of state may be thus tabulated.

Physical Action.*—Where one substance only is employed which becomes more or less completely changed into another from which the first can be reproduced at will by varying the temperature or pressure, or both; as where water becomes ice or steam, either of which will reproduce the original water by altering the temperature, raising it in the first case, and lowering it in the second; or where ammonia gas is compressed to a liquid by the simple application of a considerable degree of pressure, means being employed to prevent any alteration of temperature during the action.

Chemical Action.—1. Where one substance gives rise to two (or more) products entirely different from each other and from the original substance, by simple **decomposition** or **breaking up.**

2. Where two substances coalesce or combine to form a single product different from either, by **combination** or **synthesis.**

3. Where two substances react on one another so as to give rise to two products dissimilar amongst themselves and each different from either of the original substances. This kind of action may be further distinguished as changes of—

a. **Single displacement** (Expts. 9 and 10),
b. **Double displacement** (Expts. 11, 12, and 13),
c. **Double combination** (Expts. 242, 243, and 244);
d. **Double evolution**;

in each case the total action being capable of being regarded as the result of a series of changes of decomposition and combination occurring in succession.

* A peculiar kind of action termed **allotropic modification**, and generally classed as a chemical change, occasionally occurs. In this, one substance becomes changed by heat or other agency into something quite different from the original body, *e.g.*, oxygen and ozone.

Solution of Solids, &c., in Solvents.

Besides actions of a purely physical character (like melting ice or evaporating water) where change of state in the narrowest sense takes place, and chemical actions proper where change of state in a wider sense may ensue, as in the last experiment, there are other actions known where solids or gases become apparently transformed into liquids (or separate from liquids under suitable conditions), to which the terms **solution** and **separation from solution** are applied: one of the simplest natural examples of which is afforded by the deposition of **crystals** of salt from pools of sea-water as the water evaporates under a hot sun, and the dissolving again of these crystals to a briny liquid when a shower of rain brings fresh water into the pool. Such actions as this, or the parallel cases where sugar is dissolved in hot tea, or where spirits are mixed with water (solution of liquid in liquid), are generally spoken of as **physical** changes, no essential alteration being brought about in the nature of the dissolved body, which can be recovered unchanged from the solution by appropriate means, *e.g.*, evaporation of the solution and formation of crystals. In a wider sense, however, the term "solution" includes not only actions of this class but also analogous actions where a body becomes dissolved, a **chemical** change simultaneously taking place (as when solid zinc is dissolved in dilute hydrochloric acid, Expt. 10), so that the original body dissolved cannot be regained unchanged by such simple processes as evaporation, &c. It will be convenient to study in detail these various kinds of action somewhat systematically.

---

§ 2. Physical Changes of State due to Heat and Pressure and not accompanied by Chemical Action

CHAPTER II.

Fusion of Solids and Solidification of Fluids.

When a solid substance is sufficiently heated, it *melts* or *fuses;* that is, it loses its tough, rigid texture and acquires the peculiar property of liquids, viz., that of running and flowing spontaneously in all directions unless prevented by means of a hollow containing vessel. In some cases, before complete liquefaction is brought about, the solid passes through an intermediate stage, becoming pasty; but this is not the general rule. Conversely, most liquids when sufficiently cooled down lose their liquidity and become solid masses, as water does when it freezes to ice. Some liquids, such as *quicksilver* (otherwise termed *mercury*), require the cold of the Arctic Regions to make them freeze, and some few, such as *alcohol* (spirits of wine), do not thoroughly solidify until the most intense degrees of cold possible are attained.

In the same sort of way some solids require a considerable amount of heat to be applied before fusion or melting happens; whilst others melt so readily that they have to be cooled down artificially in order to keep them solid. Thus a block of ice melts to liquid water on the application of but an inconsiderable amount of heat, comparatively speaking, whilst a mass of lead or zinc must be heated nearly red hot before it melts, and platinum requires the very highest temperature attainable in special furnaces.

Many useful arts depend on the kind of operation known as "casting," based on the property of substances to become liquid when heated and to solidify or freeze again on cooling. This is carried out very much in the same way as in the experiment above described for making a wax lemon (Expt. 5). A mould of suitable material is prepared, hollow or otherwise according to the shape of the object to be cast, and the melted matter to be cast poured into the mould until it is filled. After cooling the solidified matter is extracted from the mould, and dressed down and cleansed and polished on the surface if required. The large gas and water pipes laid down in the streets are thus cast of molten iron; brass handles, door knockers, and similar objects are frequently similarly prepared from melted brass; and plate-glass windows are made by melting glass and then pouring it out on a flat metal table so that it solidifies in a horizontal sheet on cooling; in fact, innumerable objects and articles of everyday use are fashioned by means of this casting process. Rough iron objects are generally cast in moulds of sand, not wet but slightly damped so as just to adhere together. The heat required to melt cast iron is too great to enable this substance to be easily experimented with; but lead can be readily employed for the purpose.

Expt. 15. To cast Lead into Sticks.—Get some sand in a pot or box and damp it very slightly, well intermixing it: the quantity of water used should be the smallest amount possible, only just sufficient to prevent the sand from falling in and filling up the hole left when a pencil is rammed down into the sand (previously gently pressed down in the box) and then cautiously withdrawn;

if too much moisture is present there is a liability to the sudden formation of steam when the melted lead is poured into the hole, which will blow the metal violently out of the hole and perhaps cause a severe burn: terrible accidents sometimes happen in this way in foundries and casting-houses when due care is not taken to avoid the cause.

The lead may be melted in a clay crucible on the fire or over a gas or spirit-lamp; but a more convenient arrangement is a ladle made for the purpose, consisting of an iron basin with a spout for pouring, and a long iron bar attached as a handle; this handle generally keeps sufficiently cool at the far end to prevent burning the hands, even though the bowl be full of nearly red-hot lead. The scum should be removed from the melted lead by skimming with an old knife or a piece of cardboard; the freshly skimmed surface will be perfectly bright and silvery, but whilst you watch it, it tarnishes and loses its lustre and becomes covered with a film often showing the colours of the rainbow, whence the term **iridescence.** The film thus formed results from the combination of the lead with the gas *oxygen* contained in the air forming *oxide of lead*, and is due to a chemical change termed **oxidation**, which will be examined more closely hereafter (Chapter XIV.); most ordinary metals behave in much the same way on heating, whether they become fused or not in the process; thus a polished plate of copper held horizontally over a Bunsen or spirit-lamp flame so as to heat the centre of it will develop rings of iridescent colours owing to the analogous formation of *oxide of copper;* similarly a polished steel knife blade treated after the same fashion will give iridescent films of *oxide of iron;* in these two instances the metal remains solid, whilst with lead as above, and bismuth (Expt. 18), the metal is melted by the heat.

The skimmed lead should be allowed to cool until it is nearly on the point of setting or solidifying, and should then be carefully poured into the holes in the sand which form the moulds. After a few minutes the lead will have solidified and can be withdrawn in sticks or rods from the sand, preferably by means of a pair of tongs, as it may be still too hot to handle comfortably. Zinc, tin, and several other metals may be cast into sticks in just the same way.

Smoother castings may be obtained by using polished iron moulds sold for the purpose of casting metals, &c., into sticks, made in two halves like the wax-lemon moulds in Expt. 5. Bullets are cast by means of two hemispherical moulds held together like the blades of a pair of scissors.

Expt. 16. To Granulate Zinc.—Melt some lumps of zinc in

a crucible or ladle and then pour the melted metal into a tub of cold water from as great a height as possible, conveniently by standing on a step ladder. As the melted zinc comes in contact with the water it becomes suddenly chilled, and forms thin pieces often of peculiar shape, and much more bulky than the original lumps of metal. Finally the granulated zinc is drained from the water, dried on a towel and by exposure to the air in a thin layer on a newspaper, and bottled for future use in various chemical experiments, more especially those where hydrogen gas is required (Expts. 10, 104, and subsequent ones).

Tin, lead, and various other metals can be granulated in the same way.

Expt. 17. To make Explosive Glass Drops.—In order to melt glass a sound crucible, capable of standing a high temperature, is indispensable, as well as a fire with a good draught so as to heat the crucible sufficiently. If the glass is thoroughly melted and then dropped into water it frequently does not fly about and granulate as melted metal does, but the drops solidify into solid pear-shaped lumps with long thin tails. Glass thus quickly cooled possesses the peculiar property that the pear-shaped end will stand a pretty heavy blow without breaking; but if the thin end be broken off or even scratched, the whole crumbles to powder. At the instrument dealers these glass drops may be purchased under the name of "Prince Rupert's drops," it being supposed that Prince Rupert first observed the peculiarity of glass thus treated.

Of late years glass articles have been sold said to be "toughened" by rapidly cooling them in water, or better in oil, after they have been worked into shape whilst hot; such toughened glass will often resist a wonderful amount of hard usage without breaking; but sometimes a slight scratch will cause the whole article to shiver to fragments like a Prince Rupert's drop; for which reason "toughened glass" has not come so extensively into use for household and other purposes as might have been expected from its diminished fragility in ordinary wear and tear.

Expt. 18. To Crystallise Metals by Melting.—If you try to break across a properly cast rod of lead prepared as in Expt. 15, you will find that it will bend sooner than break, lead being a soft metal not brittle when cast; but if you try to bend a thin rod of zinc cast in the same way you will find that the rod will snap in two sooner than bend, because cast zinc is not soft and flexible, but somewhat brittle. On examining the broken surfaces of the fractured rod you will notice that they are bright and shining, cast zinc being what is termed **crystalline.** Some kinds of metals crystallise very finely when they are melted and allowed to cool slowly; *bismuth* is one of the best for this purpose. Melt a few ounces of bismuth in a crucible or ladle, and then let it stand until the surface just begins to become solid, forming a crust. With an iron skewer or thick knitting-needle poke two holes in this crust on opposite sides near the edges, and then tilt the crucible so that the melted metal that has not yet solidified may run out from the interior through one hole, air entering by the other one. When the crucible and what is left in it is cool enough to touch, you will find that the crust can be broken away, and will form a sort of metallic cake, studded on the under surface with well-shaped crystals of bismuth, which often show a beautiful **iridescence** like the melted lead in Expt. 15, and for the same cause, viz., that they are coated over with a thin film of "oxide of bismuth," owing to the action of the oxygen of the air on the hot metal. Sometimes the bottom and sides of the crucible are similarly coated with crystals.

The term "crystal" is derived from the Greek κρυσταλλος, meaning originally *ice* or what is now called *quartz* (*rock-crystal*), and implying something transparent; whence the phrase "as clear as crystal." But it was early noticed that when water freezes to ice in thin films (for example, when moisture condenses inside the window of a warm room and becomes frozen by intense cold outside), or when it becomes snow, that the solid formed is bounded by plane surfaces regularly arranged in accordance with definite geometrical laws; and that the same remark as to shape applies to much of the quartz or rock-crystal occurring in nature; accordingly the terms "crystal" and "crystalline" by and by became extended to mean all substances so shaped, whether clear and transparent or not.

Cast iron is crystalline in structure; when broken across it shows a "grain." Some kinds are very fine grained, especially when cast in small quantities at a time; other kinds when fractured show a pretty large grain, being made up of crystals ¼ inch or more in length. A peculiar variety termed *Spiegeleisen*, largely used in the manufacture of certain kinds of steel, sometimes shows crystals some inches in length, the broken surface exhibiting large brilliant flat planes; the name is derived from this, being the German for *mirror-iron*.

Expt. 19. To Crystallise Sulphur.—Melt some roll brimstone at a gentle heat, taking care not to set it on fire (Expt. 8); let it cool until a crust forms, and then treat it as the bismuth in the last experiment. The cooled mass finally obtained will be seen to be made up of fine needle-shaped crystals.

Another way of crystallising sulphur is by dissolving it in

certain fluids, such as *carbon disulphide*, and allowing them to evaporate. This kind of action will be described later on (Chapter V.). Many solid substances can be thus crystallised by dissolving them in an appropriate solvent liquid and removing this by evaporation, or by making a hot solution and allowing it to cool; such operations are extremely common in chemical manufactures as well as in laboratory experiments.

Expt. 20. To make Teaspoons that Melt when used.—The alloy above described (Expt. 7) under the name of "Rose's fusible metal" will, as already stated, melt in boiling water, so that if a hollow plaster mould of a teaspoon be made and fusible metal cast therein, a teaspoon will be obtained which will melt if very hot water be poured into it, or if boiling water be stirred with it. If the water be not quite boiling, however, as is pretty sure to be the case when tea from a teapot is used, in all probability the heat will not be quite sufficient to melt the spoon. By melting the alloy and adding to it a small quantity of quicksilver, it is possible to make a mixture which, though solid at the ordinary temperature, will melt in water very much below the boiling-point. Another variety of easily-fusible alloy, known as "Wood's alloy," is made by melting together—

Bismuth, ...	7 to 8 parts.
Lead, ...	4 ,,
Tin, ...	2 ,,
Cadmium, ...	1 to 2 ,,

This mixture melts at about 70° *centigrade* and Rose's at about 98°, the temperature of boiling water being 100° and that of melting ice zero; on this scale an ordinary warm room has a temperature of from 15° to 20° C., whilst the temperature of the blood is about 37°. Another kind of scale for measuring temperatures, called *Fahrenheit*, is sometimes used, and on this scale water boils at 212° and freezes at 32° (*vide* Expt. 28).

It appears to be a pretty general rule that when two different kinds of solids are mixed together the mixture melts more easily than would be expected from the temperatures at which each of the solids melts separately; more especially is this the case with mixtures of metals. Thus ordinary *plumber's solder* is made of lead and tin mixed in proportions differing somewhat according to the purpose for which the solder is wanted, the coarser kinds containing more lead and the finer kinds more tin; several of these mixtures melt at temperatures not high enough to cause either tin or lead alone to fuse, and it is this property chiefly which enables the mixture to be used as a *solder* to cement together gas and water pipes and for such like purposes.

In a somewhat similar way when solid snow or scraped ice is mixed with salt the two solids melt together as it were, and form liquid brine; the process of salting the streets is sometimes used in winter time to prevent the roadways becoming too slippery from snow for horses to keep their foothold. As, however, the snow is thus converted into briny slush, which becomes intensely chilled during the operation (Chapter XXI.), the process is of somewhat doubtful benefit to pedestrians who find their boots saturated with very cold salt-water which will not dry readily. This peculiarity of snow and salt to become colder than ordinary ice when they melt together is made use of in the preparation of freezing mixtures for making ices and various other purposes. By putting a thermometer first in some snow alone, and then in a mixture of snow and salt, you will at once see that the latter is much colder.

Expt. 21. To Freeze Water by the Cold produced by means of Snow and Salt.—Get two or three pounds of snow or scraped ice in a pail, preferably of wood, and then mix with it about one-third or one-half of its weight of powdered salt by means of a stout stick. Now place a tin can holding a tumblerful of drinking water (as cold as possible to begin with) in the snow-salt slush so as to immerse it nearly to the top without allowing any of the brine to run into the clean water. Cover the whole up well with a blanket and leave it to itself for a few minutes or a quarter of an hour; at the end of that time the drinking water inside the can will be more or less completely frozen hard to a block of ice, which may be extracted by removing the can from the freezing mixture and gently warming the metal for a minute before the fire or by dipping the can in warm water: this thaws the ice where it touches the can, and so loosens the block of ice. In this, as in all similar cases, the reason why the mixture of snow and salt, &c., becomes cold is that during melting heat is rendered "latent" or disappears (*vide* Chapter XXI.).

Expt. 22. To make Ices.—In order to make ordinary confectioner's ices the arrangement employed is in substance the same as that just described, with this difference that instead of using plain water, water flavoured with lemon and sugar or mixed with cream and fruit juice, &c., is employed; and that instead of leaving the liquid at rest to freeze into a block, it is kept continually agitated so that the ice forms in little fragments, converting the whole ultimately into a sort of thick paste. Very eatable ices may be made with the simple appliances of a wooden pail to hold the snow and salt, a good sized tin can to hold the liquid to be frozen, and a long spoon to keep up a continuous stirring; but specially constructed little machines with an agitator worked by a handle (fig. 14) are considerably more convenient.

Expt. 23. To Freeze Mercury.—Many substances can be used instead of salt to produce "freezing mixtures" by mingling them with snow or crushed ice. One of the most powerful freezing mixtures of this class is obtained by mixing together 2 parts of snow and 3 of *crystallised chloride of calcium;* this is sufficiently powerful in its action to freeze mercury, if the metal and all the materials have been previously somewhat cooled down by placing them in vessels chilled by contact with ice. A depression in temperature of upwards of 40° centigrade is effected by this mixture, whereas a mixture of snow and salt (about 5 parts of the first to 2 of the second) only causes a lowering of temperature of about 20°.

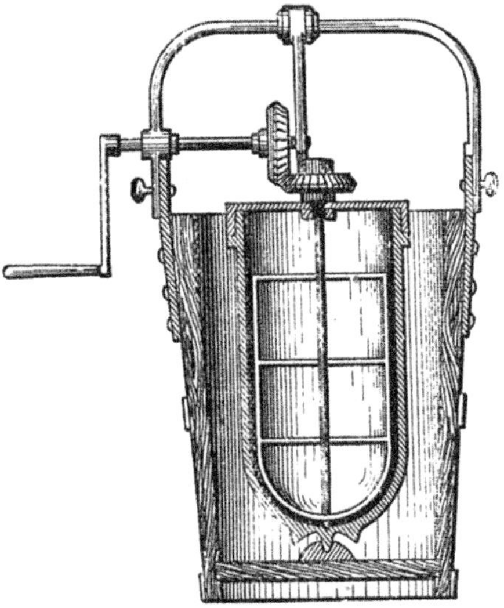

Fig. 14. Household Ice Machine.

Expt. 24. To Freeze Water without using Ice.—Many chemical substances are known which when made to dissolve in water or other fluids will generate a considerable degree of cold. *Saltpetre* (potassium nitrate) is such a substance, and has been long used in India and other hot countries as a means of cooling beverages, &c., the saltpetre and water being placed in a pail and the wine or other liquid to be cooled being immersed therein in bottles or other convenient receptacles. A better "frigorific agent" is the salt termed *nitrate of ammonium;* by taking a quantity of this in powder and stirring it with about its own weight of water, sufficient cold is often obtained to freeze water, especially if the salt and the water used have been all previously cooled down as much as possible. Instead of nitrate of ammonium a mixture of powdered saltpetre and *salammoniac* (chloride of ammonium) may be used; a lowering of temperature of upwards of 20° centigrade is produced in either case. A still more effective mixture, producing a depression of temperature of about 30°

centigrade, is obtained when the crystallised solid known as Glauber's salt, or *sulphate of sodium*, is powdered and stirred up with liquid *hydrochloric acid* (a corrosive mineral acid which must be carefully handled to avoid spilling and consequent damage). A thin glass cylinder of water immersed in a wooden pail or stoneware jar (not a metal vessel, as this would be attacked by the acid) containing a quantity of the powdered salt and acid will soon become frozen hard just as it would in snow and salt.

Many other freezing mixtures can be prepared in similar fashion, using different kinds of saline matters and other acids; for the artificial manufacture of ice these substances have long been superseded by far cheaper processes, mostly dependent on the production of cold by the quick evaporation of volatile fluids.

Expt. 25. To Freeze Water by the Evaporation of Ether.— Obtain a test-tube, about half an inch in diameter and 5 or 6 inches long. Wrap muslin or lampwick round the lower third, and pour into it two or three teaspoonfuls of clean water as cold as possible. Drop some ether on to the muslin so as to wet it, and blow vigorously on the wetted muslin with a pair of bellows. If a thermometer be placed in the water it will be immediately seen that this treatment cools the water; and by continuing to drop on ether and make it evaporate quickly by the bellows the water may in time be frozen to ice. The clamp-stand represented in fig. 12 may be used for holding the test-tube; but a simpler holder will answer perfectly well, consisting of a piece of wire twisted round the tube, somewhat after the fashion of the paper in fig. 11, with the ends thrust into a cork projecting out of a bottle full of water, so that the arrangement cannot be easily upset.

Very strong *acetic acid* (the acid of vinegar) may be thus frozen to an ice-like mass much more readily than water; the acid is often termed *glacial* acetic acid from this property, and its strength is judged of by the ease with which it freezes. Several other substances possess the same property; that is when tolerably pure they readily freeze or solidify on chilling to some extent; but if impure and admixed with other substances they do not become solid until they are much more chilled; so that the extra extent to which they have to be cooled before they solidify serves as a sort of measure of the amount of impurity present.

Expt. 26. To Freeze out Stearine from Olive Oil.—Many natural oils are not simple substances, but are mixtures of two or more fatty matters, one of which when separate is a fluid at ordinary temperatures and the other a solid. Genuine olive oil (not adulterated with other substances, as is often the case) is such a mixture, and when chilled for a long time (as in winter, or when placed in a freezing mixture) becomes pasty, owing to the separation in the solid form of the latter constituent, termed *stearine*. By pressing the pasty mass in several folds of blotting paper the fluid part, *oleine*, is absorbed by the paper and the stearine is obtained in the solid form, not melting at the ordinary temperature if sufficiently free from the oleine. This operation is best performed by means of a copying-press in a cold room, so as not to thaw again the solidified portion before the liquid part is separated.

This experiment affords an illustration of a kind of process largely employed in certain manufactures where solid matters analogous to stearine are thus separated by cold and pressure from more fluid impurities naturally accompanying them; such methods are chiefly used in the arts for preparing hard fatty matters for candle-making from soft oils, &c., for obtaining paraffin wax from petroleum and shale oils, spermaceti from sperm oil, and a variety of similar operations. To produce the cold requisite in such cases powerful freezing-machines are used, in most of which ether or some still more volatile fluid is made to evaporate very quickly, the evaporated liquid being recovered and used over and over again. The production of cold in such cases is primarily due to the fact that when a liquid evaporates, heat disappears or becomes **latent** (*vide* Chapter XXI.).

Expt. 27. To Freeze out Pure Ice from Sea-Water.—If common salt be dissolved in water (as in sea-water), a greater degree of cold will be required before the water will freeze than with pure water; but the ice which forms contains a much less proportion of salt than the original brine, especially if the salt water be agitated whilst freezing, so as to form small particles of ice instead of a solid block, as the former can be readily drained from the as yet unfrozen brine, whilst the latter is apt to enclose drops of strong brine here and there throughout the mass. When a quarter or a third of the water has frozen, the ice so formed should be drained on a cloth from the brine; on melting the ice the water will often be found to be sufficiently free from salt to be drinkable; but if still somewhat brackish, it may be further purified by repeating the process. In this way fresh water for cooking purposes and for drinking may be prepared in cold climates from water too salt to be otherwise used.

It is evident that the unfrozen part of the water will be richer in salt than the original brine, so that a concentration of the liquor may be effected by freezing out some of the water, thus producing the same result as if the brine had been strengthened by boiling off part of the water. In the manufacture of salt from natural brine springs where coal or other fuel is cheap (*e.g.*, in Cheshire), it is usual to boil down the brine until solid salt is left; but in cold climates where fuel is dear the freezing process is often employed instead to concentrate the brine, being much cheaper.

Expt. 28. To Construct and Graduate a Thermometer.—The ordinary thermometer consists of a piece of very fine glass tube on the end of which a bulb has been blown by softening the glass in a blow-pipe flame and blowing into the tube so as to expand the glass, much as a soap bubble is blown from soap water. To do this successfully requires much more skill than is likely to be possessed by a beginner; besides, the glass tubes ready blown can be purchased very cheaply, with the further addition of being ready filled with quicksilver and sealed up.

It is a general rule that when things are warmed they **expand** or occupy more space (*vide* Chapter XIX.); this is true of the glass bulb itself, which becomes bigger on warming; but the mercury inside expands more rapidly than the glass, so that whilst the bulb becomes absolutely bigger and would hold more, it does not become big enough to hold the enlarged volume of mercury, which therefore ascends in the glass tube (held vertically, bulb downwards) in proportion as it becomes larger in bulk as compared with the glass. In short, the quicksilver necessarily runs up the stem as the temperature rises, and retracts or sinks down again as it falls.

Suppose that you possess a thermometer ready for use but not graduated. It is found by experiment that when pure clean ice is allowed to melt and kept well stirred up with the resulting water, a thermometer placed therein always marks exactly the same point; this is called the **melting-point** or **ice-point** (sometimes, but less correctly, the **freezing-point**, as the temperature at which water freezes to ice is generally, but not always, the same stationary point as that at which ice melts to water). Again, when pure water is boiled in a metal vessel under the average pressure of the atmosphere (Expt. 32), a thermometer wholly placed in the issuing vapour also marks always the same temperature, higher than the ice-point; this higher temperature is called the **steam-point** or **boiling-point**.

The thermometer is then graduated by fixing it to a piece of wood, ivory, metal, &c., on which is engraved a scale with marks or degrees on it, so arranged that when the mercury is at the temperature of the boiling-point the top of the column of mercury inside the glass tube is just level with the graduation corresponding with that temperature; and similarly when the temperature is that of freezing water or melting ice. This is effected by marking off the two points on the material to be engraved, so as to form the scale (fig. 15), and then dividing the distance between them into

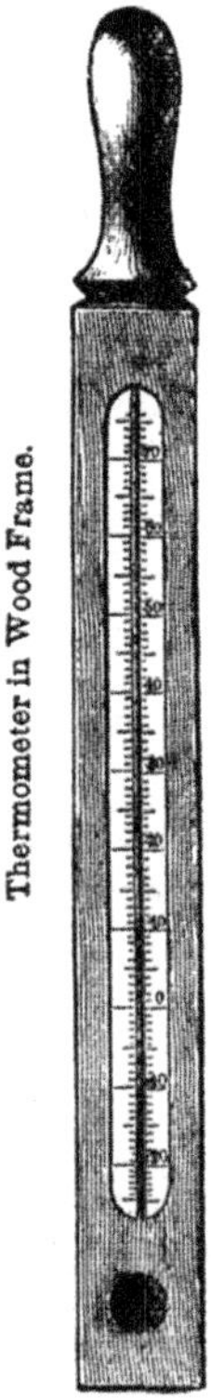

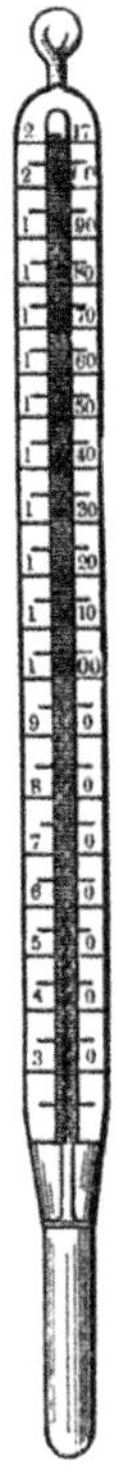

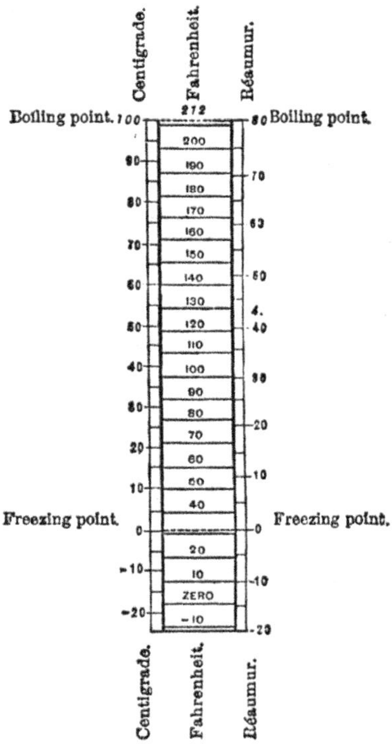

Fig. 15. Fig. 16. Fig. 17.

Fig. 15. Thermometer with attached Scale.
Fig. 16. Chemical Thermometer.
Fig. 17. Diagram showing relationship between the Centigrade, Fahrenheit, and Reaumur scales.

a number of equal parts, corresponding with the number of degrees between the freezing and boiling points. For the "centigrade" scale of temperature this distance is divided into 100 equal parts, and the graduations are also continued at the same equal intervals above and below the two fixed points; the ice-point is called 0, and temperatures below this *minus* so and so (minus ten degrees, minus fifteen degrees, &c., written $-10°$, $-15°$, and so on); the boiling-point is called 100 degrees, and temperatures above this are denoted by figures greater than 100; for these reasons this scale is called *centigrade* or "hundred-degree" scale. Degrees expressed on this scale are denoted by the letter C., thus 100° C. (boiling-point), and so on. Another form of scale is unfortunately in use in England; this is termed the *Fahrenheit* scale, after the name of its inventor, who for some inexplicable reason considered that because there are 180 geometrical degrees in a semicircle, therefore there should be the same number of steps between the freezing and boiling points; whilst to complicate the scale still more, he put his zero-point 32 Fahrenheit degrees below the ice-point; so that the ice-point becomes 32 degrees Fahrenheit (written 32° F.), and consequently the boiling-point $180 + 32 = 212$ degrees Fahrenheit (written 212° F.).

For scientific purposes the centigrade scale is far more extensively used than the Fahrenheit, but for determining the temperature of rooms, greenhouses, feverish patients, and the weather generally, the Fahrenheit scale is still chiefly used in England. In the following pages, wherever we have occasion to refer to temperature measurements in figures, the *centigrade* scale will always be employed in preference, on account of its far greater simplicity.

On some parts of the Continent a third scale known as *Reaumur's* is used. This is somewhat like the centigrade scale, the ice-point being termed 0; but the boiling-point is called 80 degrees instead of 100, and similarly throughout the whole range of the scale, so that every 5° centigrade are equal to 4° Reaumur. Fig. 17 represents the relations of the three scales to one another.

The better kind of thermometers are constructed so that the engraved scale and the glass tube cannot be separated from one another, whilst the bulb is elongated in such a fashion as not to be wider in diameter than the stem, so that the instrument can, if required, be passed through a perforated cork. Fig. 16 represents one form of chemical thermometer where the scale is a slip of ivory or paper which is placed inside a second or outer tube enclosing the stem of the actual thermometer but not the bulb, being sealed by the glass-blower to the stem just where it joins the bulb. When the graduation is finished the scale is permanently fixed to the outer tube by a drop of cement, and the tube sealed up at the upper end. A still better form is where the scale is etched on the glass stem itself (Expt. 271).

Expt. 29. To Verify the Scale of a Thermometer.—Thermometers of the cheaper sort are made in quantities at a time by the instrument makers, and far more accurately than might be anticipated considering the price; when anything like close measurements of temperature are requisite it is always essential to "verify" the thermometer indications, that is to check them and see if they are fairly exact. The most important check is to see that the ice-point is correctly recorded; for this purpose a few ounces of fresh snow or pounded pure ice are placed in a tumbler, and the thermometer plunged into the mass and used to stir it about for a few minutes, or preferably lumps of ice are placed in a perforated vessel (fig. 18), and the thermometer plunged in the ice; at the end of this time the quicksilver will have sunk to the lowest possible level under the circumstances, and by carefully looking at it it will be at once seen whether the mercury stands exactly at the zero-point or not. If not, a note should be made of the point at which it does stand; suppose this to be 2°, then it follows that the thermometer reads 2° too high at that part of the scale. If, on the other hand, the reading were −1°, the thermometer would obviously be erroneous in the other direction, reading 1° too low.

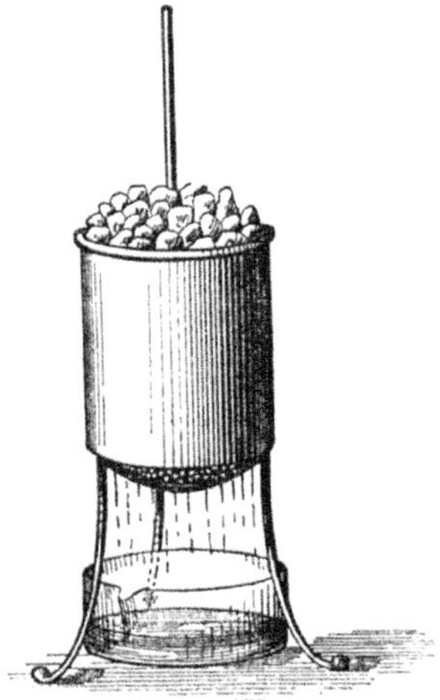

Fig. 18. Verification of Ice-Point.

The boiling-point may be similarly checked by immersing the thermometer in the steam given off by boiling distilled water when the barometer is at the average height, about 30 inches. The thermometer should be suspended inside a sufficiently wide glass tube placed vertically over the vessel of boiling water (fig. 19), so that the bulb is not immersed in the water, whilst the whole stem is thoroughly heated by the steam which must issue freely at the top. In this way the error of the reading at the boiling-point is determined. For most ordinary purposes an error at either part of the scale not exceeding one or two degrees is of little consequence; but for many kinds of scientific observations a much greater degree of accuracy than this is indispensable.

Expt. 30. Circumstances modifying the Freezing-Point of Water.—If on a cold winter's day some pure distilled water be placed in the open air (or in a sufficiently cold room) and kept stirred up by means of a thermometer placed therein, it will be noticed that ice begins to form when the thermometer marks exactly the same point as that where it stands when immersed in melting snow, *i.e.*, exactly at zero if the thermometer be accurately graduated. If, however, sea-water be used, or if some common salt, Epsom salts, sugar, or other solid soluble matter be dissolved in the water, it will be found that ice will not be formed until the temperature falls perceptibly below zero. As shown in Expt. 27, the ice thus produced contains but little of the dissolved salt, &c., as compared with the water from which it froze.

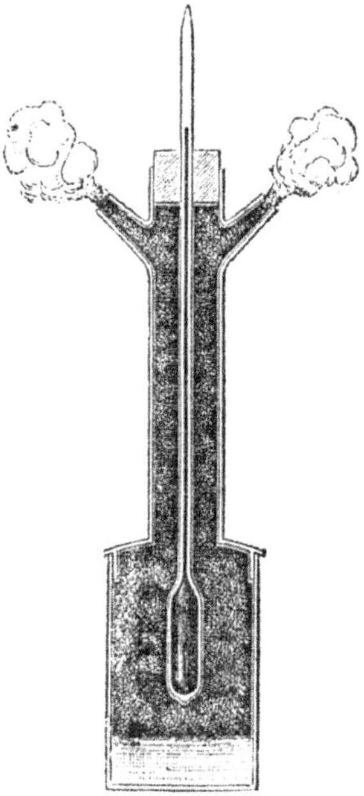

Fig. 19. Verification of Boiling-Point.

If water be allowed to stand perfectly at rest in an atmosphere a good deal below freezing-point, it sometimes happens that the whole can be cooled down several degrees below zero before any ice forms; but the least degree of agitation, such as the shaking of a passing cart or train, entirely prevents this. If water cooled down in quietude below its normal freezing-point be stirred, some of it at once turns to ice and the temperature of the remainder at once rises to zero. This property of liquids being capable of cooling below their melting-points without solidifying under certain circumstances is called **superfusion**, and is closely connected with the analogous peculiarity termed **supersaturation** exhibited by solutions under similar conditions (Chapter V.).

Water subjected to great pressure requires more chilling to freeze it than when not compressed; conversely ice strongly pressed begins to melt on the surface, but the water thus formed instantly becomes ice again on relieving the pressure. In consequence a quantity of snow or of crushed ice fragments can be cemented together into a solid block by simple compression and relaxation of the pressure. This property is termed **regelation**, and is one of the principal causes forming *glaciers*, which are loose white opaque snow at the top and clear solid transparent ice at the bottom: it is difficult to illustrate it without a powerful press, but to some extent the action may be noticed when a handful of snow is strongly squeezed in the hands to form a snowball; the separate crystals of snow become coherent together, and the mass tends to become somewhat translucent. With the aid of a blacksmith's vice or other screw press enabling powerful pressure to be developed, it is possible to squeeze a quantity of loose snow in a mould or a pair of dies, so as to convert it into a shaped solid block of ice, which will be almost perfectly clear if the snow were quite clean and the pressure powerful enough, as for example when a hydraulic press is used.

It is remarkable that whilst water is affected by pressure in such a way that the freezing-point is *lowered* by increase of temperature, the opposite is the case with many other substances. The following simple rule connects the two kinds of action. If the solid substance is lighter than the same body when melted (as is the case with ice), increased pressure lowers the melting-point, and *vice versâ.* This is due to the circumstance that in the case of ice and such like bodies, the tendency of pressure being to squeeze the mass into a smaller bulk, an increase of pressure will tend to produce the same result as fusion (*i.e.*, shrinkage in bulk); so that the greater the pressure the less heat is required to produce melting; whilst in the case of bodies that are heavier in the solid state than when melted, increase of pressure tends to produce the opposite effect to that produced by fusion (*i.e.*, in this case, increase in bulk); so that now the greater the pressure the more heat will be requisite to overcome the effect of the pressure and produce melting.

Welding.

Many solids when heated become soft and plastic before they actually melt: in this condition two lumps of hot substance may be squeezed together and will cohere to a solid mass just as two lumps of soft clay or putty may be squeezed together into one. Metals exhibiting this property are said to **weld** together, wrought iron and platinum furnishing two of the best examples. In the manufacture of iron articles two pieces of metal are frequently fastened together by this process, on which a large part of the art of the "blacksmith," or worker of wrought iron, depends. The rods or other pieces of iron to be welded are heated nearly white hot in a "forge" or cinder-fire urged by bellows; and if applied together and skilfully hammered whilst *sufficiently hot* (*at a* "welding heat") they cohere perfectly. Sometimes a little sand or other silicious matter is sprinkled over the surface, the object of which is to form a coating of fusible matter over the surface of the metal; this is forced out by the pressure, leaving clean surfaces of metal close together, which thus adhere firmly.

Expt. 31. To join Leaden Pipes by Soldering.—When a plumber wants to join two leaden water-pipes together he frequently makes use of a somewhat analogous property possessed by the mixture of metals known as "plumber's solder" (mostly lead and tin); this mixture when pretty hot is quite fluid, but on cooling a little becomes pasty before it finally solidifies; in this pasty condition the solder is spread round the two ends of the pipes applied together, one end being slightly widened out to a trumpet shape by means of an appropriate tool, and the other shaved down so as to be capable of insertion into the mouth thus formed. The portions of the lead intended to have the solder applied to them are brightened by scraping so as to enable the solder to stick firmly; and the pasty solder is manipulated with a greased piece of cloth so as to form ultimately a "wiped joint." Although the knack of being able to manipulate hot pasty solder in this way is by no means indispensable in the performance of chemical experiments, yet the ability to make joints of the kind is sometimes very useful. In the construction of apparatus intended for the preparation of gases of various kinds, a similar mode of making leaky corks and joints, &c., air-tight is frequently used, some plastic material not too hot being usually employed, such as melted sealing-wax that has cooled sufficiently to bear touching, or a stiff paste of linseed meal and a little water, or ordinary dough of flour and water, clay, and similar pasty cementing materials. Workers in glass take advantage of the pasty condition assumed by molten glass as it partially cools to shape it into bottles and numberless other articles; and methods of making small glass vessels, &c., for experimental use are constantly adopted in laboratories, the heat being usually applied by lamps, where the flame is urged by bellows forming a "blow-pipe" flame. Glass tubing, when not too large and thick, may be easily softened and bent in an ordinary gas flame (Expt. 48).

CHAPTER III.

Boiling of Liquids and Condensation of Vapours.

When water is placed in a kettle on the fire it is obvious that the heat of the fire mainly reaches the water through the bottom of the kettle. Water and many other liquids when thus heated from below exhibit the phenomenon of **boiling** or **ebullition**; bubbles of steam or vapour form at the bottom of the kettle and pass upwards, breaking on the surface and keeping up a continuous agitation. This is more easily seen when water is boiled over a spirit-lamp in a thin glass flask.

When the kettle boils vigorously, in ordinary language steam is said to issue from the spout; on looking at the spout it will be seen that half an inch or more of the space beyond the spout is perfectly clear and transparent, after which a little mist is visible, and further off still a thick cloud of mist. What is properly called **steam** is not the visible mist, to which, however, the term is often applied; the true steam is the *clear transparent vapour* forming the column first issuing from the spout, as imperceptible to the eye as the air itself. Only when this becomes partly chilled by mixing with the cooler air does it become visible, and then only because it alters its state and **condenses**, *i.e.*, becomes water again, but in droplets so small that they float about in the air forming a fog. In short, the visible mist is condensed steam, or liquid water in fine spray, resulting from the condensation on cooling of invisible steam.

What is true of water is equally true of all liquids that boil; the vapours given off are transparent, and whilst hot are usually wholly invisible (some few substances form coloured vapours), and condense again to visible droplets of the original liquids on cooling.

Some liquids boil far more readily than others; thus whilst water boils at a temperature of 100° centigrade, strong alcohol boils 20° or more lower; ether will boil with the heat of a hot summer's day, whilst many liquids are known to the chemist which will boil below zero centigrade. On the other hand, mercury does not boil till a temperature of 360° is attained; whilst melted zinc requires a pretty bright red heat to make it boil, and other substances only boil at the very highest temperatures obtainable in the most powerful furnaces or by means of intense electric currents.

Whether a substance boil at a high or a low temperature, however, the vapour or steam given off on boiling possesses the same kind of physical texture as atmospheric air; indeed this latter, like all other kinds of "gases" obtained by chemical means, is simply the steam or vapour of highly volatile liquids so constituted that they boil at extremely low temperatures.

Expt. 32. Circumstances modifying the Boiling-Point.—If the atmospheric pressure be lessened, water boils more readily, so that a thermometer placed in boiling water indicates a lower temperature when the barometer is low than it does when the barometer marks a higher pressure. An instrument for measuring the heights of mountains is based on this principle; the higher one ascends the lower is the temperature marked by a delicate thermometer placed in water made to boil by means of a spirit-lamp. The influence of alteration of pressure on the boiling-point of water is far greater than that on the melting-point of ice; a diminution of not far from 1° centigrade is brought about in the boiling-point for every 1000 feet that we ascend.

When salt or other soluble solid substance is dissolved in water the temperature indicated by a thermometer placed in the solution when boiling is always *higher* than that indicated similarly with pure water; but, what is curious, the vapour or steam that comes off *is of the same temperature in both cases.* This may be verified by means of the apparatus (fig. 19); whether the flask contain pure water or pretty strong brine, the thermometer will indicate the same temperature in the steam; but if the bulb be lowered so as to dip into the boiling fluid, it will be at once obvious that a higher temperature is indicated with the brine than with the plain water.

It results from the opposite effects of diminished pressure and presence of dissolved solid matter on the boiling-point of water that people who live at high elevations above the sea (for example at Quito, several thousand feet above the sea-level) have habitually such a rarefied atmosphere surrounding them that the temperature of pure water boiling in an ordinary saucepan is not sufficiently high to cook food in the same way as it would be done at the sea-level; so that, in consequence, to raise the temperature sufficiently to boil eggs, for example, a large quantity of salt must be added to the water.

The same result is attained by confining the steam of the boiling water by using a closed boiler or *digester* (furnished with a safety-valve) instead of an open pan; in this way the pressure is increased artificially as required, and the temperature of the boiling water proportionately raised. In a high-pressure steam-boiler (such as is used for generating motive power in factories, locomotive engines, and the like) the water has to be heated a long way above 100° before it actually boils; thus when the pressure amounts to 5 atmospheres, or about 70 lbs. per square inch, the boiling-point of water is raised to about 152° instead of 100°. Experiments on the increased temperature required to boil water in a confined space cannot be safely made without specially strong apparatus provided with a safety-valve; if you attempt to boil water in corked glass vessels you may be badly scalded by their bursting under the pressure. A strong tin can filled one-third with water and made to boil over a lamp may be used to illustrate the fact of increased pressure being produced by confining the steam; a cork placed in the mouth (not jammed in too hard) whilst gently boiling will in a few seconds be driven out with a "pop" by the pressure. If the cork be too firmly fixed, the pressure may very likely burst the can instead of projecting the cork, scattering scalding water around.

Expt. 33. To make a Model Rotating Machine working by Steam Pressure.—When water is boiled in a vessel provided with a narrow pipe through which the steam can issue, a continuous current of vapour will pass off through the jet as long as the water boils. If a small model windmill be placed in the current of steam it will be turned round just as it would by wind. Several toys based on this principle are sold—the "Dancing Nigger" is one of the kind; the jet of steam from a small metal boiler heated by a spirit-lamp impinges on the floatboards of an arrangement like a waterwheel and so makes this revolve, thereby setting in motion a sort of table under the feet of a jointed figure of a negro supported by a wire, and so shaking the figure and causing it to jump and dance about in a comical way.

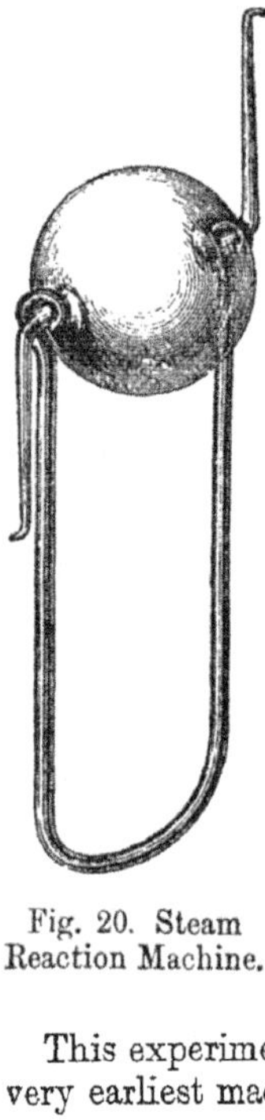

Fig. 20. Steam Reaction Machine.

Expt. 34. Steam Reaction Machine.—Fig. 20 represents another simple form of rotating machine, where the motive power is the pressure of steam generated inside a glass bulb by heating with a lamp and escaping therefrom by means of two curved jets. It requires a good deal of expertness in glass-blowing to blow the bulb and jets successfully; but the arrangement may be purchased cheaply at the instrument makers. To fill the bulb, it is held by means of the wire support so that one of the jets is undermost and dips into water which is thus sucked into the bulb by means of the other jet, conveniently by slipping a bit of fine india-rubber tube over the nozzle and sucking through it. By employing lavender water, or ordinary water scented with a few drops of eau-de-Cologne, &c., the arrangement may be used as a perfume vaporiser to scent an apartment.

This experiment is especially interesting from the fact that the very earliest machine ever made whereby motion was produced by steam pressure, the first steam-engine in fact, was constructed on this principle about 120 B.C. by Hero of Alexandria, and termed the *Æolipyle*. Fig. 20A represents a form of this engine.

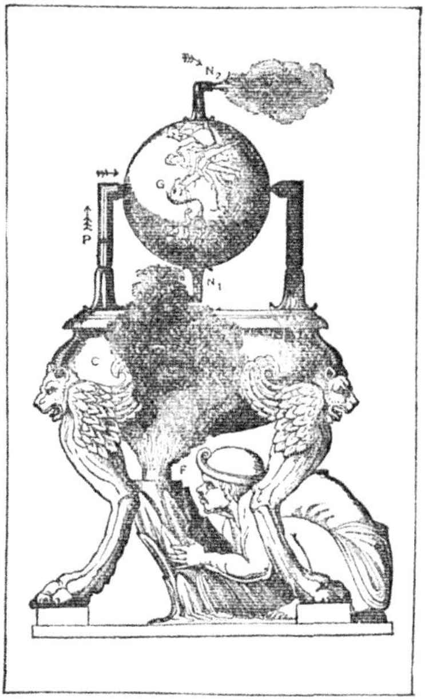

Fig. 20A. Hero's Engine.—The Æolipyle—*F*. Fire; *C*. Cauldron for Water: *P*. Pipe for Steam; *G*. Globe; N^1, N^2 Nozzles, Steam Exhaust (from Jamieson's *Steam Engine*).

Expt. 35. To make Water Boil by pouring Cold Water over the Vessel containing it.—This most remarkable paradox results in consequence of the principle that water boils at a lower temperature when the pressure upon it is diminished. Obtain a strong glass flask and an india-rubber cork fitting it tightly. Half fill the flask with water and support it on a retort-stand (fig. 6), and boil it over the gas or spirit-lamp flame, taking care that the cork is *not* in the neck, so as to give free vent to the steam. After a few minutes, when the steam is issuing vigorously, remove the lamp, and at the same moment cork up the boiling water. Take hold of the flask with a cloth and turn it upside down, supporting it in the ring of the retort-stand (fig. 21). On pouring water slowly on to the upper end of the flask the water in the lower part will commence boiling vigorously, soon stopping if the supply of cold water is discontinued, but again commencing to boil on pouring on more water. If the flask be of good glass it will not crack under the operation, which may be repeated at pleasure, heating up the water to boiling again over the flame (after removing the stopper) when it has become too nearly cold to boil any more under the influence of cold water.

The principle involved in this experiment is used in preparing a number of articles of food (condensed milk, sugar, extract of beef, &c.), medicinal extracts, and such like substances, where it is necessary to boil down to a small bulk at as low a temperature as possible to avoid damage by heat. In order to do this a powerful pump is connected with the boiler so as to pump out the vapour as fast as it is formed, and always maintain in the boiler only a very small degree of pressure. This arrangement is usually termed a *Vacuum Pan.*

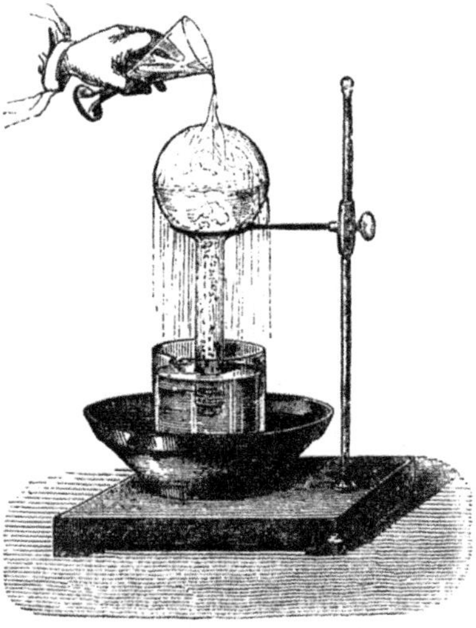

Fig. 21. Water made to boil by pouring Cold Water on Flask.

By rapidly removing the vapour of water as fast as it is evolved, it is possible to chill the water by its own evaporation so much that ice forms on the surface.

Expt. 36. To make Distilled Water.—For small quantities, obtain a stoppered glass retort holding about a pint (fig. 22), half fill it with water, place it on a stand, and heat it to boiling over a gas or spirit-lamp. The steam produced passes over into the neck of the retort, and there condenses to drops which run down into a long-necked flask which acts as a receiver. In order to obtain a more extended cooling surface the receiver may conveniently be placed in a basin of water and its upper surface kept moist by means of a piece of cloth, &c., laid over it and dipping into the water in the basin; or better still, as indicated in fig. 22, a small stream of water is allowed to trickle over the top of the receiver by means of a flexible tube connected with a water tap. A yet more effective method is to interpose a water-jacketted condenser (often called a *Liebig's Condenser*) between the retort and the vessel into which the condensed water drops; fig. 23 indicates such an arrangement of retort and condenser, the water in the jacket being continually renewed by cold water which runs in at the lower end of the condenser from a convenient vessel (or from the water-main by means of a flexible tube), whilst the heated water issues from the upper end and flows into another receptacle (or directly into the sink). A large glass flask or tin can may be conveniently substituted for a glass retort, the mouth being fitted with a perforated cork through which a piece of metal or glass tubing is passed, bent so as to convey the steam to the condenser (fig. 24).

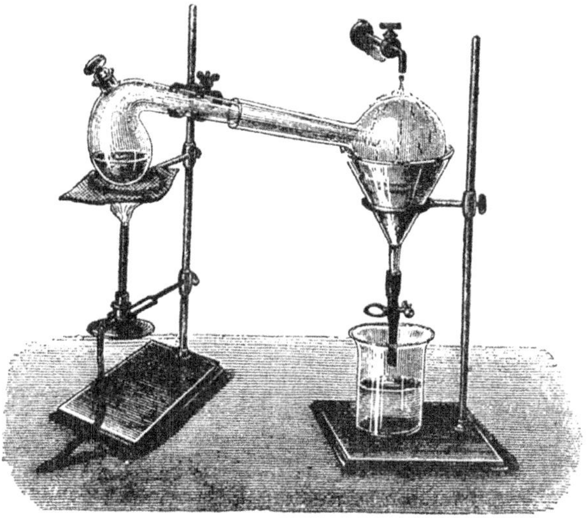

Fig. 22. Distillation by means of Retort.

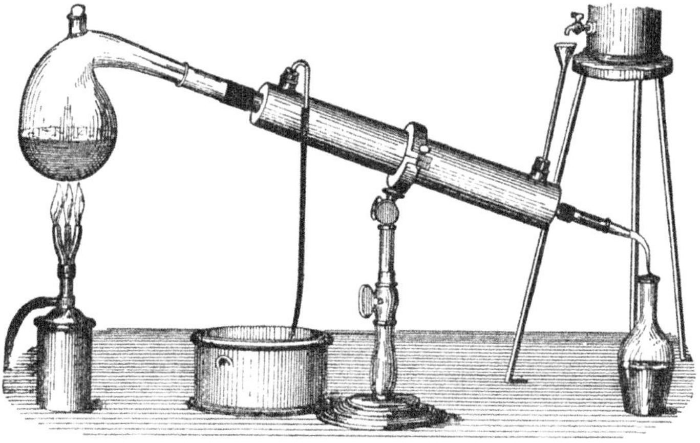

Fig. 23. Distillation with Liebig's Condenser.

In the manufacture of alcohol and other distilled products stills are employed frequently of enormous size, but acting in just the same way, the vapour evolved being condensed in large worm tubes, cooled by being placed in a tank of water. Fig. 25 represents such

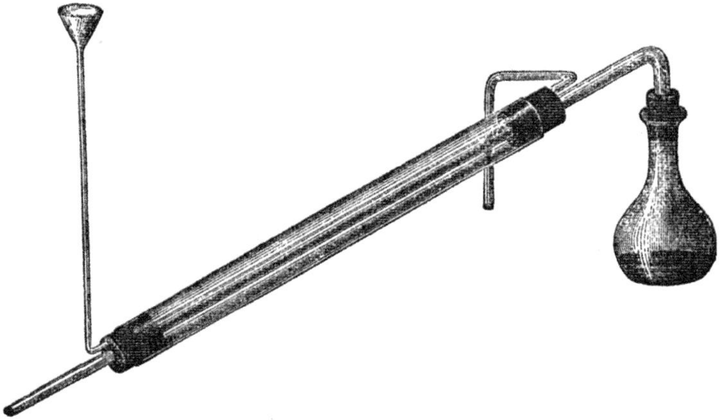

Fig. 24. Flask and Condenser.

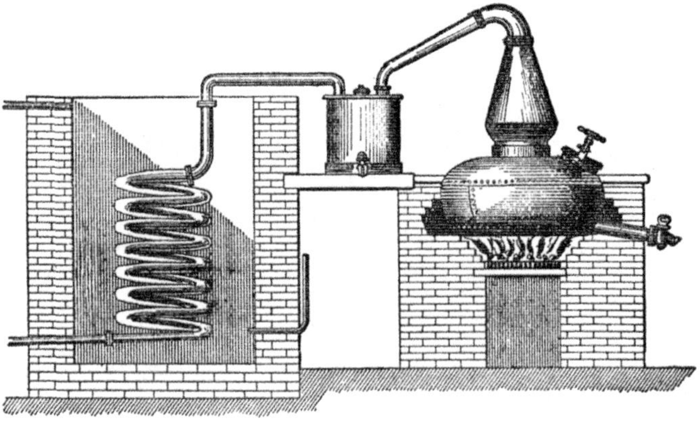

Fig. 25. Still and Worm.

an apparatus, an intermediate vessel being placed between the still and worm to catch any liquid that may spirt or froth over.

In all these cases any solid saline matter previously dissolved in the water is left behind in the retort or still, and pure water obtained after condensation.

Expt. 37. To prove that Fresh Water can be obtained from Salt Water by Distillation.—Place in the retort as above a tea-spoonful of common salt and half fill the retort with water, and warm until the salt is all dissolved. Pour out a few drops into a wine-glass and add a little solution of *nitrate of silver;* the two colourless clear fluids will at once become turbid (just as in Expt. 12, No. 1 glass), a flocculent white solid precipitate being formed which by and by falls to the bottom of the glass. This solid matter is due to the chemical action of the salt on the nitrate of silver by double decomposition; if pure water be used instead of salt solution in another wine-glass, the addition of the nitrate of silver will of course produce no result and the liquid will remain clear; so that the nitrate of silver thus serves as a chemical **test** or **reagent** for salt, giving rise to the formation of a **precipitate** if salt be there, but not if it be absent.

Now distil off some of the water, taking care to avoid frothing over of the brine in the retort, which would contaminate the distilled portion; on testing the distilled fluid with nitrate of silver it will remain clear, showing that the salt has not passed over with the steam but is left behind. This chemical test is far more delicate than the taste, and will easily detect small quantities of salt quite imperceptible to the most sensitive palate; it is, in fact, so delicate that if the hands be dipped in some distilled water and a few drops of nitrate of silver be then added, the liquid will become more or less cloudy, owing to the formation of a small amount of precipitate, due to the circumstance that salt (sodium chloride) is contained in the perspiration, so that a minute quantity becomes dissolved in the water from contact with the skin.

This method of preparing fresh water from sea-water is largely used aboard ships, especially steamers. Boilers fed with sea-water will produce steam which on condensation is free from the salt and other impurities of the sea-water, and can be used for cooking directly. For drinking purposes this distilled sea-water is unpleasantly flat, like boiled water, and for the same reason, viz., that the air and other gases present which give the fresh taste to ordinary spring water, have been removed by the heating. Consequently it is usual to aerate and freshen up the distilled water before using it for drinking in order to improve its taste; but for boiling vegetables, making tea and coffee, &c., this treatment is not necessary. In certain tropical cities, where a supply of good natural water is difficult to obtain all the year round owing to the comparatively short period when rain falls and the difficulty of storing it in tanks, this method of obtaining drinkable water from sea-water or saline springs is of considerable importance.

In default of any other apparatus, distilled water may be obtained from impure or saline water by the aid of an ordinary kettle; this should be tilted slightly so as to bring the spout to a little higher elevation, and the lid made as tight as possible by means of a little clay or putty; a piece of clean "Compo" or other metal tubing should be fixed to the spout by jamming a cork therein, the central part of which has been cut away so as to allow the tube just to pass through. The tube should be cooled as far as possible by wet cloths, &c., to promote the condensation of the steam.

Expt. 38. To Extract Spirit from Wine or Beer.—The chief intoxicating ingredient in fermented liquors is the substance termed *alcohol*, present in such fluids to very varying extents, according to their nature and quality, but usually to the extent of from 3 to 6 per cent. or more in ordinary beer, and somewhat larger proportions in wines, especially the stronger ones like port and sherry. Ordinary home-made gingerbeer, prepared by means of yeast from sugar, water, lemon-juice, ginger, &c. (Expt. 190), contains quite enough alcohol to be quite as exhilarating a beverage as many kinds of light ale. From any of these fluids weak alcohol may be extracted by distilling in a retort or larger vessel according to the quantity of liquid available. When the boiling has been continued so long that the distilled fluid measures one-fourth to one-third as much as the liquid originally employed, practically all the alcohol present has passed over along with some water; the reason being that alcohol has a lower boiling-point and is more easily vaporised than water, so that when a mixture of the two is boiled the steam coming off contains more alcohol-vapour relatively to the water-vapour, and therefore yields on condensation a liquor stronger in spirit than the original fluid. By this treatment all solid matters present in the beer, &c., used are left behind in the still; by repeating the distillation and collecting the first third that distils a much stronger spirit is obtained, practically all the alcohol originally present being thus concentrated into perhaps one-tenth of the original bulk.

Ordinary spirits, whisky, gin, rum, brandy, &c., are prepared in very similar fashion, the flavour depending on the nature of the fermentable matters used to produce the weak spirit in the first place, and on the kind of flavouring materials (juniper berries, &c.) used in certain stages of the process subsequently.

Expt. 39. To make Anhydrous Alcohol.—It is impossible to bring about a complete separation of water and alcohol by distillation alone, the strongest spirit thus obtainable always containing a little water which requires treatment with some chemical reagent to ensure its complete removal. In order to obtain nearly *anhydrous* or "absolute" alcohol (alcohol free from admixture with water) from wine or beer, the tolerably concentrated liquor obtained by two or three successive distillations as in the last experiment, collecting the first third or so each time, is placed in a bottle with a quantity of solid *carbonate of potassium* (purified pearl ashes). This substance has a strong attraction for water, and dissolves freely therein, forming a solution with which alcohol will not mix; so that after a while the contents of the bottle separate into two layers of liquid floating one on the other like oil on water; the top layer is alcohol almost wholly free from water, which can be poured off into another bottle and preserved, whilst the heavier liquid is a solution of the carbonate of potassium in the water withdrawn from the spirit.

Another way in which absolute alcohol can be prepared from spirit already of considerable strength is by putting some lumps of freshly burnt quicklime in a glass flask, and then pouring the spirit on to them and corking up the flask for twenty-four hours; a cork and piece of bent tube are then affixed to the flask and the spirit distilled off and condensed by means of a "Liebig's condenser" (fig. 24), the heat being applied not by directly heating the flask with a lamp but by placing it in a vessel of water which is gradually heated to boiling; this contrivance is termed a *water-bath*, and is identical with the arrangement used in an ordinary gluepot and certain cooking utensils to avoid overheating (*vide* Expt. 89).

If the spirit used in this experiment is not pretty strong to commence with, the lime is apt to get very hot and may burst the flask; the quicklime chemically combines with the water present, becoming "slaked" in so doing (Expt. 240), whilst the alcohol remains unchanged; so that on applying heat the alcohol rises in vapour whilst the water remains behind combined with the lime.

Expt. 40. To Strengthen Weak Spirit by means of a Bladder. —Animal membranes, such as bladders, possess a remarkable property, viz., that whereas they are readily porous towards water they will not let alcohol pass through them at all easily. In consequence, if a sheep's bladder be nearly filled with a mixture of spirit with so much water that the mixture will not burn in a spoon (about twice as much water as spirit will generally suffice), and tightly tied up and hung up for a few days in a warm room, it will be found that the contents of the bladder have considerably diminished in bulk by the passage of the water through the bladder and its evaporation, but the strength of the liquor is so much increased by the operation that the spirit will now readily take fire in a spoon. Almost the whole of the water present can thus be in time removed, leaving nearly absolute alcohol in the bladder.

Membranes of different kinds are frequently employed to effect a separation of substances from one another when we have mixed together bodies which will pass through the membranes with different degrees of ease (Chapter VIII.).

Absolute alcohol obtained by any of these processes is a highly inflammable fluid, which readily takes fire if spilt about near a flame. The "methylated spirit" usually burnt in spirit-lamps consists of alcohol not quite absolute (but nearly so), to which a certain amount of a peculiar ill-flavoured spirit-like substance obtained from wood by distillation is added in order to prevent the liquid being used for drinking. The excise duty on potable spirits is extremely high in Great Britain, far above the actual value of the alcohol present, and this prevents the use of pure spirit in this country for many manufactures that can be carried on more cheaply in other countries where the duty is lower; to diminish the injury to trade thus effected a much smaller duty is levied on "methylated spirit," which, although too badly-flavoured to be drunk as a beverage by most people, is nevertheless applicable for varnish-making and many other trade purposes almost as well as pure spirit.

In case of alcohol or methylated spirit being spilt about and taking fire (as for example by upsetting a spirit-lamp), it should be remembered that pouring water on the flames is far less effective than smothering them out by throwing over them a piece of carpet, an old blanket or coat, or any *woollen* fabric that may be handy: cotton and linen burn so much more readily than wool that they are much less serviceable for extinguishing flames, unless they have been previously wetted; a wet towel, however, will often suffice to put out a small fire caused by some accident if applied in the very early stages. Lamentable accidents sometimes happen through ladies' muslin dresses and the like inflammable clothing being set on fire; if in such cases the persons on fire have sufficient presence of mind to lay themselves down and roll over on the carpet whilst calling for assistance, they will often escape serious injury; any one coming to help should fling a coat, woollen curtain, or carpet, &c., over the burning dress and so smother out the flames. Unfortunately, persons on fire are apt to "lose their heads" and run about shrieking instead of lying down; this fans the flames and makes them burn far more fiercely, and so the unfortunate sufferers are far more severely injured than need have been the case had they had the presence of mind to lie down and roll over trying to extinguish the flame.

Expt. 41. To Extinguish Burning Spirit or Ether.—Get a small saucepan and warm it slightly and then pour into it a teaspoonful of strong alcohol or methylated spirit, or, better still, of ether: the warmth of the pan will quickly vaporise the ether or spirit, so that by applying a light to the mouth of the saucepan a large flame will be produced. Put the flaming pan on a table where there is nothing near that can be set on fire, and then quickly put over the pan a wet towel or piece of blanket; the flame will be extinguished just as a candle is when an extinguisher is dropped over it. Precisely the same result will be produced if a little strong spirit be spilt over a *brick* or *stone floor* and set on fire and a wet blanket cautiously dropped over the flame.

Experiments of this sort *must on no account be tried in an ordinary room with a wooden floor*, as the burning spirit is liable to run down under the boards through the crevices. Builders often leave shavings lying about under the flooring when putting up houses, and should such shavings be set on fire by the burning spirit much damage may be done. Even in cases where there is fire under the floor, however, it may often be extinguished by covering up the spots where smoke issues from the crevices and pouring water on the floor so as to smother the fire and at the same time quench it with water.

One of the most effective appliances for extinguishing small fires is an ordinary siphon of soda water or lemonade, the valve being opened so as to squirt the effervescing fluid on the flaming object. Here it is not only the water which acts as a quenching agent, but also the gas used as aerating material; this keeps the ordinary air away from the burning matter, and thus tends to put out the flame just as a wet blanket would do. The fire extinguishing appliances termed "extincteurs" work on much the same principle; they contain water and certain chemicals set in action by striking a knob on the machine; these chemicals generate a quantity of gas which forces out the water in a jet and also acts like the gas of the soda-water siphon, keeping off air from the flames. "Hand grenades" for extinguishing fires are glass bottles filled with water containing dissolved therein some saline matter which tends to prevent substances burning readily when the solution is sprinkled over them; so that at the very commencement of a fire, if one or more of these bottles are broken in the flames, the water and saline matter jointly tend to put out the fire, or at least to prevent it spreading.

Expt. 42. To render Muslin Incombustible.—This property

of certain saline substances to interfere with the free burning of easily combustible bodies treated therewith is often utilised for the purpose of making such substances more or less *fireproof*. Prepare some ordinary laundress's starch paste by rubbing up solid starch to a cream with a little water in a basin with a spoon, and then pouring *boiling* water into the basin, well stirring the while. Muslins, &c., starched with this will flame readily if fired when quite dry; but if a little *alum*, or, better still, the salt termed *tungstate of sodium*, be mixed with the starch paste, it will be found that muslins starched with the mixture will no longer take fire and blaze up when a light is put to them; if held in a candle flame they will char where the flame touches them, but the burning portion will not spread around. The most effective way of showing this is to make two small wire-frames, shaped like the wicker dress-stands used by ladies, so that when draped, one with prepared muslin and one with ordinary starched material, they may represent two miniature ball dresses, something like dressed dolls. On putting a light to the first nothing remarkable happens, but on doing the same to the second the muslin takes fire and flames up, and in a few seconds only the bare wire-stand is left. Of course this experiment must not be tried in a room where there are curtains or other furniture that might be easily set on fire should burning fragments of muslin fly about in the air.

Vaporisation of Liquids without Ebullition or Boiling.

In order to convert a liquid into vapour it is not necessary to boil it from beneath. By holding a hot plate of iron over water in a shallow saucer the water is quickly **evaporated** and dried up without visible ebullition; in the same kind of way many other fluids will gradually form vapour and disappear by simple exposure to air at the ordinary temperature. When the pavement dries after a shower, or wet clothes after washing, the water gradually assumes the condition of vapour which becomes intermingled with the air, and is removed by winds and currents. In colder regions of the air the vapour thus present in consequence of evaporation from the sea and land becomes more or less condensed just as the steam from a tea-kettle; mist or cloud thus first results, consisting of minute drops or **vesicles** of liquid water floating about in the air. When these run together or **coalesce** so as to form larger drops which are heavy enough to fall quickly, **rain** results; if the drops fall through a very cold stratum of air so as to freeze before they reach the earth, **hail** is produced; whilst if the mist is chilled to the freezing point under certain conditions the minute drops freeze together so as to form feathery crystals or **snowflakes.**

This condensation of mist or liquid water in small particles from moist air on cooling is due to the fact that air (like all other gases) is capable of taking up in an invisible form a certain quantity, but no more, of water-vapour; the higher the temperature the more water can be thus dissolved, as it were, in the air. Suppose then that air containing as much water-vapour as it can take up at the particular temperature existing at the moment be somewhat cooled, the surplus amount of vapour present over and above the quantity sufficient to "saturate" the air at the lower temperature, becomes condensed to liquid water, making its appearance either as mist suspended in the air or as **dew** on the surface of the vessel containing the air; and this is the explanation of the formation of dew in nature. On a clear night the surface of the earth cools down spontaneously by the process called **radiation** (Chapter XXII.), and consequently any moist air coming in contact with objects thus cooled becomes also chilled and deposits its surplus moisture as dew on the cold object. Accordingly those things that radiate heat best become most chilled during the night and most copiously bathed with dew by condensation of moisture from the air; whilst objects prevented from cooling so much by radiation by a cover of any kind are correspondingly less bedewed. If the chilling action is sufficiently great to freeze the moisture thus deposited, **hoarfrost** results instead of liquid dew.

Expt. 43. To cause Artificial Dew.—If we examine our breath on a warm summer's day nothing is visible; but in colder weather it is obviously misty, for the reason that, as the moist air exhaled from the lungs becomes mixed with the cold external air and cooled down, the temperature is not sufficient to enable the total air to retain permanently vaporous all the moisture present, so that the surplus condenses as liquid water mist-particles. If we breathe on a cool window-pane the moisture condenses thereon as visible dew; similarly on putting a cool chimney over a burning lamp or gas jet a deposition of dew becomes visible for a short time until the chimney becomes hot; in this case the moisture comes in virtue of the chemical changes produced during the burning of the lamp or gas (Chapter XIV.).

Fill a flask or glass decanter with ice-cold water or with fragments of ice, and then bring it into an ordinary sitting-room, preferably one where many people are congregated; immediately the glass will be dimmed with dew, which after a few minutes will roll down as good sized drops.

Expt. 44. To cause Artifici l Hoarfrost.—Repeat the preced ing experiment, filling the decanter with a freezing mixture of snow and salt, or sulphate of sodium and hydrochloric acid (Expt. 21 and 24), instead of ice-cold water; the moisture deposited on the cold glass will be frozen to hoarfrost as fast as it forms, so that after a few minutes a perceptible quantity can be scraped off the glass with a knife.

Expt. 45. *To Determine the Dew-Point.*—The **dew-point** is that temperature to which moist air must be cooled in order that it will just begin to deposit dew; the greater the degree of cooling requisite to produce this, the further removed is the air from being **saturated** with moisture. Obtain a test-tube about $\frac{3}{4}$ inch in diameter and 6 inches long with a cork fitting tightly into it. By means of a "rat-tail file" bore three smooth holes side by side through the cork. Through one pass a piece of quill glass tubing about 7 or 8 inches long, so that it may reach almost to the bottom of the test-tube; through the second pass a shorter similar piece about 2 inches long, and through the third the stem of a thermometer, preferably one with the scale engraved on the glass, and with a cylindrical bulb no wider than the stem, which is $\frac{1}{4}$ inch thick or so (fig. 26). (Thermometers are specially made of this form for chemical purposes, so that they can be passed through corks, &c.) Instead of a cork an india-rubber stopper ready perforated with the three holes can be purchased. The thermometer must descend so as nearly to reach the bottom of the test-tube. Attach a long piece of thin india-rubber tubing to the projecting short piece of glass tube, pour ether into the test-tube so as to fill it about an inch deep, fix it in a holder obtainable at the instrument dealers, and insert the cork with the longer tube and thermometer bulb dipping into the ether.

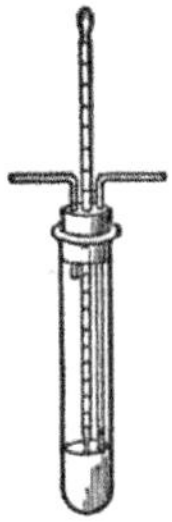

Fig. 26. Dew-point Hygrometer.

Now stand a short distance off and suck the air out of the test-tube by means of the india-rubber tube; if the cork fits sufficiently tightly air will bubble through the ether passing in through the longer tube from the outside air. This will make the ether evaporate quickly, the vapour being carried off by the current of air; in consequence the ether will be cooled down (as in Expt. 25) on account of heat being rendered latent during evaporation (Chapter XXI.), and the thermometer will gradually fall. By and by you will notice that the lower part of the tube becomes dim from the formation of dew; the temperature indicated by the thermometer when this just begins to form is the **dew-point**. To be very accurate, stop sucking the air, when the dew will begin to clear off again, the thermometer slightly rising. The average of the two readings, first when the dew began to form and second when it just clears off, may be taken as the true dew-point. For instance, suppose that the temperature of the room was 18° centigrade to begin with, and that the thermometer showed 10° when the dew first became visible and 10°·5 when the glass became clear again, then the true dew-point would be the average of 10° and 10°·5 = 10°·25 centigrade. This shows that the air of the room was not saturated with moisture at the temperature of 18°, but that as much moisture was present as would saturate it at the temperature of 10°·25.

In order to avoid the taste of the ether-vapour being sucked into the mouth, a contrivance called an *aspirator* may be conveniently used; this may be simply made from a tin can with a tap soldered in near the bottom (fig. 27), and a tightly-fitting cork in the neck through which passes a short piece of glass tubing on to which the india-rubber tube is slipped; the cork is taken out and the can filled with water; the cork is then fixed in its place and the tap

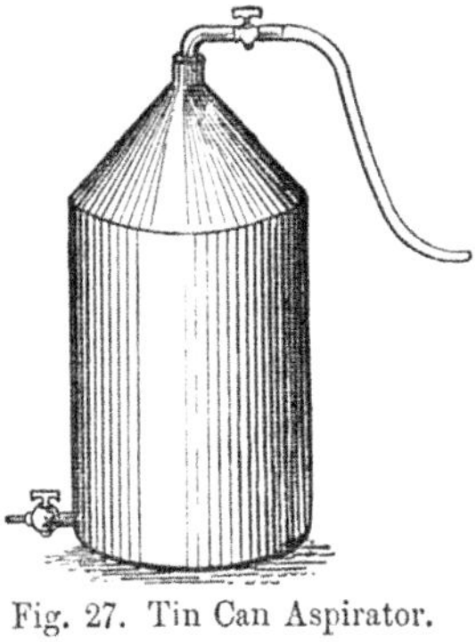

Fig. 27. Tin Can Aspirator.

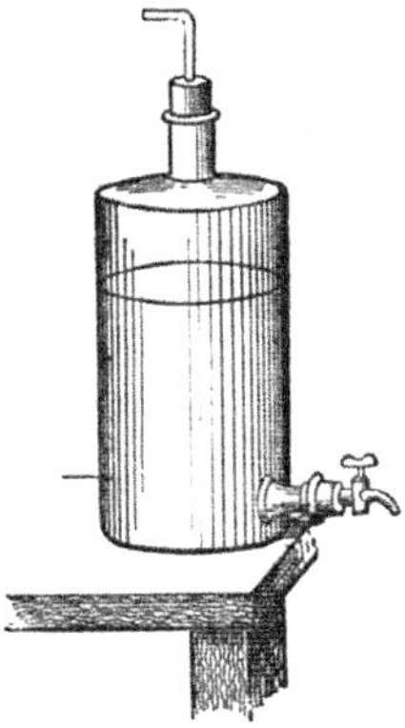

Fig. 28. Glass Aspirator.

turned on, when as the water runs out air will be sucked in in its place. This contrivance is often used in laboratories to produce a current of air through tubes, &c. Fig. 28 represents a glass aspirator consisting of a bottle with a second neck near the bottom into which a tap is fitted by means of a cork.

The cork of the aspirator ought to fit air-tight; if this is the case, on tightly pinching the india-rubber tube the water will cease to flow. If the cork is not quite air-tight it may be made so by means of sealing-wax melted over it, or with a little putty or linseed-meal paste smeared over it so as to fill up leaks and crevices. The same remark applies to the cork in the test-tube. The instrument thus made is termed "Regnault's Hygrometer," or apparatus for measuring the amount of moisture in the air. In order to see the point where dew first forms and finally clears off more readily, the bottom of the glass test-tube should be silvered outside, or fitted into a closely-fitting thin polished silver case like a thimble.

India-rubber stoppers, perforated with one, two, or more holes (fig. 29) may be purchased at the instrument dealers; these generally fit much better than ordinary corks, and are more easily made air-tight. Some little practice is requisite in order to pass a glass tube through the holes; the tube should not be so wide as to require any great amount of force to pass it through with a screwing motion, the tube being moistened with water; nor so narrow as to be leaky when the apparatus is used. With thin glass tubes, the glass may break in the hand, giving a nasty cut if incautiously forced in a direction inclined to its length. The cut-off ends of the glass tube should be smoothed with a file so as not to cut the rubber (*vide* Expt. 48).

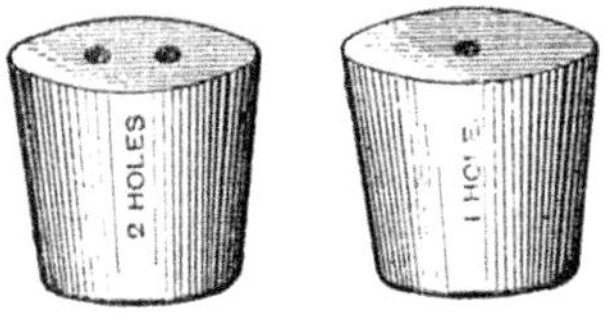

Fig. 29. India-rubber Stoppers.

Instead of india-rubber perforated stoppers, tubed caps (fig. 30) may often be conveniently employed.

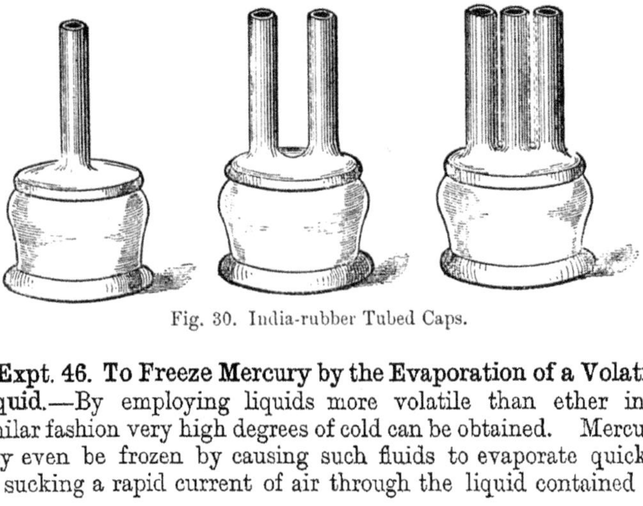
Fig. 30. India-rubber Tubed Caps.

Expt. 46. To Freeze Mercury by the Evaporation of a Volatile Liquid.—By employing liquids more volatile than ether in a similar fashion very high degrees of cold can be obtained. Mercury may even be frozen by causing such fluids to evaporate quickly by sucking a rapid current of air through the liquid contained in a tube similar to that of Regnault's hygrometer with some mercury at the bottom. Fig. 31 represents a tube arranged for this purpose, the liquid being *sulphur dioxide* (the gas produced on burning brimstone), which at ordinary warm temperatures is a gas, but condenses to a liquid on chilling below 0°. The lowest obtainable temperatures produced by the most powerful frigorific agents are obtained in somewhat similar fashions (Chapter XXI.).

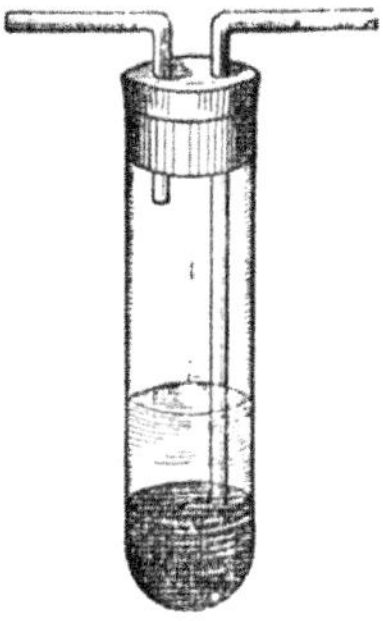
Fig. 31. Tube Arrangement for freezing Mercury.

Expt. 47. To Solder Tin and other Metals together.—The tap mentioned in the last experiment is fixed to the tin can by boring a hole in the latter, applying the tap in position thereto, and then cementing the two together by means of a little melted solder held by a hot tool usually called a "soldering iron," although it is really made of a lump of copper riveted to an iron bar to support it, and provided with a wooden handle. The construction of home-made apparatus of metal is greatly facilitated by knowing how to use this soldering tool; the chief points to notice are that all the surfaces to be soldered must be clean and bright; if dirty or rusty they must be filed or scraped bright. The "tongue" of the soldering tool (the copper "bit") must not be heated too strongly; if this should occur the solder adhering to it will probably be burnt off it and it will not then take up any more melted solder; the tip should then be filed whilst hot, dipped for a moment in a solution of *chloride of zinc*, got by dissolving a fragment of zinc in a teaspoonful of hydrochloric acid (whereby hydrogen gas is copiously evolved, Expt. 13), and then applied to a strip of solder; a little of this melts and sticks to the hot copper, wetting it with melted solder just as a cold lump of copper would be wetted with water if dipped therein. It generally facilitates the sticking of the solder to the scraped surfaces to be stuck together if they are brushed over with this chloride of zinc solution just before applying the tool; powdered resin or salammoniac is sometimes used instead of chloride of zinc solution with the same results.

To make a neat join with a soldering tool requires some practice and dexterity, but a little patience will soon enable you to manage the tool sufficiently well to make a serviceable, if not a very elegant, connection. The tool is usually heated in an ordinary fire, taking care not to make it quite red hot; a very convenient way of heating a small-sized tool is to support it on a tripod or three-legged stand of iron (fig. 7) and place a good sized Bunsen gas-burner under the copper part so that the flame laps well round it. Specially constructed gas-burners for heating soldering tools in workshops, &c., are to be bought tolerably cheaply. As a rule a spirit-lamp will not give sufficient heat, except for very small tools not applicable for good sized joints.

Expt. 48. To Bend and Cut Glass Tubing.—Another matter highly useful in the construction of apparatus is the cutting to length and bending of glass tubes. For tubes not much bigger than a stout quill (the kind mostly used) the cutting is effected simply by drawing across the tube the edge of a triangular file (which may be slightly wetted) so as to make a scratch; by pressing the tube with both hands as though you were trying to bend it at the scratched part, it will generally break clean across at the scratch (fig. 32); it is always safe to grasp the tube by means of a handkerchief or towel whilst doing this in case it should not break properly, so as to avoid possibly cutting the fingers. The

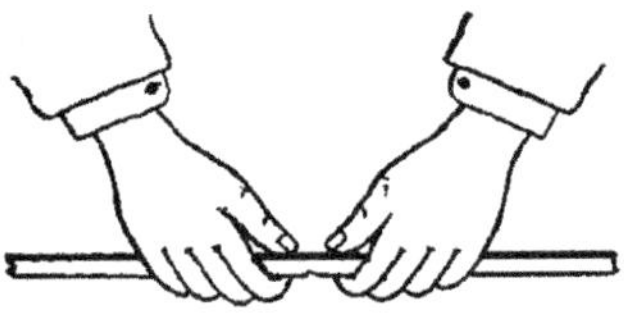

Fig. 32. Cutting Glass Tube.

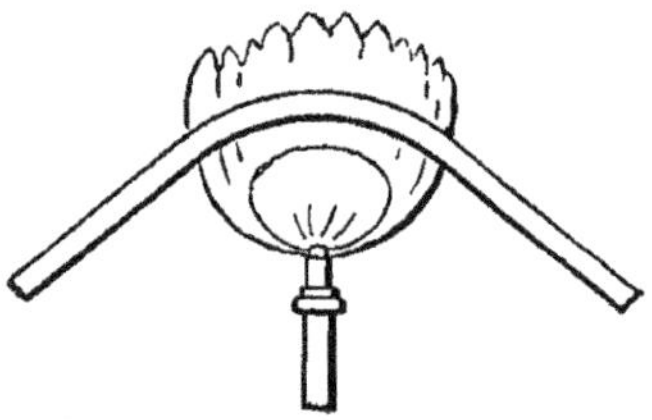

Fig. 33. Bending Glass Tube.

edges of the cut parts should be smoothed down slightly with the file, otherwise they are apt to cut and injure corks when passed through perforations therein.

In order to bend the tubing, the best way is to hold the tube horizontally in a "bat's wing" ordinary flat-flame luminous gas-burner (fig. 33) (preferable to a spirit-lamp or Bunsen gas-lamp, either of which may, however, be employed in default of a bat's wing burner), slowly turning it round so as to heat it equally; in two or three minutes a length of some 3 inches is thus heated sufficiently to form a neat regular bend on pressing the tube in the proper direction. After removal from the flame, when the bend is complete the glass should be held in the air a while by hand or a clampholder to cool; if laid on a table whilst hot the table will be burnt and very probably the glass will crack at the bend from the comparatively sudden chilling. When quite cold the film of lampblack or soot sticking to the glass is wiped off and the bend is complete. Care must be taken not to attempt to bend the tube before it is thoroughly hot and soft enough, otherwise it is likely to snap, or at least to give an ugly and misshapen bend; also not to draw the ends asunder, as this will thin the glass at the bend and make it more fragile there. If, on the other hand, the tube is wanted to be tapered off instead of bent, as is sometimes the case, the glass is heated till soft and then the ends pulled asunder by a slow steady motion, the drawn-out part being cut by means of a file when cold. In the same way thin tubes may be drawn out from thicker ones if required.

The process of making glass tubes in the first instance is carried out in much the same way. A glass-blower collects a lump of hot soft nearly molten glass at the end of a hollow "blow-pipe," and blows into this so as to swell out the glass into a sort of enormously thick bubble; another workman sticks a second rod on to the end of the bubble by means of a little melted glass adhering to the rod, and the two then rapidly move asunder so as to pull out the bubble into a long comparatively thin tube; the glass cools quickly as it is pulled out, so that when the long tube is complete the glass is sufficiently nearly solid to retain its shape.

Like most other operations in shaping molten glass, a considerable degree of practice and dexterity is here requisite to obtain a good result.

CHAPTER IV.

Direct Passage from Solid to Vaporous State and Vice Versâ.

When a vapour is suddenly chilled to a temperature not only below the boiling-point of the body in the liquid condition, but also below the freezing-point of the substance, the vapour often condenses directly into the solid condition without ever becoming visibly liquefied. Thus on a cold autumn or winter's night, the aqueous vapour in the air becomes chilled by contact with cold bodies and condenses as hoarfrost (Expt. 44) or particles of solid ice; whilst snow is similarly producible when a current of warm moist air meets with another current of intensely cold air, the condensed aqueous vapour making its appearance as a crystallised solid instead of liquid particles of mist or rain.

Many other substances behave in similar fashion; thus in the case of brimstone or sulphur, when heated to boiling, if the vapour is allowed to condense on a surface heated above the melting-point of sulphur (about 111° C.) it condenses as liquid brimstone which runs down, just as water would do on distillation. If the surface is below this temperature, the sulphur condenses as small solid crystals, just like hoarfrost; whilst if sulphur vapour is quickly chilled, it solidifies to "flowers of sulphur," partly crystallised like fine snow and partly consisting of minute spherical particles or *utricles*, resembling miniature hail.

Sublimation.

Some few solids on heating pass directly into the vaporous condition, and, *vice versâ*, their vapours condensing directly to solids without liquefying previously. When a solid is heated so that its vapours condense again in a different part of the apparatus, the process is termed **sublimation**, being to the solid exactly what distillation is to a liquid.

Expt. 49. To sublime Salammoniac and Camphor.—Place in a dry test-tube a few grains of *salammoniac* (chloride of ammonium) and apply heat; the solid substance will soon disappear from the hot bottom of the test-tube, but nearer the mouth a ring of white condensed solid matter will become visible; this solid is sublimed salammoniac.

Repeat this experiment, using a few grains of *camphor;* very similar results will be observed. When camphor is kept in half-empty bottles it generally sublimes to some extent spontaneously in the bottles, so that the upper part becomes covered over with particles of sublimed camphor often well crystallised. It is a curious fact, and one for which no clear explanation is known, that the camphor which thus sublimes usually deposits chiefly on the side of the bottle which faces the light, and comparatively little on the other side. You can easily verify this by corking up a little camphor in a bottle, and leaving it for some weeks on a shelf, where as much light falls on the bottle as possible, the camphor bottles in druggists' shops often show this peculiarity, which is not due simply to one side of the bottle being cooler than the other, as might perhaps be supposed.

Expt. 50. To sublime Iodine.—Repeat the previous experiments with a few fragments of *iodine* instead of salammoniac or camphor; you will notice that the hot part of the tube becomes filled with a beautiful deep violet-tinted vapour, which cools in the upper part of the tube to little black shining crystals. Iodine is one of the comparatively few substances that form coloured vapours when heated, as well as belonging to the comparatively small class of solids that sublime without previous fusion; its name is derived from the former circumstance (ἰώδης = Greek for violet-coloured).

Expt. 51. To sublime Carbonate of Ammonium.—Repeat the previous experiment with some solid carbonate of ammonium; sublimation will occur as before, and the upper part of the test-tube will become covered with minute crystals. The vapour given off in this experiment is *alkaline*, so that if a red litmus test-paper be moistened and placed in the mouth of the tube it will be turned blue (Expts. 142, 203).

Carbonate of ammonium constitutes the basis of ordinary smelling salts; a drop or two of oil of bergamot, or some analogous sweet-smelling essential oil, is added to the coarsely-powdered salt, so that the vapours given off consist not only of ammonia, but also of essential oil, thus modifying the odour. A little quicklime added to the mixture makes the production of ammoniacal vapours much more vigorous.

Expt. 52. To separate different kinds of Solids by Sublimation.—Just as water may be separated from saline matters by distillation (Expts. 36 and 37), so may volatile solids like salammoniac, camphor, and iodine be separated from non-volatile solid matter by sublimation.

Mix intimately together a little fine sand and salammoniac, place the mixture in a test-tube, and apply heat; a sublimate of pure salammoniac will be obtained, whilst the sand will remain unchanged at the bottom of the tube.

Mix a little common salt (dry and in fine powder) with some camphor, and heat the mixture gently in a watch-glass covered over with a second one, the two being placed mouth to mouth, taking care not to crack them by too strong a heat. By and by the camphor will have all sublimed into the upper glass to which it will stick, whilst the salt will remain in the lower one. You can prove that the lower glass contains salt by dissolving the contents in distilled water (after cooling) and adding a little nitrate of silver solution as a test (Expt. 37); if you shake up the sublimed camphor with water, and test this fluid in the same way, you will find that no salt is present therein, or at most only a trace due to the mechanical carrying away of minute particles of salt by the camphor-vapour whilst escaping, just as air might carry dust with it.

Expt. 53. Preparation of Crystals by Sublimation.—Many substances when sublimed deposit on the side of the subliming vessel as crystals; but generally the crystals are but small, and not easily examined without a magnifying glass. Some few bodies, however, form large feathery crystals when properly heated. *Benzoic acid* is one of the best examples of this, when prepared by cautiously heating the natural product *gum benzoin*, an exudation from the bark of a peculiar tree (*Styrax benzoin*), occurring in Sumatra, Siam, and other Eastern hot countries, something like the sticky gum often visible flowing from the bark of plum and peach trees in this country, but harder.

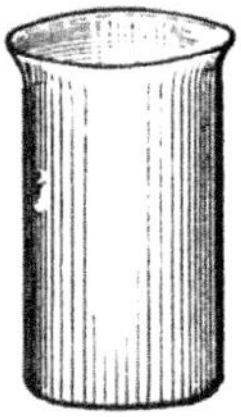

Fig. 34. Glass Beaker.

Procure two small glass shades or beakers (fig. 34) of different sizes, so that one will slide within the other, and cover over the mouth of the smaller one with two or three folds of blotting-paper tied on like the cover of a jampot, having previously placed in it an ounce or two of coarsely-powdered gum benzoin. Now slide the wider shade over the other one, and place the lower one in a small saucepan or tin-pot full of sand nearly up to the level of the lower edge of the upper shade. Cautiously heat the "sand-bath" thus formed by means of a lamp; the benzoic acid contained in the gum will be gradually volatilised along with small quantities of tarry or oily matter; the latter will mostly be absorbed by the paper, but the benzoic acid vapour will pass through the pores of the paper, and condense in the upper shade to beautiful light feathery crystalline plates of peculiar odour. When the operation is over these can easily be removed and bottled for use; the sublimed benzoic acid thus prepared is employed in pharmacy and for other purposes.

Expt. 54. Frosted Trees.—Fix in a glass shade or cylindrical jar a twig or small branch of any garden shrub, and then place the jar mouth downwards over an iron plate on which some benzoic acid is thrown, and quickly heat the plate, so as to cause the acid to sublime; the crystals will deposit on the leaves and stem of the branch, giving a pretty frosted appearance. It is usually best to heat the plate strongly first, then throw on the benzoic acid, and immediately place the jar over it to catch the vapours.

§ 3. Changes of State due to Solution not accompanied by Chemical Action.

CHAPTER V.

SOLUTION OF SOLIDS IN LIQUID SOLVENTS AND SEPARATION OF SOLIDS FROM SOLUTION.

If common salt or sugar be shaken up with water it gradually disappears, becoming **dissolved** in the water, which becomes converted into brine or syrup (as the case may be), acquiring the saline taste of the salt or the sweet taste of the sugar; the fluids thus obtained are **solutions** of sugar or salt, or more strictly speaking, "watery" or "aqueous" solutions, for other fluids besides water may be used as "solvents," or *menstrua, i.e.*, means to dissolve solids; thus "camphorated spirits" is a solution of camphor in alcohol, obtained by putting a little camphor into a bottle, half filling with strong spirit, and allowing to stand with occasional shaking. Similarly a bit of butter may be dissolved in ether, chloroform, or turpentine, or a few grains of tallow may be dissolved in benzene or benzoline; or sulphur, or phosphorus in carbon disulphide (Expts. 19 and 237).

As a general rule, a given quantity of water or other solvent will dissolve more of a particular substance when hot than when cold. If you boil some water in a flask, and gradually keep adding to it little by little alum, sugar, or other soluble matter (preferably in tolerably fine powder), you will at first see that the powder dissolves rapidly, but by and by it ceases to disappear so quickly, and the liquid must be boiled awhile before the last portion added will dissolve; finally a point is reached when the liquid cannot dissolve any more, when it is said to form a *saturated solution.*

Expt. 55. To obtain Crystals from a Saturated Solution by further Evaporation. — Dissolve common salt in boiling water little by little until the water will not take up any more. Now pour the hot liquid into an *evaporating basin*, that is, a round-bottomed circular dish (fig. 35), made of porcelain, so as to stand heat, obtainable at the instrument dealers. Place it on a tripod stand and heat it; the water now gradually evaporates, and as,

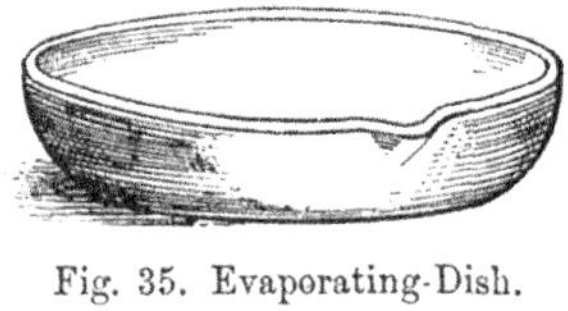

Fig. 35. Evaporating-Dish.

in consequence, there is not enough water left to dissolve all the salt present, some of the salt separates in the solid form as small crystals. If too large a flame be used, the liquid will probably "bump" violently. To avoid this, the flame should be small, and the liquid kept continually stirred with a piece of glass rod. Table salt is manufactured in this way from brine obtained naturally from springs in the salt districts, or made by dissolving rock salt in water; the liquor is allowed to settle awhile to clarify it, and is then evaporated down in pans by fire. The solution first becomes saturated by the driving off of the water, and then salt deposits in crystals. If the evaporating brine is kept well agitated, the crystals formed are small, and make *table salt;* but if the evaporation goes on more slowly and gently, the crystals are larger. *Bay salt*, used for curing fish, &c., is thus prepared in large crystals $\frac{1}{4}$ to $\frac{1}{2}$ inch across. The salt as it separates is "fished" or raked out of the brine pan by means of a ladle perforated with holes forming a strainer, or by a sort of rake, so that the solid particles of salt are retained whilst the liquid brine runs away.

It is particularly noticeable that in this operation various impurities soluble in water, and contained in the brine along with the true salt, become separated from the salt, the latter being obtained pure in the solid form, whilst the former remain dissolved in the remaining liquid or "mother-liquor."

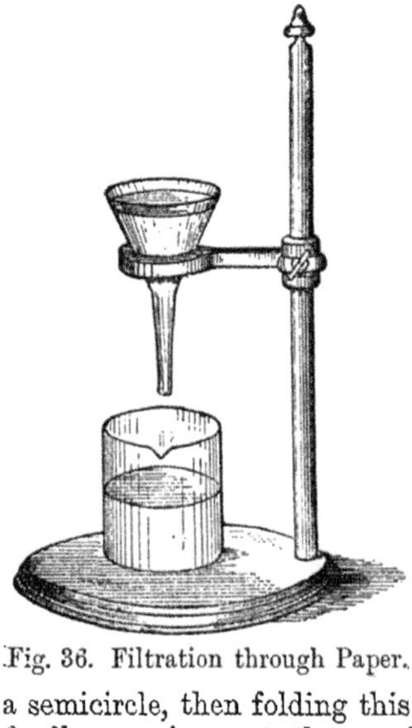

Fig. 36. Filtration through Paper.

Expt. 56. To prepare Salt from Sea-Water.—Obtain a quart or so of sea-water, and boil it down in an evaporating basin holding half a pint, adding more sea-water as the liquid diminishes in bulk by evaporation. By and by the liquor will become so strong in salt that solid crystals of salt will begin to form. When some little quantity has thus separated, let the whole cool, and separate the crystals of salt from the mother-liquor by placing the whole in a glass funnel in which a blotting-paper "filter" is placed (fig. 36). This filter is prepared by cutting a piece of blotting-paper into a circle, and folding this double, so as to form a semicircle, then folding this again so as to form a quadrant, and finally opening out the quadrant into a hollow cone, which is placed inside the funnel; the mixture of salt crystals and mother-liquor is poured into the hollow paper cone, when the liquor runs slowly through the pores of the paper, and drops out clear and bright at the end of the funnel, whilst the crystals of salt remain upon the paper. When all the liquid has run away, the crystals may be scraped off the paper on to a plate, and left to dry in the air. Fig. 37 represents a method of plaiting or folding the paper which furnishes a filter, through which the fluid passes much more rapidly than through the plain cone.

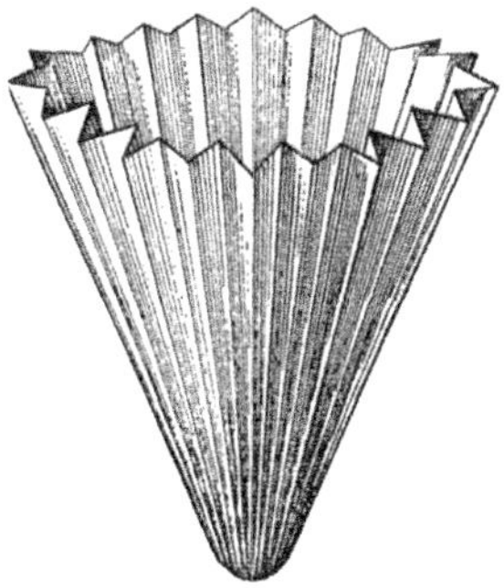

Fig. 37. Plaited Filter

Sea-water contains many substances dissolved in it besides common salt; these communicate the bitter nauseous taste observable in sea-water in addition to the saline taste of the salt. These impurities and other substances for the most part remain dissolved in the mother-liquor, so that the salt crystals separated as above are free from the bitter taste, and only taste of pure salt.

In some countries salt is largely manufactured from sea-water by penning up a large quantity of water in a shallow reservoir; the heat of the sun gradually evaporates the water, so that by and by a very strong brine is produced. If this is now boiled down by artificial heat with continuous stirring, fine crystals of salt separate, which are "fished" out, slightly sprinkled with water to wash away the small quantity of adhering mother-liquor, and finally dried for table use. If, on the other hand, the evaporation is continued by the natural heat of the sun in the reservoir, larger crystals of *bay salt* are formed.

On a hot summer's day this process may be seen going on, on a small scale, among the rocks at low water at the seaside; shallow puddles of sea-water left in the morning as the tide goes down often become sufficiently evaporated by the heat of the sun before the tide rises again to give perceptible films of crystallised salt at their edges; or they may even dry up almost entirely, giving a distinct layer of crystals of salt in the hollow where the water originally lodged. In all probability natural rock salt was produced in byegone years in this sort of way.

The Dead Sea is a lake into which river water flows (the Jordan), but which has no outlet, all the water brought in by the river becoming evaporated by the heat of the sun. As the river water contains more or less salt, the result is that during many centuries past salt has been accumulating in the lake, which is now a very strong briny liquor quite undrinkable, like sea-water. In the shallow parts of the lake the sun evaporates the water so quickly that crystals of salt are continually forming, and sinking to the bottom, forming a layer of solid salt.

In many parts of America a similar action has gone on with lakes into which streams once flowed containing other forms of saline matter besides common salt; and the dried-up beds of these lakes are consequently valuable deposits of the saline matters thus left.

The substances termed *borax* and *nitrate of sodium* (or Chili saltpetre) are obtained in large quantities from sources of this description.

The "mother-liquors" from which salt has separated by crystallisation, obtained by evaporating down natural brine or sea-water, are sometimes used medicinally, as the substances communicating the bitter taste are, generally speaking, strongly purgative. "Friedrichshall Wasser" (water from the spring at Friedrichshall), and "Kreutznach Bittern" (bitter mother-liquors from the Kreutznach spring) are fluids of this kind. The mother-liquors of the Cheshire salt-works contain substances rendering them valuable for the cure of rheumatism, &c., when used as baths; and the same remark applies to many other saline springs. The comparatively scarce substances *iodine* and *bromine* are often contained to a considerable extent (combined with other substances) in such mother-liquors, which are consequently valuable as a source of these chemicals.

Expt. 57. To separate Common Salt from other Soluble Substances by Crystallisation.—Take a couple of ounces of common salt and dissolve it in a sufficient quantity of hot water; add to the liquid a few drops of a solution of "perchloride of iron," otherwise termed *ferric chloride*, and also a few drops of diluted *hydrochloric acid*. You thus obtain a compound solution containing chiefly salt, but also small quantities of other soluble matters. Obtain some blue *litmus test-papers;* these are made by soaking the substance called litmus (a colouring matter derived from a vegetable substance termed the *litmus lichen*) in spirit, so as to obtain a "tincture" thereof, *i.e.*, a solution in alcohol of those constituents soluble in that medium; a blue or purple fluid results, in which fine blotting-paper is dipped and dried. The coloured paper cut into slips can be used as a **test** for hydrochloric acid, which, like all other acids, possesses the peculiar property of turning this particular blue colouring matter *red*, even when only present in very small proportions. Consequently, if you dip a blue-litmus test-paper in the compound solution prepared as above directed, the paper will be reddened by the acid present. Also obtain a little of the yellow salt termed "yellow prussiate of potash" (more accurately *ferrocyanide of potassium*), and dissolve it in water: the solution or test-liquor thus obtained *serves as a test for iron*, for when mixed with a liquid containing iron in solution (such as the solution of perchloride of iron employed as above) double decomposition takes place, and a precipitate is formed, consisting of a beautiful blue substance termed *prussian blue;* so that if a little of this test-liquor be added to a portion of your compound solution (of salt, hydrochloric acid, and perchloride of iron) in a wine glass, the liquor will become blue like blue ink, depositing on standing a blue powder consisting of precipitated prussian blue.

Now boil down the compound solution to a small bulk in an evaporating basin, so as to obtain a considerable proportion of the salt in the form of separated crystals; let the whole cool, and then filter these crystals as directed in the last experiment. Sprinkle a few drops of water over the crystals, so as to wash out most of the adhering mother-liquor, and then dissolve the crystals in pure distilled water by the aid of heat. On testing the solution thus obtained with the litmus test-paper, it will be found that the paper is either not reddened at all, or only very slightly, in comparison with the effect produced by the original compound solution; and on testing it with the ferrocyanide of potassium test-liquor, either no blue colour at all will result or only a comparatively faint one. On the other hand, the mother-liquor that has run through the filter will give a stronger acid reaction with the litmus paper than the original compound solution, and will react more powerfully with the ferrocyanide test-liquor producing prussian blue, for the obvious reason that the boiling down has concentrated the fluid. It is thus evident that the salt crystals that formed have been nearly, if not completely, separated from the hydrochloric acid and perchloride of iron. By boiling down the solution of these crystals obtained as above, until it deposits crystals for the second time, and collecting them on a second clean paper-filter as before, you will obtain a more complete separation, if the first batch of crystals were not quite pure.

This process of crystallisation and re-crystallisation (or dissolving the first crystals in fresh water and crystallising afresh) is most extensively employed for the purification of chemical products manufactured for various purposes in the arts, as being at once the cheapest and most effective way of separating impurities.

Expt. 58. To make Sugar Candy.—When a saturated solution made boiling is allowed to cool somewhat the same thing happens as when moist air is chilled; just as the cooler air is incapable of retaining intermixed with it all the water-vapour at first present, so that the surplus deposits as liquid water, forming mist or dew (Chapter III.); so a hot saturated solution, when allowed to cool, allows some of the dissolved solid to reappear in the solid form, generally in the form of crystals of larger or smaller size according to circumstances; these crystals have usually a tendency to deposit on any foreign solid body placed in the cooling fluid, so as to coat it over with little crystallised masses.

Boil some water in a small saucepan, and add sugar to it as long as any dissolves, taking care not to burn the syrup. Lay some pieces of stick across the saucepan with bits of string hanging down from them so as to dip into the syrup; or put thin bits of stick into the pan, so as to stand nearly upright. Now leave the whole at rest for 24 hours; at the end of this time the strings and immersed sticks will be coated over with crystals of sugar, forming sugar candy; by leaving these in the syrup some days longer, the crystals will gradually grow larger.

Sugar that has been overheated, or brownish imperfectly purified sugar (containing treacle, &c.), generally requires a longer time to deposit crystals than pure white loaf-sugar, as the products of the action of heat on the sugar, and the other impurities, have the power of stopping more or less the separation of solid crystals from the solution. Crystallised fruits are made by steeping dried fruits in syrup, so that sugar crystals form all over them.

Fig. 38. Alum Crystals.

Expt. 59. To make Crystallised Baskets.—Very pretty objects may be made by placing wicker-work baskets or skeleton figures, such as a crown made of wire lapped over with cotton or wick, in a saturated solution of alum in hot water and allowing to stand for some days; the alum separates in well-defined crystals all over the surface, the crystals so formed growing larger the longer they are left immersed, because the water slowly evaporates. Fig. 38

represents a block of alum crystals slowly deposited from a large bulk of fluid. Many other substances may be used instead of alum to produce the same kind of effect; thus the salt called *sulphate of copper* will coat the objects with beautiful blue crystals, and *ferrocyanide of potassium* (the salt used in Expt. 57 as a test for iron) with pretty yellow ones.

Expt. 60. To frost Glass Windows.—A peculiar frosted appearance is often given to windows and other glass objects by brushing over the glass a hot solution of some salt that will crystallise easily, such as *Epsom salts*, or *alum;* as the solution cools the crystals form, giving a frosted appearance to the glass; and for the same reason, glass frosted in winter time is covered over with fine ice crystals (Expts. 18 and 44). If a little size be added to the liquid, the crystals are less easily rubbed off when dry; whilst a little whiting (tinted to fancy with prussian blue or other pigment) makes the frosting more opaque, preventing people from seeing into the room from outside.

Expt. 61. To obtain a Colourless Solution of a Yellow Substance and Yellow-coloured Crystals from a White Solution.—As a general rule, if a colourless substance is dissolved in a colourless solvent, unless some chemical change takes place (Chapter X.), a colourless solution results; and conversely, if a substance crystallise from a colourless solution, the crystals are generally also colourless; whilst if a coloured substance be dissolved in water or other appropriate solvent, the solution is usually of the same colour as the solid substance dissolved. Thus blue copper sulphate, when dissolved in water, forms a blue solution; yellow ferrocyanide of potassium produces a yellow fluid; whilst colourless common salt, alum, and loaf sugar give rise to colourless solutions in each case. There are, however, some exceptions to these general rules; thus *iodide of lead* is a yellow solid,* but on boiling this with the right quantity of water it dissolves, forming a clear colourless fluid. On allowing this fluid to cool again, small shining bright yellow crystals of solid iodide of lead form, like minute spangles in the liquid, these redissolve on heating to a colourless fluid, and reappear again on cooling, the changes being capable of alternate repetition as often as you please, provided that a little water be added from time to time to replace that lost by evaporation during the heating.

* Iodide of lead is easily obtained by putting in a flask some clear filtered solution of nitrate of lead or sugar of lead, and then adding some clear solution of iodide of potassium; a chemical change takes place, one result of which is that iodide of lead is formed as a solid substance, gradually subsiding as a yellow "precipitate" Add some distilled water and boil, adding more water if the whole of the yellow precipitate does not dissolve at first on boiling. Compare Expt. 11.

Expt. 62. To obtain differently Coloured Solutions by dissolving the same body in different Solvents.—Another remarkable kind of exception to the above mentioned general rule is noticed when differently tinted fluids are obtained by employing different liquids to dissolve the same solid, no chemical change taking place during the action.* Place in a small bottle a few small particles of *iodine* (Expt. 50), and dissolve a few crystals of *iodide of potassium* in a little water in another glass; pour the colourless solution of iodide of potassium into the bottle and shake up; the iodine will soon dissolve, forming a liquid of a yellowish brown tint if sufficiently strong, and of lighter shades, somewhat resembling sherry of various kinds, when weaker. (Similarly, strong spirit of wine dissolves iodine, forming a "tincture" of iodine, or alcoholic solution of iodine, of much the same range of colour as before.) Now add a teaspoonful of chloroform to the watery fluid, shake up, and then allow to stand a few minutes; the chloroform will soon settle at the bottom of the bottle, being a liquid which will not readily dissolve in or mix with water (Chapter VIII.); but before doing so it dissolves a large portion of the iodine, and in consequence acquires a beautiful *violet* colour; so that you can obtain two differently coloured solutions of iodine in the same bottle—one a solution in chloroform, of violet hue; the other in watery iodide of potassium, of brownish yellow tint.

Expt. 63. Circumstances modifying the Rate at which Crystals form from Solutions. Supersaturated Solutions.—Sometimes when a hot saturated solution of a crystallisable substance is cooled slowly and without agitation, it does not deposit crystals all at once, but still remains clear and perfectly fluid, even though there is more substance dissolved in the liquid than could be taken up by the solvent at the temperature of the room by simply agitating it with the crystallisable body. Sugar is one of numerous substances exhibiting this property, to which the term **Supersaturation** is applied. A very strong hot syrup, made by dissolving pure sugar in boiling water, may often be cooled down completely without depositing crystals; but on standing for some time the supersaturated cooled liquid gradually forms crystals of sugar candy, more especially on the surface of any foreign solid body introduced for the purpose, such as the strings and sticks used in Expt. 58.

The salt termed *sulphate of sodium* (or *Glauber's salt*) is a par-

* When chemical changes take place during solution, as described later on, all sorts of alterations in colour may take place under different circumstances; thus the element called chromium (χρωμα = colour) derives its name from the fact that by chemical action it can be converted into bodies of most diverse colours, forming solutions of almost all the colours of the rainbow.

ticularly good example of a substance apt to form supersaturated solutions. If this salt be added in powder little by little to boiling water contained in a clean glass flask until the water will dissolve no more, and the liquid be then allowed to cool gradually and without any shaking or agitation, all chance of dust falling in being excluded by placing in the mouth of the flask a lump of clean cotton wool, it will generally happen that no crystallisation at all takes place during cooling, the contents of the flask being perfectly fluid even when completely cool. If the cotton wool stopper be now taken out of the neck it will often happen that a particle of dust will fall in which will cause the formation of crystals round the particle, which thus acts as a "nucleus" or centre round which solid matter concretes. The crystallisation thus commenced rapidly spreads throughout the whole of the liquid, which thus becomes almost entirely solidified in the course of a minute or so, so that the flask can be turned upside down without any fluid escaping. If no dust fall in on removing the stopper, the crystallisation can be started by throwing in a fragment of wood or stone, or introducing a stick or a glass rod; if a thermometer be thus employed, it will be seen that the mercury rises very perceptibly whilst the crystallisation progresses.

This experiment in fact is the converse of producing cold by freezing mixtures of snow and salt, or by dissolving saltpetre, &c., in water (Expts. 21 and 24); here a liquid becomes solidified, and heat is developed, so that the mass becomes warmer than at first in consequence of the action; with snow and salt, the two solids become a liquid brine, and heat is absorbed, making the whole colder than at first. Just in the same way, when saltpetre is dissolved in water, or when sulphate of sodium is stirred up with hydrochloric acid, so as to produce a chilling action, the solid matter employed becomes practically liquefied, and therefore cold results.

Expt. 64. Illustration of Comparative Solubility in different Solvents.—It is most frequently observed that whilst a given substance may be freely soluble in one kind of solvent, it dissolves much less readily, or even not at all, in another kind. For example, common salt dissolves readily in water; but if added to alcohol, especially if pretty strong, very little indeed is dissolved. Many other substances (such as *sulphate of copper*) behave in the same way; so that if some alcohol be added to a cold saturated solution of the substance in water and the whole well shaken together, a quantity of crystallised substance will often be thrown out of solution, because the weak spirit resulting from the mixture of the alcohol and water present is incapable of permanently dissolving all the solid matter, and therefore the surplus separates in crystals. Mixed fluids such as these often show the phenomenon of supersaturation to some extent; crystals not appearing all at once on mixing strong alcohol with the watery solution of the dissolved body, but depositing on standing, especially if the fluid is stirred well.

Some solids are more easily dissolved by strong alcohol than by water or weak alcohol, so that by adding water to a concentrated alcoholic solution of the substance a "precipitation" of crystals is brought about, because the resulting weak spirit is unable to dissolve the whole. Camphor forms a good illustration of this; if some crushed camphor be placed in a bottle with a little nearly absolute alcohol, and the bottle corked and shaken from time to time, a "tincture of camphor" or "camphorated spirit" is produced as the camphor dissolves; if a few drops of this be poured into a little water, and the whole shaken up, the camphor separates again as a crystalline solid.

In the manufacture of ordinary hard washing-soap, fatty and oily matters of various kinds (tallow, lard, palm oil, olive oil, cocoanut oil, and such like) are boiled with solution of caustic soda, whereby a chemical change is produced, the sweet liquid termed *glycerine* being formed, together with soap (Expt. 276); both of these products dissolve in hot water. In order to separate the soap from the water and glycerine, advantage is taken of the fact that soap will not dissolve easily in brine; so that by adding some strong brine to the liquor (or by throwing a little solid salt in, which comes to the same thing, as the salt dissolves and forms brine), the soap is rendered insoluble (just as the camphor is thrown out of solution in alcohol by adding water), and separates as a soft pasty mass, which floats up to the top of the brine; this is then skimmed off whilst hot, and poured into large moulds, in which it concretes on cooling to blocks of solid soap, which are then cut up into bars and tablets. The glycerine is not thrown out of solution in this way, but is retained dissolved in the brine, from which it is separated by boiling down and allowing the salt to crystallise out, thus leaving impure glycerine, which is subsequently purified by distillation and other processes (Expt. 277).

Expt. 65. To separate Soap from Solution.—Cut up a quarter of an ounce of ordinary yellow soap into small fragments, and place them in a flask with a cupful of distilled water; heat the whole gently without boiling, with occasional shaking; by and by the soap will dissolve up to a liquid, which will be more or less clear according to the nature of the soap. Now throw in a tablespoonful of common table salt in powder and stir or shake; the salt will dissolve in the water, and the soap will separate in soft pasty flakes.

Expt. 66. To separate a Mixture of Sugar and Sand by means of a Solvent.—The property possessed by certain liquids of readily dissolving substances which are almost insoluble in other liquids is often made use of for the purpose of separating substances when mixed together, not only in carrying out analytical operations in the laboratory, but also in various manufactures; *e.g.*, the preparation of many kinds of dyeing materials. Sometimes sugar is adulterated with sand; such an admixture is easily recognised by treating the mixture with hot water, when the sugar dissolves, leaving the sand unaltered.

Expt. 67. To separate Salt and Clay.—In the same way salt and dried clay or brick-dust may be separated. Mix a tea-spoonful of each in a mortar, and then pour a cupful of warm water on the powder. The salt will dissolve, whilst the brick-dust will be unaltered. By filtering the fluid through a paper filter supported by a funnel (Expt. 56), a clear solution of salt will be obtained, in which the presence of salt may be recognised by testing with nitrate of silver (Expt. 37), or by evaporating down (Expt. 55).

Expt. 68. To separate three Solids by two different Solvents.—Take some *camphor*, a little of the blue salt termed *sulphate of copper*, and some fine white *sand*, and rub the whole together in a mortar until a uniform bluish powder is obtained. Put some of this into a bottle with some strong alcohol, and cork the bottle and shake it up a few times; let the undissolved matter subside, and pour it off the clear spirit; if not sufficiently clear it may be passed through a paper filter (Expt. 56), and the clear filtered liquid set aside in a basin; by and by the spirit will evaporate, and the camphor will be left; or water may be added to the filtered spirit, when the camphor will be precipitated in crystals. Repeat the treatment with alcohol once or twice, to make sure that *all* the camphor is dissolved. Next pour some warm water into the bottle and shake up; allow to subside, and pour off the bluish watery fluid, filtering it if not quite clear; pour more water into the bottle, and repeat the process until the water no longer takes any blue colour. You will now have the sand just as at first, wholly undissolved by either of the two solvents successively applied; the camphor has been dissolved out by the first treatment with spirit, and the sulphate of copper is obtained in the solid form by evaporating down the blue watery solution until the water is driven off.

Solubility of Liquids in Solids.

It is a matter of everyday observation, that substances of a porous nature will absorb fluids; thus a sponge, a towel, a piece of unglazed earthenware, or a lump of charcoal, will each absorb a considerable quantity of water or other analogous fluid without allowing it to drain completely away by gravitation on standing. This phenomenon, however, is not exactly a case of solubility, inasmuch as the action only takes place **on the surface** of the minute cavities or pores with which such substances are permeated, being in fact simply a variety of the action of minute tubes in taking up fluids termed **capillarity**, which again is only one result of what is known as **surface action** (Chapter XVII.). Certain solid substances, however, exist, which although so compact and destitute of porosity that no gas or liquid can be forced through them by simple pressure (unlike such things as a porous earthenware jar, which will allow water to pass slowly through it by weight alone), will nevertheless gradually allow certain liquids applied to the outside of a lump of the solid to pass inwards, and become disseminated throughout the whole mass. Thus solid articles of gold and silver allow liquid mercury to creep into their interior should this fluid wet their outsides; one result of which is that the precious metal becomes considerably brittle.

Expt. 69. To penetrate Gold by Mercury.—Wet a small gold coin with mercury (by well rubbing), and allow it to stand some time; it will become so brittle that it may easily be broken in half; and if the fractured surfaces are examined, it will be seen that the mercury has crept inwards, and become distributed throughout the entire mass, the action much resembling the passage of water into a mass of blotting paper, excepting that with the blotting paper the pores are relatively large, and will admit any fluid capable of wetting the paper; whereas the gold, if porous at all to mercury in the same sense that unglazed pottery is to water, is only porous to that fluid, and will not allow other fluids to pass into the interior of a solid lump, even though these fluids are freely capable of adhering to, or wetting the exterior surfaces of the lump of gold.

Silver is similarly penetrated by mercury, though not quite so readily as gold.

Caution.—In certain experiments, where fluid mercury is used (*e.g.*, collecting gases, as in Expt. 280), it is well to take off one's watch and chain, rings, &c.; otherwise these may be seriously damaged by the action of little drops and splashes of mercury accidentally coming in contact with them.

Compact india-rubber, although not porous to gases or liquids in the ordinary sense, is capable of absorbing or dissolving small quantities of certain fluids (*e.g.*, benzene), without in any way losing its solidity, the fluids penetrating the entire mass in much the same way as the mercury does the gold in the above experiment. When somewhat larger proportions of the same fluids are employed, the rubber becomes more or less softened and disintegrated, and with pretty large quantities entirely dissolved; the solutions thus prepared are employed for rendering textile fabrics waterproof, by impregnating them with the liquid, and allowing the solvent to evaporate, thus leaving the fabrics coated over with a film of rubber. Such goods are often said to be *Macintoshed*, from the name of the inventor of the process.

Gold treated with large quantities of mercury is acted on in just the same way, a semisolid or liquid *amalgam* (or solution of gold in mercury) being formed; such a solution is sometimes used for gilding steel and other metals, the amalgam being applied to the surface of the metal to be gilded, and the mercury then volatilised by heat, leaving a film of solid gold on the surface of the other metal, which is then brightened or "burnished" by rubbing with a hard substance. This method of superficially coating with gold is termed **water-gilding** (not because water is employed, but because the gold is virtually rendered fluid by the mercury), and is quite a different process from **electro-gilding**, which has of late years largely superseded the mercurial process, partly because it is cheaper, and partly because the water-gilding process is a somewhat dangerous and unhealthy trade, owing to the poisonous action of mercury vapours on the workmen when inhaled.

Expt. 70. To absorb Water by a Solid.—Place a piece of hard glue (impure gelatine) in a cupful of cold water, and allow it to stand for some time; the water will gradually pass into the solid glue, causing it to swell considerably and increase in weight; in time, especially with hot water, the glue will be entirely disintegrated and apparently dissolved; on allowing the hot liquid to cool and stand, the water will gradually evaporate, and the mass will first thicken and set to a jelly (**gelatinise**), and then the jelly will slowly **indurate** or dry up.

Lumps of gum and many other substances possess the same property; animal membranes, such as bladders, intestines, sausage skins, and the like, are constituted of substances closely akin to gelatine, and behave in much the same way. Substances of glue-like characters are often spoken of as **colloids**, and exhibit certain remarkable differences from crystallisable solids (*vide* Expt. 93).

CHAPTER VI.

Solution of Gases in Liquid Solvents and Separation of Gases from Solution, no Chemical Action taking place.

In some of the previous experiments we have seen that besides ordinary air (which is obviously different in many respects from coal-gas) a number of substances can be obtained by chemical action resembling air and coal-gas in the one respect of physical texture, but otherwise considerably different not only from these two substances, but also from each other. Thus hydrogen (Expt. 10), and sulphuretted hydrogen (Expt. 13), obtainable as above described, are specimens of a much larger number of similarly-textured substances obtained by means of various chemical changes, all of which are generically grouped together as **Gases**.

Strictly speaking, all gases are really the steams or vapours given off from so many highly volatile liquids, just as actual steam or water vapour (Chapter III.) is given off on boiling water, and differing from ether vapour as regards volatility in much the same way as ether vapour differs from water vapour, or water vapour from mercury vapour; just as by chilling any one of these vapours the substance resumes the liquid condition, so may any one of the gases known to chemists be condensed to liquids by sufficiently lowering the temperature; but in some cases the amount of chilling required is so great that it was only within a few years that scientific appliances became sufficiently perfected to enable this to be done; previously to that, although it was inferred by analogy that all gases could be condensed to fluids under proper conditions, still some half dozen (of which hydrogen was one) had resisted all attempts to liquefy them, and were accordingly known as the "incondensible gases." At present no such distinction is recognised; and a "gas" simply means a vapour requiring a lower temperature to condense it to a liquid than the ordinary atmospheric temperature in temperate climates. Sulphur dioxide, indeed, liquefies at but little lower than the freezing point of water; and both this gas and another comparatively easily liquefiable one, ammonia, are employed for various purposes in the arts in the liquid state.*

* Principally for freezing machines, in which they are employed in somewhat the same sort of way as in Expt. 46.

When a gas and a liquid are brought together, the liquid generally dissolves more or less of the gas, forming a solution thereof; several such solutions have already been used in the experiments above described, for example, solution of *hydrochloric acid gas* (Expt. 10) and solution of *sulphuretted hydrogen* (Expt. 12). Ordinary aerated water (soda-water) is a solution of *carbon dioxide* or *carbonic acid gas* in water. When the cork of the bottle is taken out, the liquid effervesces violently; this is because the solution has been made under the influence of a considerable degree of pressure, which enables the water to dissolve much more of the gas than it otherwise could do. The bottle is filled with the fluid solution and corked by a peculiar kind of machine whilst this pressure is maintained, so that the liquid in the corked bottle is still under pressure; for which reason, soda-water bottles sometimes burst, especially when kept in a warm place, which tends to increase the internal pressure. When the cork is extracted the pressure is relieved, and as there is more gas present than the water can permanently dissolve at the ordinary atmospheric pressure, the surplus escapes as gas with effervescence, just as a solution of a solid when saturated hot deposits some of the solid in crystals on cooling (Chapter V.), and just as air saturated with moisture deposits some of that moisture as dew on cooling (Expt. 43). Precisely the same remarks apply to aerated water kept in siphons; instead of a cork, these are fitted with a glass tube arranged like a feeding-bottle, with a valve at the top worked by a handle or trigger (fig 39). On pressing the trigger the valve is opened, and the pressure inside forces out the liquid, which passes up the tube and through the valve and so out of the nozzle.

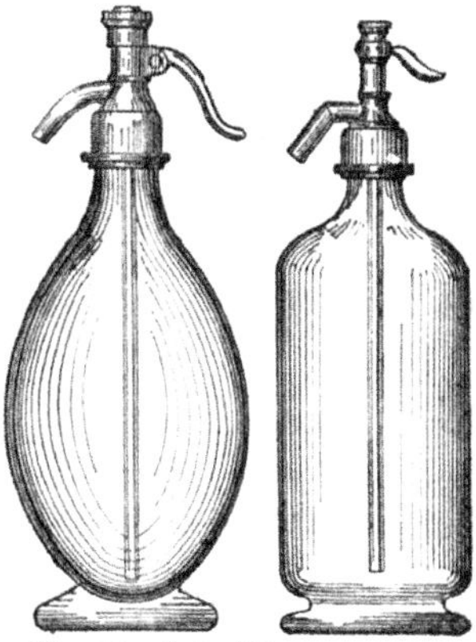

Fig. 39. Soda-Water Syphon.

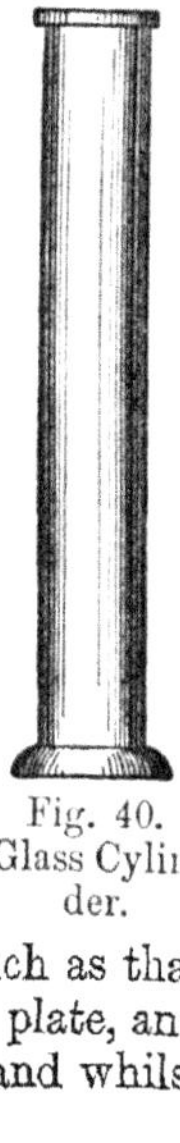

Fig. 40. Glass Cylinder.

Expt. 71. To show that Soda-Water Gas extinguishes a Candle.—Get a large tumbler or a wide-mouthed quart bottle, or better still, a glass cylinder, such as that represented in fig. 40, capable of being covered by a glass plate, and holding about a quart. Uncork a bottle of soda-water, and whilst it is effervescing pour the liquid into the cylinder so as to fill it one-third or one-quarter full; or similarly, partly fill it from a siphon. The carbon dioxide gas escaping from the solution will displace the air from the upper part of the cylinder, so that by lowering into it a lighted taper tied to a wire or string as in fig. 41, the light will be extinguished, because carbonic acid gas does not possess the power of keeping up the chemical actions effected by the air during the process of burning.

Fig. 41. Taper.

Bottled beer, zoedone, champagne, sparkling Moselle, aerated lemonade, and similar effervescing drinks, evolve gas on being poured out, for exactly the same reason as soda-water; they all consist of fluids (mostly water containing more or less alcohol, sugar, and flavouring matters, &c., dissolved therein), in which more carbon dioxide is also dissolved under pressure in the bottle than the liquid can retain dissolved when the pressure is relieved by uncorking. In the case of many such liquids the aeration is artificially conducted in a machine similar to that used for soda-water; and much inferior or spurious sparkling wine is thus manufactured. Genuine champagne, like bottled beer, on the other hand, becomes impregnated with gas in virtue of chemical changes (fermentation, Expt. 189) going on inside the bottle after the liquid has been introduced and the bottle securely corked up, no aerating machine at all being used; but the end result is the same so far as producing an effervescent fluid is concerned.

Expt. 72. To make Aerated Water for Home Use.—An instrument called a "gasogene" is often employed for the production of aerated water in small quantities for table purposes (fig. 42). This consists of a large hour-glass shaped vessel of thick glass with a piece of wide porcelain or glass tubing cemented into the narrow part joining the two bulbs, A and B, in such a way that liquid will stand in the upper bulb without running down into the lower one.

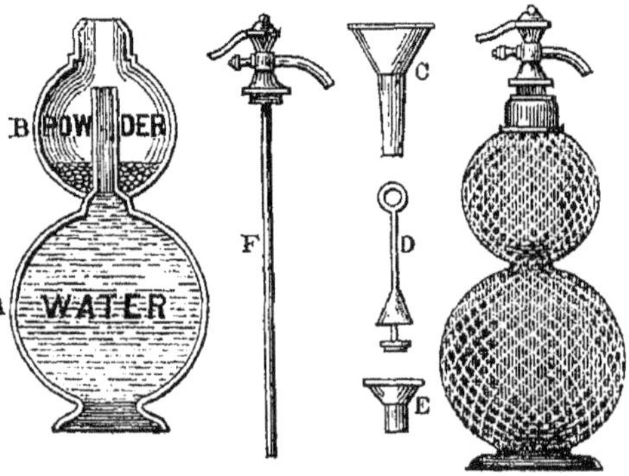

Fig. 42. Gasogene.

A thinner piece of glass tubing, F, passes through this short wide tube so as to reach down to the bottom of the lower bulb; this glass tube is surmounted with a trigger valve and jet exactly like that of a soda-water siphon, and is made to screw air-tight into the top of the upper bulb. To charge the machine this screw is undone, and the thin glass tube with valve attached, F, lifted out. A long funnel, C, provided for the purpose is inserted with the narrow end inside the porcelain wide tube, so that by pouring drinking water through the funnel this lower bulb is filled with water. The funnel is now removed, and a shorter one, E, substituted, and the upper part of the porcelain tube loosely covered with a conical piece of metal introduced through the shallow funnel E, by a wire D. Some crystals of tartaric acid and a proportionate quantity of "bicarbonate of soda" are now poured in through the shallow funnel, the quantity depending on the size of the gasogene; the cone prevents these substances from passing down through the porcelain tube into the water, so they remain in the upper bulb, B (which must be empty of fluid). The cone, D, is now removed, and the thin glass tube and valve, F, introduced in position, and screwed up tight. The gasogene is then turned on its side, so that some water flows from A (which was at first the lower bulb) into the other, B, through the porcelain tube. As soon as the water comes in contact with the solid tartaric acid and bicarbonate of soda, it begins to dissolve these substances, with the result of setting up a chemical action that does not ensue as long as no water is present; the gasogene being now placed erect, the effect of this chemical change is to cause carbon dioxide gas to be evolved, which gradually passes up the porcelain tube, and is ultimately dissolved by the water. A considerable degree of pressure is thus generated, the effect of which is that when the trigger valve is opened the water in the lower bulb is forced up the narrow pipe, and escapes at the jet just as the aerated water from a siphon, effervescing in precisely the same way.*

By placing a teaspoonful of fruit syrup, lemon juice in which sugar has been dissolved, syrup of ginger, or such like flavouring matter, in the tumbler in which the aerated water is received, delicious summer drinks are obtainable, especially if a little crushed clean ice is also added to cool the beverage.

* The portable fire-extinguishing machine known as an "Extincteur" acts on precisely the same principle; by striking a knob on the outside certain chemicals become mixed together with water inside in such a fashion as to generate gas, and set up sufficient pressure to force out the effervescing water in a stream through a flexible delivery tube attached on turning on the tap connected therewith.

Expt. 73. To make Sherbet (Lemon Kali).—In the gasogene above described, the substances employed to generate the gas are kept separate from the aerated water produced, the former being in the upper bulb, and the latter in the lower one; but when the presence of these substances in the water drunk is not objectionable, a much more simple method suffices to prepare an effervescing beverage. Thus a mixture of powdered *tartaric* or *citric acid*, *bicarbonate of soda*, and *sugar* may be made and kept in a dry bottle without change; but on adding a teaspoonful to a tumbler of cold water and stirring up, a vigorous effervescence ensues, and a sort of aerated lemonade results. Such mixtures, flavoured with a little essence of lemon or ginger, &c., are largely used for summer beverages, under the names of *sherbet*, *lemon-kali*, &c., but are apt to exert a medicinal action, owing to the presence of the substance formed by the action of the acid on the bicarbonate of soda (*tartrate* or *citrate of soda*). *Seidlitz powders*, and various patent medicines that effervesce when thrown into water, are essentially mixtures of this kind, together with some powdered saline matter, such as *Epsom salts* (sulphate of magnesia) added to produce purgative or other medicinal action.

Many natural spring waters are effervescent, on account of chemical changes going on underground, in virtue of which the water becomes charged with carbonic acid under some degree of pressure (owing to the depth), just as in a gasogene. When much mineral matter is present in the water, a medicinal effervescent water results; many such springs exist in different parts of the world, to which invalids and people out of health resort in large numbers for treatment. When only a little mineral matter is present, the water is often bottled for use as an ordinary beverage; *Apollinaris water* and many similar natural slightly effervescent table waters, are fluids of this kind.

Water impregnated strongly with carbonic acid is capable of dissolving certain substances insoluble in plain water; much of the lime giving "hardness" to natural water is dissolved in this way. When iron is taken into solution in such water, a "chalybeate" spring is formed; when very much lime is present, this water often forms a calcareous deposit on all objects with which it comes into contact, owing to some of the lime being thrown out of solution (in the form of carbonate of lime), when the water is exposed to the air. What are known as "petrifying springs" mostly act in this way; a bird's nest, a twig, ferns and flowers, baskets, and similar objects, when exposed to such water, become gradually covered over with a stony coating, and are apparently converted into stone or "petrified." A well-known well of this description is at Knaresborough near Harrogate, in Yorkshire, and other similar ones in Derbyshire, and many places in Continental Europe and elsewhere.

Carbonic acid is by no means the only gas contained in natural waters; sulphuretted hydrogen (Expts. 13 and 14) is sometimes present, as in some of the Harrogate and many Continental springs, giving to the water a strong smell of rotten eggs. Waters of this description are extensively employed in the cure of certain complaints, partly by drinking the water, partly by bathing therein. In some places, mostly in volcanic districts, springs exist containing sulphurous acid (Expts. 117, 161, and 238); whilst combustible gases, somewhat akin to ordinary coal gas, are given off from deep wells in the American petroleum districts and elsewhere, sometimes in such quantity as to be capable of being largely used as fuel for raising steam, smelting metals, and such like manufacturing purposes.

Circumstances modifying the Solubility of Gases in Liquids.

It does not at all follow that because a particular liquid, such as water, will dissolve one gas freely, that therefore it will act similarly with a different gas; just as water will dissolve salt or sugar in considerable quantity, but will fail entirely to dissolve chalk or sand, so is each particular gas possessed of a different power of dissolving in water; whilst just as a particular substance, such as camphor, will hardly dissolve in water at all, but is freely soluble in a different solvent, alcohol (Expt. 64), so a gas will often dissolve readily in one kind of liquid, but nothing like so freely in another. For example, sulphuretted hydrogen dissolves pretty readily in cold water, communicating to it its peculiar nauseous smell and flavour; but liquid petroleum and heavy coal oils will hardly dissolve the gas at all. The solubility of gases in liquids, however, is largely affected by circumstances that do not have the same effect on the solubility of solids on liquids; thus it has been already stated that when the pressure is increased the quantity of gas (carbon dioxide) that a given quantity of water can dissolve is increased too, and *vice versâ*. This is a general law, it being uniformly found that if a given quantity of gas can be dissolved in a certain quantity of any solvent at a particular pressure, twice that quantity of gas can be dissolved when the pressure is doubled, three times when trebled, four times when quadrupled, and so on; whilst if the pressure is halved or reduced to one-third or one-quarter, &c., the quantity of gas dissolved in each case is similarly reduced to one-half, one-third, one-quarter,

&c. Variations in pressure do not affect the solubility of solids in liquids to anything like the same extent.

Again, in the case of solids, the general rule is that the higher the temperature the more substance can a given quantity of solvent dissolve, although there are some exceptions to this rule (Chapter V.); but with gases the opposite is uniformly the case; the warmer the solvent, the less gas can it dissolve.

Expt. 74. To make a Cold Liquid apparently Boil without becoming perceptibly Heated.—Pour into a small flask enough strong solution of ammonia (*liquor ammoniæ fort.*, sp. gr. ·880) to fill it one-third full, and then gently warm it over a lamp; if the solution be strong enough, you will immediately see a torrent of bubbles arise in this flask precisely as though it were boiling, although the heating has hardly been continued long enough to make the flask feel warmer to the hand than at first. The bubbling arises from the fact that the liquid has really become a little warmer, and in consequence cannot dissolve as much gas as when cold, so that the surplus escapes with effervescence, much as the bubbling is caused in soda-water released from a siphon, except that in the latter case the inability of the water to retain all the gas dissolved is due to diminution of pressure, and in the former to slight increase of temperature. When water is saturated with ammonia gas at the ordinary temperature, it dissolves several hundred times its bulk of the gas, and increases somewhat in volume itself in so doing.

Expt. 75. To collect a Jar of Ammonia Gas.—The bubbles escaping in the last experiment consist of the gas *ammonia*, which possesses a very strong odour, and produces a choking sensation when inhaled in too large a proportion along with air; it is not, however, directly poisonous, except through the irritation it thus causes, and when largely diluted with air is refreshing and gently stimulating, being the essential constituent in the vapours given off by "spirits of hartshorn" (somewhat diluted solution of ammonia) and ordinary "smelling salts," Expt. 51; *sal volatile*, a medicinal stimulant, also smells of ammonia, being in point of fact essentially a solution of ammonia in alcohol. Ammonia gas is lighter than air; consequently, in order to collect some of the gas for experiments, it suffices to fix in the flask used in the previous experiment a cork with a round hole perforated therein, through which passes tightly a piece of quill glass tubing a foot or so in length; or to close the flask with a tubulated india-rubber cap through which the glass pipe passes (fig. 43). By holding a narrow jar (or piece of wide glass-tubing with the end sealed up by the blowpipe or tightly corked), mouth downwards, over the tube, any gas escaping therefrom will be collected in the jar, displacing the heavier air; the flask may be conveniently supported by a clamp, such as is shown in fig. 44, fixed on to a retort stand (fig. 6), whilst the jar should be held by the hand or another clamp, so that the tube passes up inside it as far as possible. After the liquid has been warmed in the flask for a minute or so, the strong current of ammonia gas passing upwards through the tube will have pretty thoroughly displaced all the air in the jar, which may then be corked up with a well-greased cork or stopper, and set on one side for use when several jars have been similarly collected.

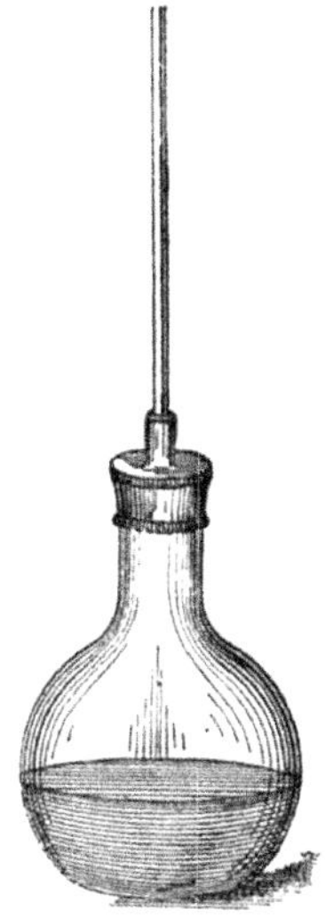

Fig. 43. Flask and Delivery Tube for Ammonia.

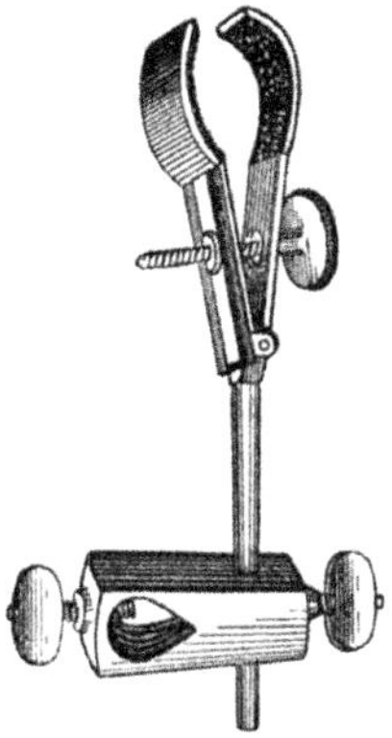

Fig. 44. Clamp.

Expt. 76. To show the Solubility of Ammonia Gas in Water or Alcohol.—Prepare a jar or narrow closed tube full of ammonia gas as just directed, and, without closing it with a cork, carefully lift it off the delivery tube from which the gas escapes, and quickly set it mouth downwards in a tumbler of water; the water will soon begin to dissolve the ammonia gas, the effect of which is that the atmospheric pressure will force the water up in the tube until it nearly fills it. If little or no residual air is mixed with the ammonia gas in the tube, the water will rise almost to the very top, leaving only a small bubble undissolved; but if much air is mixed with the ammonia, the water will rise to a proportionately less extent. This experiment illustrates the different degree of facility with which water dissolves ammonia gas and ordinary air, the former being readily dissolved, whilst the latter is scarcely affected. Instead of water, alcohol (methylated spirit) may be used with much the same results.

Expt. 77. To make a Miniature Fountain.—The easy solubility of ammonia in water and the tendency of the atmospheric pressure to force water into a partial vacuum caused by this solution or absorption, causing a fountain to play, may be thus shown.

Get a stout flask or bottle of white glass; provide a well-fitting cork to the mouth, perforated with a round hole through which passes, air-tight, a piece of glass tube 10 or 12 inches long, and drawn out at the end inside the bottle, so as to make a moderately fine jet. The flask should be quite dry inside, and then filled with ammonia gas, as in Expt. 75. When it is judged that practically all air has been displaced, the flask is lifted off the delivery pipe and the cork inserted in its mouth; the cork and tube should fit perfectly air-tight, which is best ensured by using a perforated india-rubber bung, or by means of sealing-wax, or thoroughly greasing with tallow, &c. Place the flask so that it shall be supported by the ring of a retort stand (fig. 6), with the projecting tube dipping some inches into a vessel of cold water, and then pour a little cold water over the flask. This will cause the gas inside to cool a little, and consequently contract (Chapter XIX.); the external atmospheric pressure therefore forces a little water up the pipe to the drawn-out jet. As soon as this water comes in contact with the ammonia gas it dissolves some of it, and consequently produces a partial vacuum inside the flask,

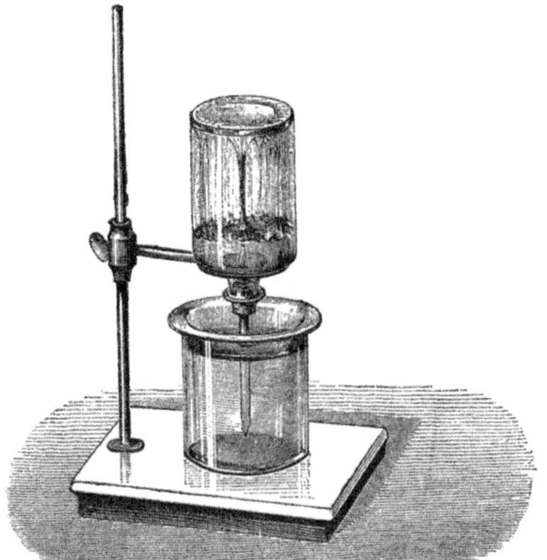

Fig. 45. Ammonia Fountain.

the result of which is that the atmospheric pressure forces the water up the pipe with some considerable degree of violence, producing a vigorous miniature fountain (fig. 45). If the flask is large and not sufficiently strong, it may happen that it will be crushed in and broken by the pressure of the atmosphere upon it, when the counterbalancing internal pressure is greatly diminished by the rapid absorption of the ammonia gas by the water.

For other experiments in which ammonia gas or solution takes a part, *vide* Nos. 79, 114, 124, 138, 145, 160, 198, 214, &c.

Expt. 78. Absorption of Ammonia by Ice.—The tendency of water to absorb ammonia gas is so strong that it is exhibited even when the water is frozen solid; the absorption of the gas by the outer surface of the ice develops sufficient heat to thaw the next layer underneath, and the water so produced absorbs more ammonia, evolving more heat, and so the action goes on until all the gas is absorbed, and all the ice melted, if the two are in suitable proportions.

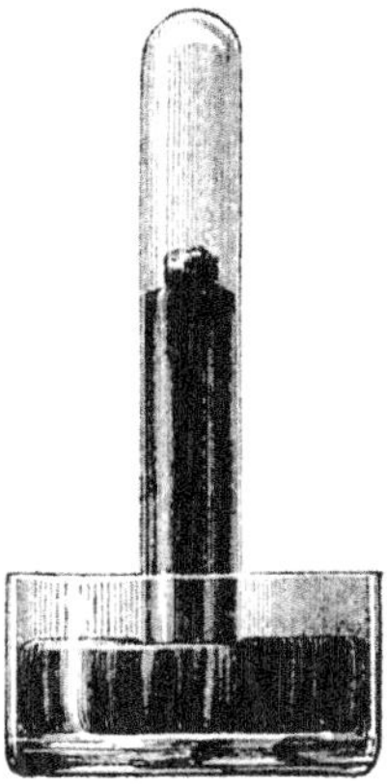

Fig. 46.

Collect a jar of ammonia, as in Expt. 75, and throw into it a lump of ice the size of a small walnut; the ice will be soon melted, and a solution of ammonia formed.

A better way of showing the same result is to float a piece of ice on the surface of a basin of quicksilver, and then place over it a narrow tube sealed at one end, and full of ammonia gas (fig. 46). The ice will absorb the ammonia, and become liquid; whilst the mercury will rise in the tube, being forced up by the atmospheric pressure.

Separation of Gases from one another by means of Solvents.

We have seen (Expts. 66, 67), that it is often possible to bring about a more or less complete separation of solids mixed together by taking advantage of the fact that one of them is readily soluble in a particular solvent which will not dissolve the other constituents of the mixture freely; similarly sulphate of copper, camphor, and sand may be separated (Expt. 68) by first treating the mixture with alcohol, which dissolves the camphor, leaving the other two undissolved, and then with water, which dissolves the sulphate of copper, leaving the sand undissolved. In similar fashion, gases when mixed together may often be separated by means of solvents. Thus Expt. 76 shows how ammonia gas and air may be separated by means of water or spirit in which the ammonia dissolves freely, and the air but little. If the experiment be repeated with a jar only partly filled with ammonia gas, and consequently containing a considerable admixture of air, it will be found that when the ammonia is all absorbed, the remaining air is practically unchanged.

Expt. 79. To separate Ammonia Gas and Air from one another. —Place a jar or bottle mouth downwards over the delivery tube attached to a flask in which a strong solution of ammonia is being warmed, as in Expt. 74, but instead of waiting some little time to allow of the displacement of the whole of the air by the ammonia gas, remove the jar after a few seconds, so that only a small portion of the air in it is displaced. Place the jar mouth downwards in a basin of water, and allow it to stand for half an hour; by this time all the ammonia in the jar will probably be absorbed by the water; to make sure, cork the jar under water with a greased cork, so that some water is retained in the jar, take it out and shake up vigorously; then replace it in the basin mouth downwards, and remove the cork, so as to allow more water to enter should the shaking have produced any further diminution in bulk of the gas inside the jar by virtue of absorption of ammonia. Again cork, and remove the jar, and stand it on a table mouth upwards. Uncork the jar, and introduce a lighted taper at the end of a wire (fig. 41). It will be seen that the taper burns in exactly the same way as it would do in the same jar if filled with ordinary air (easily effected by filling the jar with water, and then pouring out the water so as to allow air to enter and fill the jar); thus showing that the air equally mixed with the ammonia gas is practically unchanged. If too much ammonia have not been added, it will be found also that the air left in the jar unabsorbed, after shaking up with water, will have little or no smell of ammonia, showing that all, or nearly all, of the ammonia gas in the mixture has been removed by the solvent action of the water.

If instead of water a weak solution of hydrochloric or sulphuric acid be used, an absolutely complete removal of all the ammonia will be effected, as the solvent action of the water alone will be intensified by the chemical action of the acid.

Owing to the different solubility of gases in water and other solvents, and the tendency under suitable conditions of gases to evaporate or pass off out of solution, it results that a thin stratum of liquid, or a film of water coating a porous material, or even a soap bubble, will exert a separating action on mixed gases brought into contact with the film. Thus a soap bubble filled with one kind of gas (*e.g.*, with carbon dioxide, or with ordinary coal gas), and allowed to stand in the air for some time, will cause a considerable alteration in the composition of the gas inside to take place, because the constituents of the external air will dissolve in the moisture of the film of the bubble, and will then evaporate again inside the bubble; whilst the gas contained inside the bubble will also dissolve in the moisture, and will similarly evaporate into the external atmosphere; so that a continual passage of the original gas from within outwards, and of atmospheric air from the outside to within, will be brought about. Thus a soap bubble (made of soap solution of such a kind as to give a bubble permanent for some hours Chapter XVIII.), and filled with coal gas, will be at first light enough to ascend in the air like a balloon, and capable of carrying up with it a small attached car or weight (Expt. 307); but after some time so much air will have passed in, and so much of the lighter gas will have passed out, that the bubble and car will no longer be light enough to float in the air, and will sink again instead of tending to ascend.

Expt. 80. Passage of Gases through Moist Membranes.—The following experiment affords another illustration of the same kind of action. Obtain a small bladder, immerse it in water till thoroughly moist and softened, and then blow it up so as to fill it not quite full of air, using a pair of bellows for the purpose, and not the lungs. Tie the neck up tightly, and then suspend the bladder in a large jar of carbon dioxide (see Expt. 100); this gas will dissolve in the moisture of the membrane, and evaporate again into the air inside more rapidly than the air will pass outwards by the reverse action; accordingly the bladder will gradually become more and more distended, and if nearly full at first, and free from punctures, may ultimately burst by the increased pressure inside.

Certain solids possess the power of exerting an action of this kind on certain liquids (as the bladder in Expt. 40, where alcohol and water are thus partially separated), and are consequently capable of being used for the purpose of separating mixed fluids. Some solids exert an analogous action on gases, and will consequently separate these gases from mixtures of them with other gases; thus, india-rubber possesses the property of absorbing and parting with oxygen in this way, and is consequently sometimes employed to separate oxygen from nitrogen in the air (*vide* Chapter VII.).

Supersaturated Solutions of Gases.

When aerated water is drawn from a siphon or gasogene, or poured out from a bottle, a vigorous effervescence takes place at first, but soon subsides, so that no further bubbling is visible; if, however, the liquid be well stirred, or if a rough object (such as a crust of bread or lump of sugar or stone) be dropped in, a further disengagement of bubbles of gas takes place; similarly a glass of sparkling wine that has stood awhile, and apparently become "still," can be made to recommence bubbling by stirring it up with a spoon or crust of bread. This arises from the circumstance that the surplus gas originally contained in the aerated water or wine does not all escape on first pouring out, but a portion is retained for awhile, forming a "supersaturated" solution, just as a supersaturated solution of a solid may be prepared by dissolving in a hot fluid more solid than it can permanently retain dissolved when cold, and then cooling without shaking (Expt. 63). In each case the surplus dissolved matter gradually separates, if the solution is left to itself, whilst the separation is brought about far more rapidly by shaking or stirring up, and especially by contact with portions of foreign solid matters that can act as a nucleus. Most gaseous solutions behave in the same way; whilst if the solution is not supersaturated to begin with, it can often be made so by cautiously warming it; the warmer fluid being unable to retain as much gas dissolved as it could when cooler, becomes supersaturated for the time being, so that on stirring it up, or bringing foreign solid matter in contact with it, bubbles of gas are disengaged.

Many natural spring waters issue from the soil comparatively cool, and these contain dissolved no more gas (usually carbonic acid) than suffices to saturate them, or not quite so much; but if a tumbler of such water be allowed to stand a few minutes on a warm summer's day, bubbles of gas will begin to form here and there on the glass, especially at the points where it is a little roughened; owing to the slight rise of temperature brought about by contact with the warmer air, the liquid is unable to retain in solution all the gas that was completely dissolved whilst cooler, so that it first becomes slightly supersaturated as regards the higher temperature attained, and then allows the surplus dissolved gas to escape in bubbles. The peculiar sensation on the tongue and palate felt when the mouth is filled with effervescent water or wine, &c., is largely due to the same cause; the liquid becomes warmed in the mouth, and consequently bubbles of gas form in contact with the tongue, chiefly on the "papillæ," or little roughnesses on its surface, producing a sensation different from that due to the mere taste of the fluid. This sensation is still felt, on filling the mouth with aerated fluid that has stood just long enough to be no longer visibly effervescent; bubbles of gas form on the papillæ precisely as they do on stirring the liquid with a spoon or rough object, the effect being heightened by the warming of the liquid in the mouth, which promotes the disengagement of gas bubbles.

CHAPTER VII.

Solution of Gases in Solid Solvents and Separation of Gases from Solid Solutions, no Chemical Action taking place.

Just as liquids will absorb or dissolve gases, forming solutions thereof, so can solids to some extent absorb or **occlude** gases; and just as the solubility of a gas in a liquid depends on the nature of the gas and on that of the liquid, and also on the temperature and pressure and such like considerations, so also is the quantity of gas absorbed by a solid influenced by the same circumstances. In most cases, in order to extract the gas dissolved by or occluded in a solid, it is requisite to heat it highly in a vessel whilst the pressure is largely diminished by exhausting by some form of air-pump; but in some instances the dissolved gas is easily expelled by heat alone.

Thus, when metals of various kinds are heated, they frequently part with gases which they have previously absorbed, either during the process of their manufacture or subsequently. Sometimes a metal will absorb a large quantity of one particular kind of gas, but only comparatively small amounts of other kinds; thus iron possesses the property of freely absorbing the gas termed *carbon oxide* or *carbon monoxide* (not the same thing as the "carbon dioxide" or carbonic acid gas used for aerating water, &c.), and the somewhat rare and costly metal *palladium* can absorb several hundred times its bulk of hydrogen gas. On heating in an exhausted vessel the absorbed gas is released, and can be pumped out with facility. India-rubber possesses the property of absorbing oxygen pretty freely, and on this property is based a method of separating oxygen from the air. Silver possesses the peculiar property of absorbing oxygen when it is very hot indeed, and considerably above its melting-point; whereas on cooling somewhat most of the dissolved gas is given off again, thus causing the molten metal to "spit" or effervesce, somewhat after the fashion of soda-water, but much less vigorously.

Expt. 81. Absorption of Hydrogen by Palladium.—One of the simplest ways of causing palladium to absorb hydrogen is to decompose acidulated water by means of a voltaic current, whereby two gases are evolved—hydrogen at one pole or electrode and oxygen at the other. If the terminals of the wires leading from the voltaic battery are made to be strips of thin palladium foil, the strip attached as the "negative" electrode, at which the hydrogen is disengaged, will absorb most of the liberated gas, and in so doing will increase somewhat in dimensions, the metal swelling as it absorbs the gas, much as water does when it absorbs ammonia (Expt. 74). When a strip of palladium foil, or a thin palladium wire, is fully charged with hydrogen in this way, a slight elevation of temperature suffices to expel a little of the occluded gas; in consequence the charged metal can be used as a candle or torch; when the end is held for a moment to a light the heat expels some of the hydrogen, which takes fire and burns; the heat thus produced causes more hydrogen to be liberated, so that the flame keeps burning for some time, the metal being clothed with a very pale flame, almost invisible in daylight, somewhat as though it had been dipped in strong alcohol, and then fired before the spirit evaporated.

Expt. 82. Palladium Eels.—A curious effect may be produced by taking advantage of the fact that palladium expands when it absorbs hydrogen. Varnish one side of each of two long narrow strips of palladium foil with shellac or other impervious varnish, and then connect them with the wires of a sufficiently powerful voltaic battery, and immerse them in water to which a few drops of sulphuric acid have been added; the water will then be decomposed by the current, and the hydrogen evolved at the negative pole absorbed by the palladium strip. As only one side of the strip is in contact with the water, the other being protected by the varnish, only one side becomes charged with hydrogen, and in consequence this side only swells; so that the strip by and by becomes curved and bent round, ultimately twisting up into a sort of coil or spiral on account of the expansion of the charged side relatively to the other one. When one strip is thus pretty fully charged, reverse the direction of the current; hydrogen will now be evolved on the surface of the second strip, which will soon begin to curl up like the first one; but simultaneously oxygen will be evolved on the surface of the first strip, and this will act chemically on the hydrogen already absorbed, removing it, and consequently discharging the gas dissolved in the metal, and causing the latter to shrink again. The effect of this is that the one strip coils up and the other uncoils, as though they were alive. When this action has gone on a while, it may be renewed in a different way by again reversing the current; the second strip, that formerly coiled up by virtue of its swelling through absorption of hydrogen, now uncoils in consequence of the shrinking through removal of hydrogen; and *vice versâ* with the other strip.

It is remarkable that hydrogen, when thus dissolved in or absorbed by palladium, becomes more chemically active than before; so that chemical changes may be produced in liquids by immersing therein palladium containing occluded hydrogen, that would not be produced by simply bringing the liquids into contact with hydrogen gas.

Expt. 83. To form Prussian Blue by means of Occluded Hydrogen.—Mix together solutions of perchloride of iron and red prussiate of potash (*ferricyanide* of potassium—not the same compound as the *ferrocyanide* of potasssium used in Expts. 57 and 59); the mixture will become dark brown sherry colour if the compounds be pure. Into this fluid insert a piece of palladium foil or wire charged with occluded hydrogen; by and by the surface of the metal will become covered with a blue film, arising partly from the action of the occluded hydrogen on the ferricyanide of potassium, converting it into ferrocyanide of potassium, which then forms prussian blue with the perchloride of iron, as in Expt. 57; and partly from an analogous action of the occluded hydrogen on the perchloride of iron converting it into another chloride of iron, which develops a blue pigment termed *Turnbull's blue*, by reacting on the ferricyanide of potassium. In either case a chemical reaction is brought about by the occluded hydrogen, which ordinary gaseous hydrogen does not effect so readily, if at all, showing that the chemical activity of the hydrogen is increased by the physical process of absorption by the palladium.

In consequence of the power possessed by solids of dissolving gases, a remarkable property is possessed by thin plates of certain kinds of solids; viz., that if on one side of the plate the air in contact therewith is continually removed by a pump, whilst the other side is kept constantly in contact with a suitable gas, this gas will dissolve in the solid, pass inwards, and finally make its way to the other side, where it will evaporate again and be removed by the pump. Thus if an india-rubber bag be filled with pieces of wood or other solid objects to prevent its collapsing, and the air be then pumped out of the interior, the outside air will slowly pass through the rubber in this kind of fashion, so that the pump will extract a small continuous stream of gases. Moreover, since the air is a mixture of two gases, oxygen and nitrogen (Expt. 148), and since the former of these dissolves in and passes through rubber more readily than the latter, it results that the air thus pumped out is much richer in oxygen than the original outside atmosphere; instead of containing only one-fifth of its bulk of pure oxygen, it will contain about two-fifths. This action is obviously analogous to the more easy passage of water than of alcohol through a bladder (Expt. 40), or to the alteration produced in the composition of a given gas contained in a soap bubble, when allowed to stand some time, the gas dissolving in the watery part of the soap film passing through, and evaporating again into the outside air, whilst air passes inwards by the reverse operation (Expt. 79); or the similar action of a moist bladder containing air when placed in a gas of carbon dioxide (Expt. 80).

If a jar of hydrogen gas be tied over, jam-pot fashion, with a thin sheet of india-rubber, the hydrogen will slowly pass outwards, and the external air inwards through the rubber; the former action goes on more rapidly than the latter, so that the total gaseous contents of the jar are lessened; accordingly, if the temperature remains the same so as not to affect the result by altering the volume of the gas in the jar, the rubber cover will be gradually pressed inwards and become concave. Conversely, if a jar filled with air and similarly tied over be allowed to stand some days inside a larger jar filled with hydrogen, the latter gas will pass inwards more rapidly than the air will pass outwards, so that the rubber will become convex, bulging outwards.

Platinum and palladium, and to some extent iron, more especially when considerably heated, possess a similar power of allowing hydrogen to pass through walls of these metals more readily than most other gases; red-hot iron is also permeable to carbon monoxide to some considerable extent. This property is well illustrated in the case of coal gas (a complex mixture of several gases, of which hydrogen is one main constituent), by enclosing a long tube of platinum or palladium inside a glass tube, which thus serves as a jacket, just as in a Liebig's condenser (fig. 23). The central part of the compound tube is heated, and a current of coal gas led through the jacket, whilst the inside of the metal tube (stopped up at the far end) is connected with a good air-pump. Under these circumstances, the hydrogen of the coal gas passes pretty freely through the metal, and is extracted in a nearly pure state by the pump, whilst the other gases do not pass through at all, or at least only to an inconsiderable extent. The action may to some extent be compared with that of a wet paper filter into which a mixture of oil and water is placed (fig. 36): if the filter be wet with water, only water passes through, and oil remains behind; whilst if the paper be previously impregnated with oil, only oil passes through, the water being now left behind (*vide* Expt. 84).

Closely akin to the phenomena of occlusion, or solution of gases in the body of solids, are various physical actions, whereby gases are strongly attracted to and condensed upon the surfaces of solids, more especially when these are of a highly porous nature. These **surface actions** will, however, be more conveniently discussed later on (Chapter XVII.).

CHAPTER VIII.

Miscibility of Liquids, or Mutual Solubility of Liquids in one another, no Chemical Action taking place.

It is a matter of every day observation, that whilst certain fluids will mix readily together, no matter in what relative proportions they are used, others will not do so, at any rate when employed in anything like equal quantities; thus spirits of wine (alcohol) and water are a pair of fluids that will dissolve in each other or intermix in all proportions; whilst olive oil and water are a pair of fluids practically quite insoluble in one another, and quite incapable of intermixture; as again are mercury and alcohol.

Many cases are known where two fluids are intermiscible in certain proportions, but not in others; this occurs when the first fluid is soluble in the second, but only slightly so; and when the second is similarly soluble in the first, but only to a limited extent. Thus, if a few drops of ether are added to a pint of water, or a few drops of water to a pint of pure (anhydrous) ether, and the whole well shaken, the smaller quantity of the one fluid will permanently dissolve in the larger quantity of the other, and one single uniform fluid or solution will in either case be obtained. But if you put into a bottle equal bulks of ether and water, and shake them up well together, and then allow the whole to stand awhile, the liquid will separate into two layers, as oil and water would do; but with this difference, that the upper layer is not pure anhydrous ether, but is ether containing as much water dissolved therein as it can possibly take up: whilst similarly, the lower layer is not plain water, but is a saturated solution of ether in water.

If quicksilver, distilled water, and pure ether be shaken up together, on standing three layers will be formed; the lowest being the quicksilver, which is *absolutely insoluble* in water, ether, and mixtures of them; the middle one, water containing a little ether in solution; and the uppermost ether containing a little dissolved water.

Expt. 84. To illustrate different Degrees of Solubility of Liquids in one another.—Place a teaspoonful of distilled water in each of three test-tubes, and then pour into one about as much olive oil, into a second a little chloroform, and into the third as much alcohol (methylated spirit will do). Close each tube tightly with the thumb or with a cork, and shake up vigorously, and then allow each tube to stand for a few minutes. The oil and water in the first tube *will entirely separate from one another*; if you prepare a paper filter (Exp. 56), and wet it thoroughly with water, and then pour on to it the mixture of oil and water, the water only will pass through, whilst the oil will remain, being incapable of running through the pores of the paper whilst these are filled with water. In the second tube the chloroform and water will also apparently separate from one another, the water, being lighter, rising to the top, and the heavier chloroform sinking to the bottom. The separation here is not complete, for the water that rises to the top contains dissolved a small quantity of chloroform, as may easily be perceived by tasting it; similarly the heavy chloroform at the bottom dissolves a minute quantity of water, although this can only be distinguished by certain chemical tests. In the third tube *no separation at all takes place*, because the alcohol and water can dissolve one another much more freely than chloroform can dissolve water, or water chloroform. Now add to the mixture of alcohol and water a teaspoonful of solid *potassium carbonate*, or "pearlash," and shake up again; after standing awhile the liquid will separate into two layers, whilst the solid pearlash will have disappeared; the lower layer is the water which has dissolved the pearlash, forming a solution of potassium carbonate; the upper layer is the alcohol which will not dissolve in or mix with solutions of potassium carbonate, although it will mix readily with water alone (compare Expt. 39). The two liquids may be conveniently obtained separate by means of a funnel provided with a tap (fig. 47); the mixed fluid is poured into the funnel, and after separation is complete first one is run off and then the other.

Fig. 47. Tap Funnel.

Expt. 85. To throw one Liquid out of Solution in another by adding a third.—We have seen, in Expt. 64, that when a solid is dissolved in a solvent (like camphor in alcohol), a portion of the solid may sometimes be thrown out of solution by adding another liquid in which the solid is less easily soluble (such as water). Exactly the same thing occurs with liquids; thus, if a little *essential oil of lemon* be dissolved in alcohol, and water be then added, the whole becomes milky, because the oil of lemon is thrown out of solution, making its appearance in the form of very small globules, which at first remain suspended in the alcohol and water, like fatty matter in milk, but by and by separate into larger drops of essential oil. Eau de Cologne and many similar perfumes behave in exactly the same way when water is mixed with them, and for the same reason, these perfumes being mostly composed of alcohol, containing certain odorous essential oils dissolved therein, giving the particular perfumes. *Absinthe* and various other liqueurs similarly turn milky on mixing with water, because they are spirituous solutions of certain essential oils and similar substances. Very coarse kinds of spirits, such as crude whisky, &c., sometimes become milky when mixed with water, because they contain oily impurities (*fusel oil*), which are similarly thrown out of solution when the spirit is diluted.

Expt. 86. Comparative Solubility and Insolubility of Melted Metals in one another.—Melt in a crucible an ounce of lead (obtainable at the plumber's), and then throw into the melted mass as much tin; the latter will immediately melt, and the two fluid metals will mix together, like the alcohol and water in the last experiment, so that no matter how long the melted mixture be kept hot and fluid, the two metals do not separate from one another.

Repeat this experiment, using *zinc* instead of tin; you will now find that the melted zinc and lead will not mix together any more than chloroform and water; if you stir the mass up well with an iron rod, and then allow it to stand, the zinc will float up to the top, and the lead will sink down to the bottom, forming two layers of liquid metals. A good way to show this is to get a clay tobacco pipe, and plug up the bottom of the bowl with a bit of clay or chalk; heat the bowl red hot in the flame of a Bunsen burner, and then pour into it the melted metals from the crucible; keep the tobacco pipe bowl hot for half an hour, and then remove the flame; when the whole is cold, break away the clay from the ingot of metal. You will find that the lower end is soft, and can be cut easily with a knife, and will mark paper, being in fact lead (containing a small quantity of zinc dissolved, just as the chloroform retains a little water); whilst the upper end is hard, and is zinc containing a little lead (as the water dissolves a little chloroform).

A still better way of showing the separation of lead from zinc is to mould a very narrow long clay crucible (Expt. 1), shaped like a test-tube, using a thick pencil rubbed with oil as a "core" or mould to shape the interior by rolling and working the clay round it. Such a tube should be slowly dried and fired by heating in the lamp flame; if any cracks form these should be filled up with clay after cooling, and the tube heated a second time, and so on until the clay test-tube is tight enough to hold the molten metal. The tube is then heated in the flame of a large Bunsen lamp like the pipe bowl, and the melted metals carefully poured in from the crucible, and kept melted for half an hour or more. Finally, after cooling, the clay is carefully broken away from the long narrow ingot or rod of metal; the lower end will be soft enough to bend and hammer out flat, but the upper end will be hard and rigid, and if forced by sufficient power will not bend, but will break off with a "crystalline fracture," *i.e.*, the broken surfaces will show a crystalline structure (*vide* Expt. 18).

Bismuth and *zinc* behave in exactly the same way as lead and zinc when melted together, the heavier bismuth separating to the bottom (dissolving some zinc in so doing), whilst the lighter zinc floats up to the top (dissolving a little bismuth at the same time). If equal weights of the two metals are melted and stirred together, and then poured into a hot clay test-tube, and kept melted for an hour, the rod of metal finally obtained after cooling will be extremely brittle, and easily broken at the lower end, owing to this consisting mostly of bismuth; whilst the other end, being mainly zinc, will be much harder, and less easily broken.

The different degrees of solubility of melted metals in one another are often utilised in metallurgical operations in a variety of ways; for example, certain lead ores, especially the one termed "galena," contain silver, which is smelted along with the lead in the process of extracting that metal from the galena, or natural sulphide of lead. In order to extract the silver, a process sometimes used is to stir up some zinc with the melted lead; the zinc rises to the top, being mostly undissolved by the lead, and carries a great part of the silver with it, because molten zinc dissolves silver more freely than does melted lead. By skimming off the melted zinc from the top, and then heating it strongly, the silver is ultimately obtained (mixed with a little lead), the zinc being volatilised and driven off by the heat.

Bromine is a liquid element of a most intolerable odour; it will dissolve in water to a slight extent, forming a brownish-orange solution. If a weak solution of bromine in water be shaken up with a little chloroform or bisulphide of carbon, the latter will dissolve most of the bromine and separate it from the water, so that after standing a heavier coloured liquid will subside to the bottom, consisting of the chloroform or bisulphide of carbon containing the bromine in solution, whilst a much lighter coloured watery fluid will rise up to the top; the action being exactly analogous to that taking place with zinc, lead, and silver, excepting that the silver solution in zinc rises to the top, whilst the bromine solution in chloroform sinks to the bottom. These actions are precisely analogous to that occurring in Expt. 62, where iodine is removed to a great extent from a watery solution by means of chloroform, the chief difference being simply that iodine is a solid, whilst bromine and molten silver are liquids.

Caution.—Bromine is very volatile, and its vapours are excessively injurious to the lungs if breathed, producing much irritation and coughing; be very careful, therefore, not to inhale the fumes.

Expt. 87. To separate two Intermixed Liquids by Evaporation.—When a liquid not easily volatilised is dissolved in another that is readily volatile, separation is easily brought about by the agency of heat, just as when a solid substance is dissolved in water or spirit, and the solvent evaporated by applying heat (Expts. 37 and 38).

Pour a few drops of olive oil into a test-tube, and then add a tablespoonful of ether and shake up; the ether will dissolve the oil, and form a solution or mixture of the two liquids. Pour this out into a small evaporating basin, and allow it to stand in a warm place for a short time, taking care that the ether does not take fire (Expt. 41); the ether will speedily evaporate, and leave the oil unchanged. If the ethereal solution be placed in a distilling apparatus furnished with a condenser (Expt. 36), the ether can be condensed and recovered for use over again.

Obtain a teaspoonful of essential oil of lemon, and add to it a wine-glassful of strong alcohol; the spirit will dissolve the essential oil, forming a fluid from which the alcohol may be evaporated or distilled off, leaving the essential oil behind. In this case, if the alcohol be recovered by condensation, it will be found that it retains the odour and flavour of lemon to some extent, as this essential oil is not altogether non-volatile, and consequently a little distils over along with the alcohol.

Expt. 88. To extract Oil from Seeds, &c.—The principles illustrated in the last experiment are applied in the arts for the extraction of oily matters from seeds and various other sources, so as to obtain materials useful in the manufacture of soap and numerous other products.

Crush a handful of linseed in a mortar, and place the coarsely powdered mass in a bottle. Pour in some ether so as to cover the powder, cork the bottle, and let the whole stand awhile; the ether will gradually penetrate into the fragments of seed and dissolve out the oily matters therein contained without affecting the rest of the vegetable matter. Now add a little more ether, shake up, and allow to stand awhile to settle; pour off the ether (through a filter, (Expt. 56), if not clear) into another bottle, and shake up the partly exhausted seeds with some more ether, repeating the operation two or three times in succession, and collecting all the ether poured off in the same bottle. In this way practically all the oily matter contained in the seeds will be dissolved out. Now place the ethereal solution of oily matter obtained in a retort or a flask provided with a cork and bent tube, and connected with a Liebig's condenser (fig. 24), and gently heat the flask so as to boil off the ether and leave the oil in the flask.

To do this safely, *the flask should not be directly heated by a lamp*, but should be placed in a pan of warm water, to which a little more hot water is now and then added so as to keep it sufficiently warm to make the ether boil; if, on the other hand, a flame is placed underneath the flask, there is not only a liability of the flask cracking, and a large mass of flame being produced through the firing of the ether, but, further, there is a possibility of the ether vapour being generated faster than it can be condensed again in the condenser, so that a quantity of uncondensed vapour streams out at the far end; this being heavy, is apt to flow along the table, and ultimately to take fire at the lamp, thereby producing a body of flame that will probably severely burn the experimenter, and possibly do other damage. When bottles of ether are unstoppered, especially in hot weather, anywhere near a light (such as a gas or candle flame) similar firing of escaping vapour is very apt to take place; and many accidents happen through the incautious opening of bottles of ether, or substances containing ether (such as the "collodion" used by photographers), too near a candle. Some very inflammable substances used for burning in lamps, such as "benzoline" and spirits of wine, are liable to produce the same danger, more especially in summer or hot climates (compare Expt. 41).

For manufacturing purposes on the large scale, ether is too costly for use as a solvent for oils, &c.; cheaper fluids, such as "benzene" (obtained from coal tar), and "bisulphide of carbon," are generally employed, the oily materials to be treated being placed in a kind of tank and the solvent poured on them, and then run away by means of a tap at the bottom of the tank into a kind of boiler heated, not by fire, but by means of a steam pipe, so as to avoid all chance of firing the inflammable fluid or its vapour. In order to economise the liquid, a series of tanks are generally arranged in such a way that the liquid that runs from the first is made to pass into the second, where it takes up more oil, and thence into the third, and so on.

Expt. 89. To separate the Fatty Matters from Milk.—Cow's milk is a very composite fluid, containing amongst other things little globules of fatty matter floating about in a watery solution of "milk sugar," "casein," and various other substances, somewhat in the same way that vesicles of water float about in the air forming mist or fog; only the water vesicles are heavier than the air, and so tend to settle down or sink, whilst the fat globules of milk are lighter than the watery fluid, and so tend to rise and float. Cream is simply the upper layer of milk, which becomes much more heavily charged with fat globules on standing, owing to this tendency to rise, whilst the lower layers become proportionately poorer.

Get a small saucepan full of boiling water, or better still the copper hemispherical "water-bath," with rings of different sizes to suit different sizes of basins, represented in fig. 48; support this on a tripod, and place on it an evaporating basin (fig. 35), so that the basin may be heated, not by direct flame, but by the steam of the boiling water; this arrangement is called a **steam bath**, and prevents any possible burning of substances placed in the basin. Pour into the basin half a small wine-glassful of milk, and let the milk evaporate slowly, stirring it now and then with a bit of glass rod. By and by all the water will be driven off, and a residue left in the dish containing the fatty matters, caseine, milk sugar, &c., originally contained in the milk. Let the dish cool completely, and then pour on the residue two or three teaspoonfuls of ether, and stir the whole up as much as possible. Now pour off the ether into a second clean basin; add more ether to the residue, and stir up again; and again pour off the ether into the second basin. If this latter be now left exposed to the air for a short time the ether will soon evaporate, and a film of fatty matter will be left. Care must be taken that the ether does not fire through its vapour coming in contact with a light (compare Expts. 41 and 88). If a little water be poured on the residue from which ether has thus dissolved the fatty matter, it will in turn dissolve out the milk sugar, which may be obtained in an impure

Fig. 48. Water Bath and Rings.

form by filtering the watery liquid and evaporating the clear liquor to dryness in a basin heated by the steam bath. What is left undissolved by both the ether and water is mainly the caseine of the milk.

The analysis of milk, in order to determine its quality and purity, is performed in very much the same way as the above experiment, except that considerable precautions are requisite in the manipulation in order to obtain exact results. A measured quantity of the milk to be examined is put in the evaporating basin first of all, this having been carefully weighed on a delicate weighing-machine or "balance" made specially for purposes of the kind. This is simply an extremely delicate pair of scales enclosed in a glass case (fig. 49), with doors and sliding glass panels to

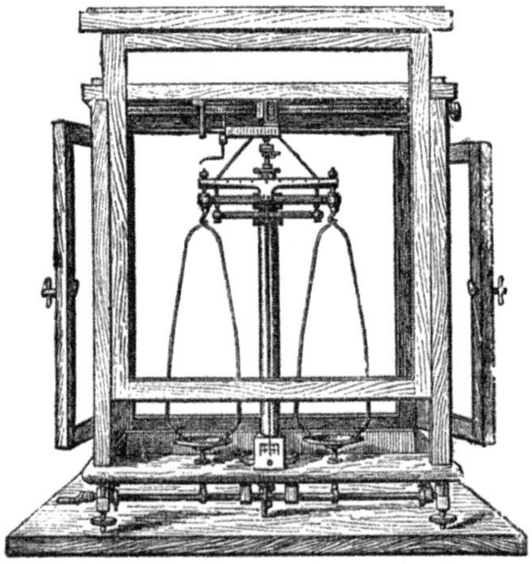

Fig. 49. Chemical Balance.

enable the weights, &c., to be put on the pans, and then all currents of air kept off by closing the doors; a mechanical arrangement prevents the balance being able to swing until a catch is released by means of a rod worked by a mill-headed button or disc, or by a lever pressed by the finger. When the water has all been driven off and the residue is quite dry, the basin is weighed again; subtracting from this weight that of the empty basin, the difference represents the quantity of "total solids" contained in the milk, viz., fatty matters, milk sugar, casein, &c., &c., all lumped together. When the residue is treated with ether, the ethereal solution of fatty matters is transferred to another weighed dish, being filtered through paper if not quite clear; after the ether has evaporated, the fatty matter left on the basin is heated on a steam bath (or in a sort of oven heated by steam) to dry it completely, and the basin is then weighed after cooling. Subtracting from this weight the weight of the second empty basin, the difference represents the weight of fatty matter contained in the milk. 100 parts of genuine cow's milk that has not been impoverished by allowing it to stand and skimming off more or less of the cream that has then risen, and that has not been adulterated by adding water, ought to yield *at least* from 2½ to 3 parts of fatty matter and from 10½ to 11½ of "total solids." Rich milks, like that of a good Alderney cow, give higher figures. According to the quantity of fatty matters found, and of the other constituents of the "total solids" present, the analyst forms an opinion as to the genuineness of the milk, and to what extent it has been skimmed or watered, should such adulteration have been made.

Spontaneous Intermixture of Liquids: Liquid Diffusion.

Liquids that are miscible together, but of different density, exhibit a remarkable behaviour when they are arranged in different layers, so that the heavier liquid is undermost, viz., that although kept perfectly at rest, and at a uniform temperature so as to avoid intermixture by setting up convection currents (Chapter XX.), the heavier liquid will gradually rise up and disseminate itself throughout the lighter one, and, conversely, the lighter liquid will sink and intermix with the heavier one. This process is termed **Diffusion.**

Expt. 90. To show the Diffusion of Salt and Water.—Fill a bottle with strong brine, and loosely cork it; lower it into a large jar full of distilled water, so that the mouth of the bottle is an inch or two below the surface of the water in the jar (fig. 50). Carefully withdraw the cork, and leave the whole to stand for a day or two; at the end of this time, more or less of the heavy brine will have risen up through the lighter water, whilst some of the lighter pure water will have descended into the bottle, weakening the brine. Carefully take out a spoonful of water from the top portion of the liquid in the jar, place it in a test-tube, and test it with *nitrate of silver* (Expt. 37). You will observe a more or less marked cloudiness, or precipitate formed, showing that the salt originally contained in the brine in the bottle has become partly diffused throughout the pure water outside the bottle.

Fig. 50. Diffusion Bottle inside Jar.

Expt. 91. Diffusion of Perchloride of Iron Solution.—Repeat the above experiment, using strong *perchloride of iron* solution instead of brine. After a day or two the water in the jar will contain enough perchloride of iron diffused into it to give the test for iron in solution described in Expt. 57; *i.e.*, if a little of the solution be taken out with a spoon, and there be added to it a few drops of solution of *ferrocyanide of potassium*, prussian blue will be produced.

Instead of operating as above described, the occurrence of diffusion may be illustrated thus. Fill a glass jar half full of distilled water, place in it a clean funnel with a long stem, so that the lower end of the funnel reaches to the bottom of the jar (fig. 51). Now carefully pour the heavier solution of salt or of perchloride of iron down the funnel; the heavier fluid will form a layer at the bottom of the vessel, on which the lighter water will float, the water rising in the jar as the heavier fluid is poured in. A sample of the water at the top of the jar taken out now and tested will not indicate the presence of any dissolved matter; but after a day or two, the same test applied to a fresh sample taken out will show that the substance dissolved in the heavier fluid has partially become diffused into the lighter water.

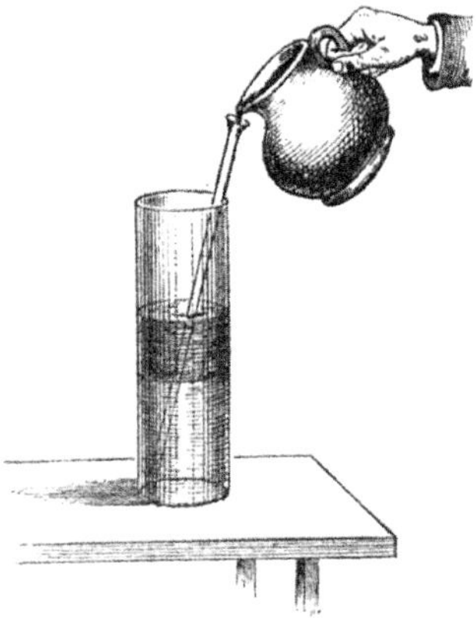

Fig. 51. Diffusion, heavier Liquid poured in by long Funnel.

Almost any solution heavier than water may thus be used: if a solution of *ferrocyanide of potassium* be poured down the funnel, the effect of diffusion after a few days will be shown by testing a sample with perchloride of iron solution, and obtaining prussian blue. A strong solution of *nitrate of lead* after diffusion has taken place will yield a fluid which will give a yellow precipitate with iodide of potassium (compare Expt. 11). A solution of *hydrochloric acid* will similarly yield a fluid capable of turning blue litmus-paper red (Expt. 57).

Intermixture of Fluids through Membranes: Dialysis and Osmosis.

In Expt. 40 we have seen that whilst water can pass pretty easily through an animal membrane (such as a bladder), alcohol cannot do so at all readily, so that if a mixture of alcohol and water is placed in a bladder and closely tied up, the water will gradually pass through the membrane and evaporate on the outside, whilst almost the whole of the spirit remains in a comparatively concentrated form. The same difference is exhibited by different aqueous solutions as compared with one another or with pure water.

Expt. 92. Illustration of Osmosis.—Thus, if a long glass tube be attached to the narrow end of a bell-shaped vessel, the mouth of which is tied over with a piece of bladder, and the whole mounted by means of a stand (fig. 52), so that the bell is immersed in the fluid contained in a larger outside vessel, it will be found that if the bell be filled with brine and the outer vessel with plain water, the salt contained in the brine will gradually pass through the membrane and produce a weak salt solution in the outer vessel; whilst, conversely, plain water will pass inwards from the outer vessel into the bladder, diluting the brine therein, and *causing the level of the fluid to ascend very considerably in the glass tube*, indicating that much more liquid passes into the bell through the bladder than in the opposite direction. Phenomena connected with the different degrees of facility of passage of dissolved solids, &c., through membranes, are referred to under the name of **dialysis**; whilst the action causing fluid to pass through in one direction, rather than in the opposite, is termed **osmosis**. Animal membranes are not indispensable to the action, as chemically prepared paper (*vegetable parchment*) and similar materials can often be used to form a *septum* or *diaphragm*, permitting dialysis and osmosis to take place through its body.

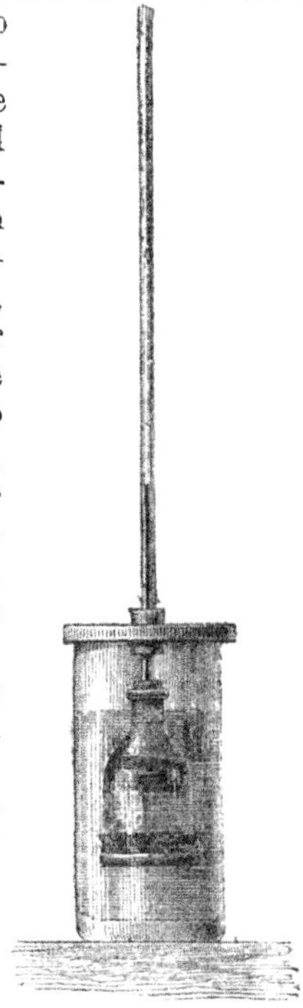

Fig. 52. Osmosis.

Substances which are readily crystallised from aqueous solution, or which can be obtained crystalline by other treatment (*e.g.*, in the case of gases, intense chilling, so as first to condense to the liquid state, and then to freeze to solids, like ice), will readily pass through such membranes, so as to become diffused through plain water on the other side of the membrane; whereas bodies of a glue-like or "colloid" character (Expt. 70), will not pass through at all, or at most only very slowly and with much difficulty. Accordingly soluble bodies are divided into two classes—**crystalloids** and **colloids**—according as they can or cannot easily undergo the process of *dialysis* or separation whilst in solution by passage through a septum—from δια, *through*, and λυσις, *separation*.

Expt. 93. To illustrate the Difference between Crystalloids and Colloids.—Dissolve a little *tannin* in water, and also prepare some thin glue or gelatine solution by digesting fragments of solid glue with warm water. Pour some of the tannin solution into the gelatine solution, and you will see that the latter becomes precipitated as a soft curdy substance, the tannin combining with the gelatine to form a substance insoluble in water. The process of making leather from skins by tanning depends on this action, the tannin contained in the oak bark, &c., used by the tanner, gradually being absorbed by the gelatinous matter of the fresh hide, and indurating it by forming an insoluble compound therewith. *A solution of tannin thus serves as a test for the presence of gelatine in solution.*

Now mix with some of the thin glue a little strong brine, and place the whole in a small bladder, or freshly-cleaned sausage skin, with one end tightly tied up. Tie a string round the neck of the bladder, and suspend the bladder by the string inside a large glass cylinder or beaker filled with distilled water. After some hours, or even one or two days, take out a sample of the water from the beaker, and divide it into two parts; to one add some tannin solution, and to the other some silver nitrate solution. If the bladder was free from visible leaks and pin-holes, no gelatine will have passed through it, so that the tannin solution will yield no visible precipitate in the first case; on the other hand, the silver nitrate solution will yield a considerable precipitate showing that the salt present has readily passed through the membrane, the salt being a crystalloid and the gelatine a colloid. The liquid remaining in the bladder will precipitate with tannin, showing that the gelatine is still there.

Expt. 94. Another Illustration.—Rub together in a cup with a tea-spoon as much solid *starch* as will lie on a sixpence and a little water, so as to make a thin pulp or creamy liquid; pour this into a cupful of boiling water, or pour the boiling water on to the starch. The solid starch will then be apparently dissolved, forming a slightly opalescent very thin gelatinous liquid. Muslins, linen, &c., soaked in such a liquid, and then dried and pressed with a hot flat-iron, become stiffened, this being the ordinary process of "starching" employed by the laundress, a little gum being sometimes added to the starch jelly to give greater stiffness and glossiness to the starched shirts, collars, &c.

Dissolve in water a few grains of *sulphate of copper*, and mix some of this solution with the starch paste; place the whole in a bladder, and suspend this in water as in the last experiment. Next day, take out some of the water, and test half of it for copper and half for starch; the former will be found, the latter not; because the copper sulphate, being a crystalloid, will have passed through the membrane, whilst the gelatinous starch, being a colloid, will not pass through it; the liquid inside the bladder, however, will readily show the presence of starch on similarly testing.

Expt. 95. To Test for Copper.—In order to test for the copper sulphate, proceed as follows. To a portion of the watery fluid add a little solution of *sulphuretted hydrogen*. The same result will ensue as when silver nitrate and sulphuretted hydrogen solutions are mixed (Expt. 12, No. 5), *i.e.*, a black precipitate will be formed by double decomposition, consisting in this case of *sulphide of copper*. Another test is, to add to a separate portion of the watery fluid a little solution of *ferrocyanide of potassium;* this will act on the copper sulphate by double decomposition, just as it does on solution of perchloride of iron (Expt. 57); but with this difference, that the precipitate produced with copper is of a peculiar brown-red or mahogany colour, instead of being blue as with iron.

Expt. 96. To Test for Starch.—The test for starch is as follows. Shake up with water a few fragments of *iodine* for some time; by and by the water will become slightly yellow in colour, a little iodine becoming dissolved; if alcohol be used instead of water, the iodine dissolves much more freely to a liquid, which is of a pale or brown sherry colour, according to the strength of the solution. To the liquid supposed to contain starch add a few drops of iodine solution; if starch is present, a blue colour will be developed, owing to the combination of the starch with the iodine. This blue *iodide of starch* possesses a somewhat remarkable property, viz., that it becomes colourless on heating. If a little of the blue liquid be boiled in a test-tube, the blue colour will disappear; but if the test-tube be allowed to cool, the blue colour will again become visible. If the test-tube be quickly cooled by immersing it in a tumbler of cold water, the blue shade will first appear at the bottom of the test-tube, because the hotter fluid tends to ascend to the top * (Chapter XX.).

* The presence of *starch* in sweetmeats, &c., may be shown by the aid of this test, thus: Pour cold water on the substances, and allow it to stand (stirring up occasionally if requisite) until the sugar is dissolved; if starch be present a sediment will be left undissolved. Pour off the liquor and transfer the sediment to a thin glass flask, and add some boiling water; this will gelatinise any starch present. Allow the whole to cool, and add a few

Instead of bladder or sausage skins immersed in water, the following arrangement may be conveniently employed for experiments in dialysis. Procure a wooden hoop, or better still the wooden drum-like cylinder of a sieve, and strain over it, tambourine fashion, a sheet of parchment paper or a flat circular piece of bladder, tying it down with string as though you were covering a jam-pot (fig. 53). Fill a basin with water, and on the surface of this float the hoop dialyser thus formed; pour the mixed solution into the tambourine, and let the whole stand for twenty-four hours: by this time the majority of the crystalloids present in the mixed fluid in the tambourine will have passed through the membrane, and become diffused through the water in the basin, whilst the colloids will be still retained in the tambourine. Instead of a hoop dialyser, the arrangement shown in fig. 54 may be used, consisting of a bell-shaped glass, with a piece of bladder tied tightly over the wide mouth, and suspended by strings inside a beaker of water.

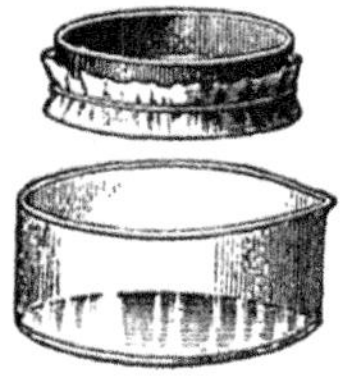

Fig. 53. Hoop Dialyser.

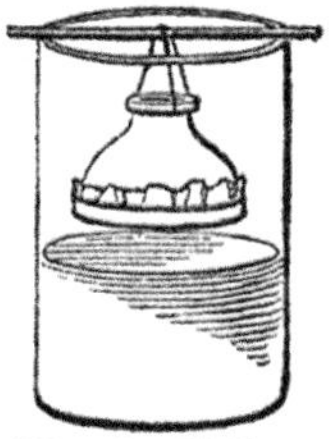

Fig. 54. Bell-Glass Dialyser.

Expt. 97. Endosmosis and Exosmosis.—If in Expt. 92, instead of brine (watery solution of common salt, or chloride of sodium), an *alkaline* fluid (Expt. 138), such as solution of *carbonate of potassium* or *carbonate of sodium*, be used, the rate of inward flow of water into the bladder by **endosmosis** is greatly increased; on the other hand, if an *acid* solution (such as highly diluted *sulphuric* or *hydrochloric acid*) be employed, the flow of fluid is reversed in direction, **exosmosis** now taking place, or passage of fluid through the membrane from within outwards, so that the level of the liquid sinks in the glass tube instead of rising. The rate of passage of the fluid is greatly influenced by the character of the membrane; the finer its pores the more rapid is the action, so that a thin fresh animal membrane will cause a less rapid flow of liquid than the same membrane partially *tanned* or acted on in analogous fashion by chemical agents (*e.g.*, chromic acid solution), which tend to

drops of solution of iodine, when, if starch be present, the blue colour of iodide of starch will appear.

Should *plaster of Paris* be present (which sometimes happens with icings of sugar on certain kinds of cakes, &c.), it will not be gelatinised by the hot water, and will subside as a white powder from the hot liquid on standing.

indurate it, and make its pores smaller. When highly porous septa are used, through which water will percolate pretty readily by its weight only (such as unglazed stoneware or porcelain vessels, *e.g.*, a terra cotta water cooling caraffe, which is always kept externally moist by the slow passage of the water through the pores, thus causing continual evaporation and consequent cooling), the phenomena become somewhat more complex, as the dissolved substances to some extent pass through uninfluenced by the walls of the passages, as they would through a series of fine-bored pipes; in such cases certain other peculiarities are noticeable, which will be more conveniently examined in connection with the subject of **Capillarity** (Chapter XVII.)

This selective passage of certain bodies through membranes and not of others, is very closely connected with many processes going on in living bodies; thus the processes of "secretion" occurring in the human body (such as the separation of urine from blood in the kidney, and the production of tears and perspiration) are actions very similar in their nature to osmotic and dialytic phenomena; whilst the processes of exhalation of carbon dioxide from the lungs, and inhalation of oxygen from the air, are again closely allied to these actions, as well as those physical processes where gases are dissolved in fluids (Chapter VI).

Alteration in Bulk on Intermixture.

When liquids are mixed together it not unfrequently happens that the bulk of the mixture is less than that of the two constituents before being intermixed; just as when a solid is dissolved in water, the bulk of the solution is usually less than the sum of the volumes of the original water and solid before being dissolved. In the act of dissolving a solid, it often happens that cold is developed (Expt. 24) although the reverse happens in certain cases where chemical action takes place, the water partly combining with the substance dissolved, as it does with quicklime in slaking (Expt. 240). When two liquids are mixed together and contraction in volume takes place, a considerable production of heat is often noticeable, even in cases where no chemical action (according to the ordinarily received views) takes place. One of the best illustrations of this is where sulphuric acid and water are intermixed.

Expt. 98. Two Half Pints do not always make a Whole Pint.—Carefully measure out in two long narrow measuring flasks two exact half pints (or other convenient volumes) of oil of vitriol and water respectively; pour the water into a large flask or stoneware jug, and then pour a little of the acid carefully into it; the liquid will get hot, and if not well cooled by immersing the body of the flask in a bucket of water, or holding it under a tap of running water, it will give off steam; gradually add the rest of the acid with continual cooling; finally, when the two liquids have been completely intermixed, and the mixture has cooled down completely, pour the mixture back into the two original measuring vessels. It will be found that, even when the greatest care is taken to avoid any loss by spilling or evolution of steam, there will be a considerable deficiency in volume, far greater than would be accounted for by the few drops of liquid remaining in the flask; thus showing that a considerable contraction has taken place.

A similar diminution in bulk is noticeable when water and alcohol and many other liquids are mixed together; also when metals are intermixed by melting to form certain alloys, the volume of mixture being less than that of the different metals used to make it.

Caution.—Whenever you are making experiments requiring the use of oil of vitriol, be very careful how you add water or watery fluids to it, or how you pour it into water, as much heat is evolved during the mixture, which under certain circumstances may evolve sufficient steam to scatter the acid about, and do serious damage to the skin, eyesight, clothing, or surrounding objects. *Never pour water into hot sulphuric acid;* and if you want to prepare diluted acid by mixing oil of vitriol with water, do not make it in a thick glass vessel (such as a tumbler), as the heat is very liable to crack the glass; a stoneware jug is much safer, pouring the acid into the water little by little, and stirring up the while with a bit of glass rod.

CHAPTER IX.

Spontaneous Intermixture of Gases without Chemical Action taking place: Gaseous Diffusion.

When two gases not capable of acting chemically upon one another are brought into contact in such a fashion that no mechanical intermixture takes place spontaneously on account of one being lighter than the other and so rising, or by stirring, &c., and none by the setting up of convective currents due to difference of temperature (Chapter XX.), they nevertheless tend to intermix, an action of diffusion being set up precisely as with liquids (Expt. 90), but in a much more marked way and far more rapidly, owing to the much greater mobility of gases. If, as is usually the case, the two gases are not of the same density, the heavier one will rise gradually and intermix with the lighter one, which will sink during the process, precisely as the liquids above described.

Expt. 99. To generate Carbon Dioxide from Chalk.—In Expt. 72 we have seen that carbon dioxide is produced when tartaric acid, carbonate of soda, and water are brought into contact with one another in a "gasogene," or even in a tumbler. Chalk and limestone are substances analogous to carbonate of soda in this respect that when they are brought into contact with solutions of acids of sufficient power, they are acted on in a similar fashion, carbon dioxide gas being evolved. This result is brought about by simply pouring some diluted hydrochloric acid upon some fragments of limestone or lumps of chalk at the bottom of a tumbler; a vigorous effervescence is produced, and the air in the tumbler soon becomes replaced by the evolved gas, which may be tested, as in Expt. 71, by lowering into the tumbler a lighted candle at the end of a wire, when the candle will be extinguished. In the manufacture of aerated water, the gas is produced by acting on chalk, &c., with a strong mineral acid (usually sulphuric acid, for the sake of cheapness), and is then pumped into the water and bottled under pressure.

Expt. 100. To produce a Current of Carbon Dioxide Gas.—In experiments with gases it is requisite to produce them inside a suitable vessel completely closed in, but furnished with an exit pipe or tube, through which the evolved gas can pass off in a current, so as to allow of its being collected in bottles or jars, &c., or otherwise treated as may be required. A convenient form of vessel, or *gas generator*, suitable for the production of gases where the materials require heating, is that indicated by fig. 55 (somewhat similar to that used for the production of ammonia, Expt. 74) the exit pipe being shortened and bent downwards; or an india-rubber tube may be attached so that the liberated gas can be led away through the flexible tube to wherever it may be required. In order to introduce more fluid, if requisite, the cork should be doubly perforated (fig. 56), or the flask closed by a two-necked cap (fig. 30), a "thistle funnel" being introduced through one orifice (passing to the bottom of the flask), and the exit pipe or "delivery tube" through the other perforation. When heat is not requisite, the thin flask may be advantageously replaced by a stout flat-bottomed bottle, such as a pickle-bottle, provided with a well-fitting doubly-perforated india-rubber bung (obtainable at the instrument dealers, fig. 29).

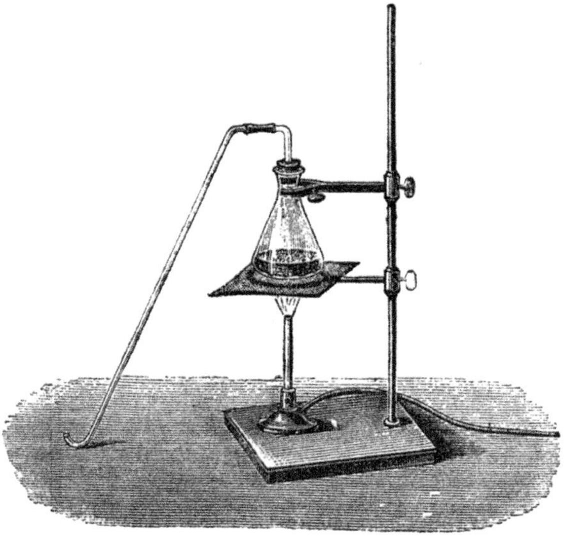

Fig. 55. Gas Generator.

Obtain some lumps (not powder) of *chalk*, or fragments of hard *limestone* or *marble*, and carefully drop them into the bottle, so as not to crack or "star" the sides; insert the cork in the tube, and if not air-tight make it so with wax, tallow, linseed meal made into a paste with a little water, or such like substances (a sound soft cork with well-bored holes of the right size, or an india-rubber bung with holes ready perforated in it, ought not to require any such "luting" to make it air-tight). Mix *strong hydrochloric acid* (solution of hydro-

Fig. 56. Gas Generator.

chloric acid gas in water, usually sold strong enough to fume visibly on taking the stopper out of the bottle) with three or four times its bulk of water, and pour a little of the diluted acid into the bottle through the funnel. As soon as the liquid comes in contact with the marble a vigorous effervescence will take place; the marble dissolves as in Expt. 99, only now, owing to the construction of the apparatus, the gas can only escape through the delivery pipe. When the evolution of gas slackens it can be renewed by pouring a little more acid in down the funnel tube.

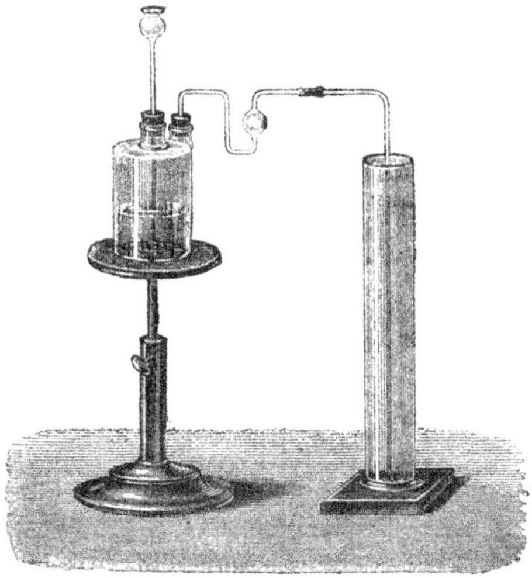

Fig. 57. Collecting Carbon Dioxide by Displacement.

Expt. 101. To illustrate the Heaviness of Carbon Dioxide.—In Expt. 75 we have seen that ammonia gas, being lighter than air, can be collected by displacement, the collecting jar being held mouth downwards; carbon dioxide, being heavier than air, can be also collected by displacement, but in this case the jar must be held mouth upwards. In order to collect the gas thus arrange the delivery tube of a carbon dioxide generator so as to pass to the bottom of a gas jar standing on the table mouth upwards (fig. 57); the air in the jar will be displaced by the heavier carbon dioxide, so that after a little while a lighted candle, lowered into the jar by means of a wire, will be extinguished, as in Expt. 71. Place a bit of lighted candle at the bottom of a second wide jar; the candle will keep on burning (unless the jar be too narrow). Pour into this jar the gaseous contents of the first jar, holding it just as though it were full of water (fig. 58); although the falling of the heavier carbon dioxide into the jar is not visible directly to the eye, yet the passage is at once rendered evident by the extinguishing of the candle.

Expt. 102. Another Illustration.—Obtain a large wide glass jar or "beaker," and half fill it with carbon dioxide from the generator; this is accomplished by passing the flexible tube to the bottom of the jar, and testing the atmosphere inside the jar with a lighted candle, removing the supply of gas when the level of the carbon dioxide is almost half-way up the jar, *i.e.*, when the candle burns freely in the upper part of the jar, but is extinguished on lowering into the lower half. Fill a bladder with air, and tie a tobacco pipe into the mouth, and then blow a soap bubble from the air in the bladder (Chapter XVIII.). Shake off the soap bubble from the pipe over the jar; the bubble, being a little heavier than air when filled with *cold* air, not with warm breath, will descend; but when it reaches the stratum of carbon dioxide in the lower part of the jar, it will fall no further, but will remain apparently floating half-way up the jar, because, although heavier than air, the soap bubble is lighter than the same bulk of carbon dioxide, and consequently behaves exactly as a cork would do if dropped into a jar half full of water; being heavier bulk for bulk than the air in the upper half of the jar, the cork falls downwards through the air; but when it meets the water, being lighter bulk for bulk than water, the cork falls no further, but floats on the surface of the water.

Fig. 58. Pouring Carbon Dioxide from one Jar to another.

Carbon dioxide gas is formed in Nature in a variety of ways, more especially though the action of heat upon limestone and similar calcareous matter underground, for which reason the gas often escapes from the soil in volcanic regions; and in consequence of the decay and oxidation of vegetable matter. Moreover, during fermentation of saccharine matter (Expt. 189), as in the brewing of beer, and the production of wine from grape juice, &c., carbon dioxide is largely formed. On account of the high density of the gas, it consequently tends to accumulate in caves and hollows in the earth, artificial wells and shafts of mines, brewer's vats, &c.; so that any one going down into such a cavity is liable to be suffocated on inhaling the "fixed air," because this gas cannot carry on the action in the lungs essential to life, which action requires the presence of sufficient free oxygen. Many fatal accidents happen yearly on account of this tendency, and the neglect of simple precautions to ascertain whether the air in a well or cave, &c., is dangerous to breathe or safe. Of these precautions the simplest is to lower down a lighted candle or lantern before making a descent; if the light continues burning, the air is not so impure as to be dangerous on account of the liability to suffocation; but if the light become extinguished, the atmosphere is unsafe to breathe, and any one descending will probably be suffocated for want of oxygen, just as the candle flame goes out on account of the same want.

When explosions take place in collieries, the air in the underground passages becomes irrespirable, because the oxygen present is consumed in burning the inflammable gas or "fire-damp" that exudes from the coal, so that a mixture of nitrogen and carbon dioxide with little or no free oxygen is produced: this suffocating gas is termed the *after-damp*, and is frequently more dangerous to life than the explosion itself.

Expt. 103. Diffusion of Carbon Dioxide into the Air.—Pass the delivery tube of a carbon dioxide generator to the bottom of a wide tumbler or pudding basin, so as to fill it with gas; a lighted candle lowered into the basin will be extinguished, as in Expt. 101. Set the basin aside on a shelf in a cupboard for twenty-four hours, and then test it again with a lighted candle; the candle will now burn freely; showing that the heavy carbon dioxide has diffused away, and the lighter air taken its place. If a very tall narrow cylinder be used instead of a tolerably wide shallow vessel, a longer time than twenty-four hours will probably be requisite before the carbon dioxide has diffused away sufficiently to enable a candle to burn.

Expt. 104. To produce Hydrogen Gas.—Place some granulated zinc (Expt. 16) in the gas generator instead of chalk (after having carefully cleaned out the bottle), and then pour diluted hydrochloric acid down the funnel, so as to cause hydrogen to be evolved (Expt. 10). Arrange the delivery tube so as to point upwards, connecting it with the generator by means of a piece of flexible tubing. Hydrogen, being lighter than air, may then be collected by placing a wide-mouthed jar bottom upwards over the issuing jet of gas (fig. 59).

Expt. 105. To illustrate the Lightness of Hydrogen.—Having collected a jar of hydrogen, place it on the table mouth downwards. Hold the jar in one hand, and a lighted taper in the other; invert the jar, and instantly apply a light; the hydrogen will take fire and burn, possibly with a slight explosion if any air is mixed with it during collection (compare Expt. 207), but without doing any damage. Refill the jar with hydrogen from the generator, and set it again on the table, but this time mouth *upwards;* after the lapse of a minute apply a lighted taper, to the jar, when nothing whatever will occur; the taper, if lowered into the jar, will burn, but no firing of gas in the jar will take place (provided sufficient time has elapsed), because all the hydrogen will have floated upwards and escaped, its place being taken by ordinary air. Balloons made of thin material, or soap bubbles, filled with hydrogen will be buoyed up and ascend in the air, because of this action (Expt. 307).

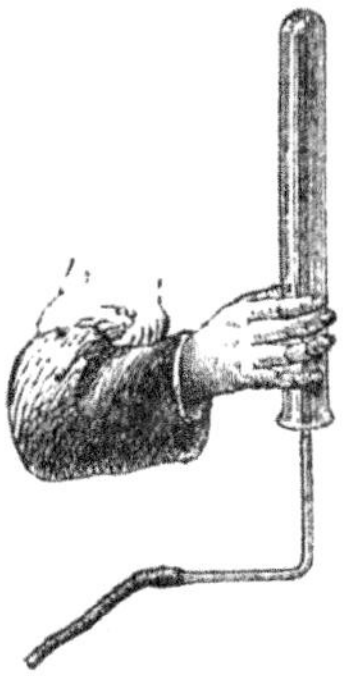

Fig. 59. Hydrogen collected by Displacement.

Expt. 106. Another Illustration.—Fill a jar with hydrogen as before; hold a second jar of the same size (as shown in fig. 60), and *pour the hydrogen upwards* from the first jar into the second, by gradually depressing the former as indicated.

Fig. 60. Hydrogen poured upwards.

After the jars have been thus held for half a minute, separate them, and apply a light to each. The gas in the upper jar will burn with a slight explosion, as some air is sure to have got mixed with it during the transfer; whilst that in the lower jar will not burn at all, again showing that the lighter hydrogen has floated upwards.

Expt. 107. Diffusion of Hydrogen into the Air.—Fill two jars with hydrogen as before, and place them on a shelf in a cupboard, each upon a tripod stand, mouth downwards, so that the air has free access to the mouths of the jars. Apply a light to the mouth of one jar, and the hydrogen will burn; if the light be applied by thrusting up a candle at the end of a wire, or a long wax taper, well into the jar, the taper will light the hydrogen at the mouth of the jar, but will be itself extinguished inside (fig. 61), because the hydrogen gas present cannot keep up the same chemical changes going on in the candle flame that will take place in the air. By cautiously lowering the candle again so as to withdraw it from the jar slowly, it may be relit by the burning hydrogen as it emerges.

Let the second jar remain at rest for twenty-four hours, and then test it by thrusting up a burning taper as before. By this time the hydrogen will have all diffused away and become replaced by air, notwithstanding that the latter is the heavier, and has to rise into the jar. If the light be applied before all the hydrogen has escaped, the mixture of air and residual hydrogen in the jar will explode with a more or less loud pop, but the explosion will not be violent enough to do any damage.

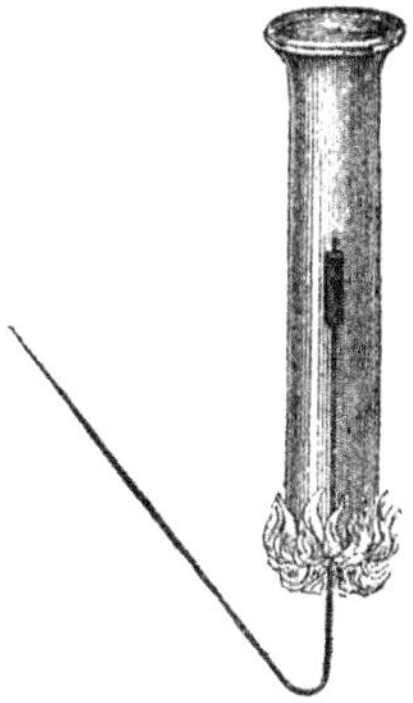

Fig. 61. Jar of Hydrogen and lighted taper.

Certain kinds of porous substances, more especially dry plaster of Paris and unglazed pottery, allow gases to diffuse through them with great facility, so that gases will pass much more readily under this influence than when forced through by pressure. In consequence it becomes possible to demonstrate that all gases do not diffuse equally rapidly under the same circumstances, but that the lighter the gas the more quickly will it diffuse.

Expt. 108. Diffusion of Hydrogen more Rapid than that of Air.—Make some plaster of Paris into a cream with water, and place it in a cup so as to fill it about a quarter of an inch deep. Dip into the cream a piece of glass tubing about a foot long and half an inch internal diameter; shortly the plaster will set, so that on lifting the tube out of the cup the end will be blocked up by a plug of plaster. Let the whole stand in the air for a day or two, and then bake the tube before a fire or in the oven so as thoroughly to dry the plaster. If the plug is sound, it will now block the tube so far completely that air will only pass into it through the plug very slowly, so that by sucking a little air out and applying the tongue to the other end (taking care first to smooth the glass with a file to avoid the chance of cutting the tongue), the tube will be

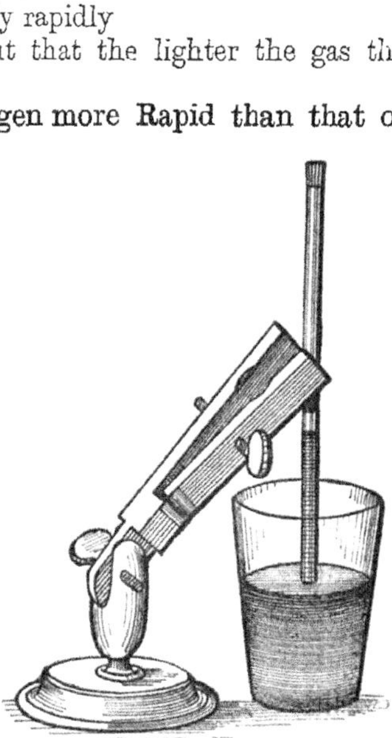

Fig. 62. Diffusion of Hydrogen.

pressed to the tongue by the outside atmospheric pressure, as a test-tube would be if similarly treated.

Fill the tube with hydrogen by displacement, just as the tube in Expt. 76 was filled with ammonia gas, and quickly place it mouth downwards in a tumbler of water; you will soon see that the water rises in the tube, so that by and by it will stand at a higher level inside than it does out (fig. 62); showing that the hydrogen gas has escaped outwards by diffusion through the porous stucco plug more rapidly than air has diffused inwards.

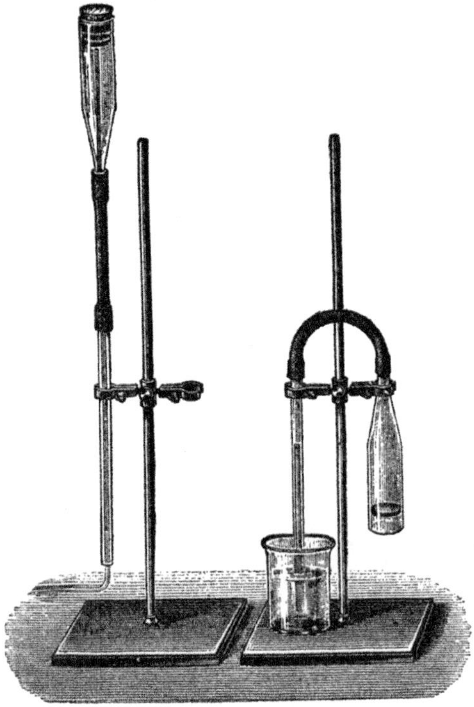

Fig. 63. Diffusion of Hydrogen.

Fig. 63 illustrates a better way of showing the same experiment; the wider end of the tapering piece of glass tube is plugged with stucco as before, and carefully dried; the delivery tube of a hydrogen generator is passed up into the diffusion tube so as to displace all air, and then withdrawn; the hydrogen reservoir at the top is then bent over as indicated by means of the connecting piece of flexible tubing, when diffusion takes place as before, notwithstanding that the hydrogen has now to pass *downwards* through the plug in opposition to gravity.

Expt. 109. Another Illustration.—Obtain a small circular porous pot, 5 or 6 inches high and 2½ or 3 inches diameter, such as is used for certain kinds of galvanic battery. Fit into the mouth of the pot an india-rubber bung with a piece of glass tube (*a*) passing tightly through it, and also passing just through a cork in one neck of a two-necked bottle at a lower level: through the other neck passes another piece of glass tube (*b*) reaching to the bottom of the bottle, and terminating upwards in a drawn-out point or jet (fig. 64). If a large jar of hydrogen be now placed over the porous pot, the gas will diffuse inwards through the sides of the pot so much more rapidly than the contained air will pass outwards, that a considerable increase of pressure results; the effect of which is to force the water with which the two-necked bottle is half filled upwards through the jet, forming a fountain. If all the junctions are properly air-tight, the stream of water may be made to rise some 8 or 10 inches in the air with properly proportioned tubes, &c.

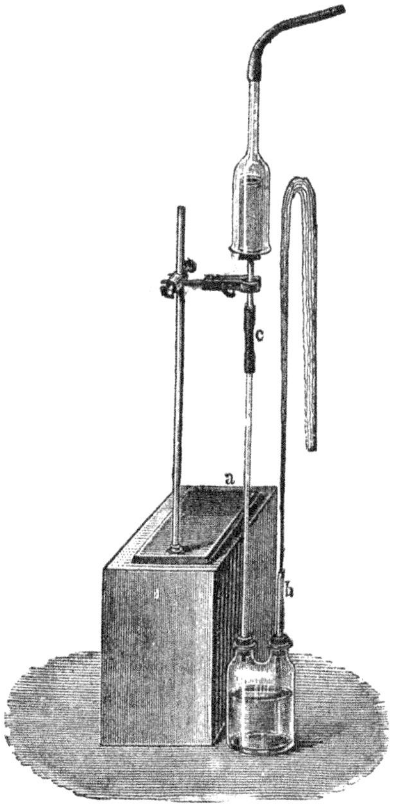

Fig. 64. Diffusion of Hydrogen.

Ordinary coal gas may be substituted for hydrogen, with somewhat inferior results.

Expt. 110. Diffusion of Carbon Dioxide less Rapid than that of Air.—By means of the flexible joint at *c* (fig. 64), the porous pot may be made to hang downwards; if now a vessel full of carbon dioxide (filled as in Expt. 101) be so placed that the pot hangs into it, air will diffuse outwards from the pot, and carbon dioxide inwards; the latter gas, however, passes much more slowly than the former, so that a diminished pressure now results inside the arrangement; in consequence, air is sucked upwards from the two-necked flask, and bubbles of air are accordingly drawn into it from the outside atmosphere through the tube *b*.

The two last experiments illustrate a general rule always observed in connection with diffusion, viz., *the lighter the gas, the more rapidly does it diffuse;** hydrogen being considerably lighter than air, diffuses much more quickly; whilst carbon dioxide, being more dense than air, diffuses less rapidly. Other experiments in connection with certain of the chemical properties of carbon dioxide and hydrogen are described subsequently (Expts. 115, 143, 144, 154, 155, 168, 207, 208, 215, 216, &c.).

---

§ 3. Solution and Separation from Solution by actions involving Chemical Changes.

CHAPTER X.

Solution of Solids in Liquids and Separation of Solids from Liquids.

It has been already explained (Chapter I.) that whilst, in the narrowest sense of the word, the term "solution" is confined to cases where the dissolved body and the solvent react upon one another so as to produce a solution from which the two substances are again separable by simple means (such as cooling, evaporation, diluting with some other fluid so as to precipitate from solution, &c.), the term is also used in a wider sense to indicate all cases where a solid disappears on treatment with a liquid, no matter whether the solid is chemically changed by the process or not; when such chemical change takes place, in general the original solid can no longer be recovered from the solution formed by any simple means, such as evaporation, &c.; what substance is thus obtained is no longer the original solid used, but is something different therefrom, being altered by the chemical action taking place. Sometimes the chemical change is not accompanied by any visible alteration, except the disappearance of the dissolved solid in the solvent fluid; sometimes a coloured solid dissolves to a colour-

* The exact stating of the law is that *the rate of diffusion is proportionate to the reciprocal of the square root of the density.* Thus the densities of the gases hydrogen, oxygen, and chlorine are nearly in the proportions of 1, 16, and 36; the square roots of which number are respectively 1, 4, and 6. Hence the rates of diffusion are in the ratio 1, $\frac{1}{4}$, and $\frac{1}{6}$; or hydrogen diffuses four times as quickly as oxygen, and six times as rapidly as chlorine.

less or differently coloured solution; and sometimes **effervescence** (or bubbling due to formation of gas which escapes) is brought about.

Expt. 111. To dissolve Magnesia by Chemical Action.—As an illustration of the first of these kinds of complex solution, where the chemical change is invisible, procure a little *calcined magnesia*, and pour over it strong *white vinegar*, little by little, stirring up well after each addition; by and by the magnesia will entirely dissolve, and in place of the white powder and the sour fluid used you will have a tolerably clear liquid of disagreeable taste, but not particularly sour (unless you have used too large a quantity of vinegar). If you put the liquid into an evaporating basin and boil it down, you will be unable to obtain from it the same calcined magnesia that you originally employed, either by crystallisation or cooling and standing, because a chemical change has taken place; *the acid of the vinegar (acetic acid) and the magnesia have so* acted on one another that both are altered into something quite different from either, viz., a compound called "acetate of magnesia," differing from calcined magnesia in being readily soluble in water, and from vinegar in not being sour.

Expt. 112. Another way of dissolving Magnesia.—Instead of acetic acid, use diluted *sulphuric acid* in quantity not quite sufficient to dissolve the whole of the magnesia; filter the solution (Expt. 56), and evaporate the clear liquid in an evaporating basin over a lamp; when it has become sufficiently concentrated, crystals will form on allowing the fluid to cool. These crystals are *sulphate of magnesia* (or *magnesium sulphate*, otherwise known as *Epsom salts*), and are obviously different from the original magnesia, as they will readily dissolve in pure water, giving a solution of somewhat disagreeable saline taste, whereas the original calcined magnesia would not dissolve at all in water, and had little or no taste of any kind. Epsom salts derive their name from being contained in the water of a spring at Epsom in such quantity that, when the water was evaporated, this salt crystallised out. Many other mineral waters contain magnesium sulphate, and owe a good deal of their medicinal efficacy to this circumstance.

Expt. 113. To dissolve Coloured Oxide of Lead or Oxide of Mercury by Chemical Action, producing a Colourless Fluid.—In the last experiment a white powder was dissolved in a colourless fluid (if white vinegar was used), yielding as result a colourless solution. *Litharge*, or *oxide of lead*, when powdered is a brownish-yellow substance; if this is treated with diluted nitric acid* like

* A few drops of strong nitric acid to a wine-glassful of water will dissolve as much litharge as will lie on a sixpence.

the magnesia and vinegar, it dissolves, forming not a yellow solution, as might be expected, but a colourless one. If the liquid is not quite clear, it should be filtered (Expt. 56); the clear filtered solution may be evaporated in a basin over a water-bath (Expt. 89) or small flame until only a little fluid is left; if this is left awhile white crystals will form, consisting of the substance called *nitrate of lead*, the nitric acid and the litharge being both altered by the chemical action taking place between them, giving rise to nitrate of lead as the product.

If you substitute *white vinegar* for diluted nitric acid in this experiment, you will finally obtain the substance called *acetate of lead*, or "sugar of lead," on account of its sweet taste. Be careful not to swallow this sweet substance in any quantity, as it is poisonous; when pickles come in contact with lead (as sometimes happens with pickle-bottles covered over with a metal capsule when the metal contains lead), sugar of lead is sometimes formed, by the action of the vinegar on the lead; so that people partaking of the pickles are made ill in consequence.

Oxide of mercury is an orange-red powder; if you substitute this for litharge, diluted nitric acid will dissolve it, forming not a red, but a colourless fluid, containing the compound termed *nitrate of mercury*, formed, as in the other cases, in virtue of the chemical action taking place between the acid solvent and the metallic oxide dissolved.

Expt. 114. To dissolve a Red Solid (Copper) in a Colourless Fluid (Ammonia Water) producing a Blue Solution.—Obtain some "spirits of hartshorn" (otherwise called solution of ammonia or *liquor ammoniæ*) and some *copper shavings* or turnings. Cram these loosely into a bottle, and pour in a teaspoonful or two of the ammonia solution, shaking it about so as to wet the copper therewith; the liquid will soon assume a beautiful blue colour, owing to the circumstance that, under the conjoined influence of the air in the bottle and the ammonia, a little of the red copper becomes converted into a soluble compound of a blue colour.

To prove that the presence of air is necessary to make the copper dissolve, fill a small bottle quite full of ammonia solution, and then drop into it some bits of clean bright copper (not dirty or dull on the surface); cork the bottle, taking care that there is no air in it, and that it is quite full of liquid. You will find that even after many hours standing the liquid does not become blue (provided the copper is quite bright and all air is excluded). Now pour out most of the liquid, leaving only a few drops, and letting air enter without allowing the bits of copper to fall out. Cork the bottle again, and shake up the contents, and you will see that the remaining ammonia solution becomes blue, now that the air too has got access to the copper.

Copper is thus dissolved in large quantity in manufacture by the simultaneous action of air and ammonia solution, the blue solution of copper being employed for preparing certain kinds of waterproof paper, canvas, and such like materials, by what is called the "Willesden process."

We have already had illustrations of the solution of solids where the chemical action taking place resulted in the evolution of gas: thus, in Expts. 99 and 104, the solution of chalk and of zinc in hydrochloric acid produced effervescence, due to the formation of carbon dioxide and of hydrogen gases, in each case respectively, and in Expt. 13 sulphuretted hydrogen gas was evolved by the action of hydrochloric acid on sulphide of iron. The following experiments further illustrate this kind of action.

Expt. 115. To produce Carbon Dioxide Gas from Barium Carbonate.—Chalk and limestone consist of *carbonate of calcium* (or carbonate of lime, both names being familiarly employed, the former being somewhat the more scientifically exact), which for present purposes may be regarded as a compound of quicklime and carbon dioxide; so that when limestone is heated, it breaks up or decomposes into carbon dioxide, which escapes as a gas, and quicklime which remains as a solid, this operation being the essential part of the manufacture of quicklime from limestone, ordinarily termed **lime burning**. Conversely, when lime is dissolved in water and carbon dioxide gas brought in contact with the solution, the lime and carbon dioxide reunite and produce carbonate of lime, chemically identical with chalk and limestone (Expt. 152). *Barium carbonate* is a mineral somewhat resembling limestone; when acted upon by hydrochloric acid, it is decomposed in a similar fashion; in the case of limestone, carbon dioxide gas is evolved, and the lime becomes transformed into a soluble substance called *chloride of calcium*, which remains dissolved in the fluid formed in the carbon dioxide generator (Expt. 99); in the case of barium carbonate, carbon dioxide is also evolved, whilst the fluid contains in solution a substance analogous to chloride of calcium, called *chloride of barium* or *barium chloride*.

Treat barium carbonate with hydrochloric acid diluted with water; when the effervescence has ceased, filter the liquid (Expt. 56), and evaporate it to a small bulk; on allowing it to cool and stand crystals of barium chloride will be formed.

In the same kind of way, by treating barium carbonate with diluted nitric acid, carbon dioxide gas will escape, and a liquid will be produced, from which crystals of *nitrate of barium* will deposit after evaporation.

Expt. 116. To dissolve Copper in Nitric Acid.—Pour some diluted nitric acid (*aquafortis*) on fragments of copper; these will rapidly dissolve, forming a blue solution, whilst a gas (*nitric oxide*) is evolved with effervescence. As with the chalk, chemical change takes place, so that metallic copper cannot be regained from the solution by simple evaporation or cooling; the gas now evolved is different from either hydrogen or carbon dioxide, one principal characteristic of it being that it forms deep red fumes when it comes in contact with the air (*vide* Expt. 223). A saline compound termed *nitrate of copper* is produced in this experiment, corresponding with the chloride of calcium formed in the previous one.

Expt. 117. Another Way of dissolving Copper.—Put a few copper filings into a test-tube, pour over them a teaspoonful of strong *sulphuric acid* (oil of vitriol), and heat cautiously with a lamp. As the acid gets hot a chemical action begins (not perceptible whilst the materials are cold); the copper gradually dissolves, and effervescence takes place; but the gas now evolved (*sulphur dioxide*) is different from either of those obtained in the last two experiments, possessing a strong smell of burning brimstone. When the contents of the test-tube have cooled, pour them slowly into a little water; a great deal of heat will be produced, and probably some steam, owing to the action of the water on the acid (Expt. 98), and a blue solution will be formed, containing the compound *sulphate of copper*, corresponding with the *nitrate of copper* formed in the last experiment.

Caution.—Be careful in mixing the sulphuric acid with water after the heating is finished. Let the whole cool, and pour the acid into the water, and not *vice versâ*. Compare Expt. 98.

The gas (sulphur dioxide) produced in this experiment in some respects resembles carbon dioxide (Expt. 99) in its chemical properties. If a little of it be collected in a jar by displacement, holding the jar mouth upwards (as in fig. 57, and using a generator like that shown in fig. 55), it will be found to extinguish a lighted candle when this is lowered into the jar at the end of a wire or string; by pouring a little water into the jar, and shaking it up vigorously, the water will dissolve the gas, forming a solution of *sulphurous acid*, such as that used in Expts. 135 and 164. If a little solution of *caustic soda* be used instead of water, the soda will combine with the gas (as with carbon dioxide, Expt. 143), forming a compound termed *sodium sulphite*; from which the gas sulphur dioxide can be reproduced by treating it as directed in

Expt. 161. When a gas is to be brought in contact with a fluid so as to be dissolved therein, as in these cases, the arrangement shown in fig. 66 (p. 139) is usually more convenient than filling a jar and then shaking up with the fluid; the gas produced in an appropriate generator being made to *bubble through* the liquid employed.

Expt. 118. To dissolve Iron by Chemical Action.—Put some iron filings into a test-tube, and then pour upon them some diluted hydrochloric acid. Effervescence will ensue, but in this case due to the escape of *hydrogen* gas, produced exactly as it was formed in Expt. 10, where zinc was used instead of iron. When the action has gone on for some time, filter the liquid in the generator through a paper filter (Expt. 56); a solution of the salt called *ferrous chloride* (not the same thing as perchloride of iron or *ferric chloride*) will then be obtained. Add to a little of the solution some solution of *ferricyanide of potassium* (*red* prussiate of potash, not the same substance as the *yellow* prussiate used in Expt. 57); a blue precipitate of a substance called *Turnbull's blue* (Expt. 83) will be at once formed, instead of a brown sherry-coloured fluid, such as would be formed with perchloride of iron and ferricyanide of potassium. If diluted sulphuric acid be substituted for hydrochloric acid in this experiment, solution of the iron with evolution of hydrogen will equally take place, producing *sulphate of iron* (ferrous sulphate) instead of chloride; by filtering the solution, evaporating and allowing to stand, this salt may be obtained in pale green crystals, which become rusty on standing moist in contact with the air for some time, owing to a further chemical change somewhat akin to the oxidation of metals (Chapter XIV.).

Caution.—Be careful how you add sulphuric acid to water in this experiment (*vide* Expt. 98).

Expt. 119. To dissolve Silver by Chemical Action.—Nitric acid has the power of dissolving most metals in the same way as copper (Expt. 116). Place a small silver coin in a vessel containing concentrated nitric acid diluted with about twice its bulk of water; the metal will be speedily attacked and dissolved. As silver coins contain a little copper as well as silver (added to harden the metal and enable it to resist wear and tear better), the solution will have a blue colour, and will contain a mixture of the nitrates of copper and silver. If, instead of a coin, a fragment of pure silver be used, a colourless solution will be obtained, containing no copper. The liquid may be cautiously evaporated to dryness, preferably on a steam bath (Expt. 89), when solid *nitrate of silver* will be obtained, which may be dissolved in water, and the solution bottled for future use.

Gold is a metal unlike most others, being entirely unaffected by nitric acid (*vide* Expt. 230).

Expt. 120. To separate Silver from Copper.—Dissolve a silver coin in nitric acid, as in the last experiment, and add to the solution some clear brine (solution of common salt or *chloride of sodium*). This substance will react by double decomposition on the nitrate of silver, as in Expt. 37, forming *chloride of silver*, which, being insoluble, will precipitate in curdy flakes. When these have subsided, the clear liquid may be poured off, and treated with iron by immersing a knife blade in the fluid, when the copper will be thrown down as a red film on the iron (Expt. 9). To obtain the silver, stir the chloride of silver up with water, let it subside, and pour away the liquid, and repeat the operation two or three times to remove all the copper solution; this is called "washing by decantation." Now pour some solution of caustic soda on the chloride of silver, and add a good-sized lump of sugar, or better, a teaspoonful of *milk sugar* (the sugar contained in milk). Keep the whole hot or boiling for some hours, and then wash the blackish powder resulting by decantation again; the chloride of silver will have become mostly transformed into *spongy metallic silver* by the treatment, so that after washing, the powder may be treated with nitric acid, which will dissolve almost the whole. The solution should be filtered to separate any chloride of silver that has not become decomposed during the treatment with soda and sugar, and may then be evaporated to dryness, furnishing *pure silver nitrate*.

Expt. 121. Separation of Solid Substances from Solutions (accompanied by Chemical Changes) as Precipitates of various Colours.—In the preceding experiments we have had instances of chemical action taking place between substances insoluble in water (oxide of magnesium, or metallic copper, &c.) and other substances soluble in water (hydrochloric acid or nitric acid), with the result of producing products of chemical change soluble in water. This kind of action may be inverted, so that by taking two substances, both soluble in water, and bringing them together, one or more different substances insoluble in water may be produced by chemical action. In such cases the insoluble product is usually said to be **precipitated**; some illustrations of this kind of action have already come before us, *e.g.*, in Expt. 37, where a white precipitate was produced by the chemical action of nitrate of silver solution on brine; and in Expt. 57, where a blue precipitate of prussian blue was formed by the chemical action of ferrocyanide of potassium solution on solution of perchloride of iron. In the first of these cases a *white* precipitate is formed on mixing two *colourless* solutions; in the second, two *pale yellow* fluids give a deep *blue* precipitate. The colour of a precipitate cannot be inferred from the colour of the solutions employed to prepare it; thus the two colourless fluids obtained by dissolving in water *iodide of potassium* and *perchloride of mercury* respectively give a *scarlet* precipitate on pouring the second into the first; if now you add more of the first fluid and stir up, the scarlet precipitate again dissolves, forming a colourless solution; on further adding more of the second fluid the scarlet precipitate reappears; and so on, alternately, as long as you please. If to the solution of perchloride of mercury you add a solution of *caustic soda*, a bright *yellow* precipitate is formed. If you mix the perchloride of mercury solution with iodide of potassium solution in such quantity as to dissolve the scarlet precipitate first formed, and then add some caustic soda solution, you obtain a nearly colourless fluid, which forms an *orange-brown* precipitate on adding to it a very small quantity of weak solution of ammonia.

A colourless solution of *sulphuretted hydrogen*, when added to a colourless solution of *tartar emetic*, gives a beautiful *orange* precipitate; if to a colourless solution of *sulphate of cadmium*, a bright *yellow* precipitate;* and if to a colourless solution of *acetate of lead*, a *black* precipitate; whilst it may also be made to form a *white* precipitate by adding it to the colourless fluid obtained by adding a little solution of ammonia to solution of *sulphate of zinc*, so as to produce a white precipitate, and then adding more ammonia to this until the liquid becomes clear again.

Expt. 122. Production of a Precipitate by simply Heating.—As already mentioned, in general solids dissolve more freely in hot water than in cold, so that it is a very unusual thing to find that a solution forms a precipitate on heating, without the occurrence of any chemical change, simply owing to the fact that more substance is present than hot water can dissolve, although the same quantity of cold water could keep it permanently dissolved. Such exceptional cases, however, sometimes occur; for instance, a clear saturated cold solution of the substance termed *butyrate of calcium* forms a precipitate, and becomes milky on heating, because this salt is more soluble in cold water than in hot, so that on heating a part is necessarily thrown out of solution, just as in ordinary cases a solution saturated hot deposits some on cooling (Chapter V.).

Cases where precipitates form on heating, owing to the occurrence of chemical changes, are much more common. For example,

* This yellow substance is the "cadmium yellow" used by artists as one of the best and most permanent yellow coloring matters known.

if a little white of egg be mixed with four or five times its bulk of water, and some of the clear solution (filtered if necessary) gently heated in a test-tube, the fluid will by and by become turbid, because the "albumin" contained in the white of egg is affected by heat chemically, and converted into a somewhat different substance not soluble in water. White of egg and other albuminous fluids (like blood deprived of the clot and red substance) are used in the arts to clarify sugar, syrups, jellies, and the like; as the albumin solidifies on heating the liquid, it catches up and entangles small suspended particles, and enables the liquid to be strained or filtered much clearer and brighter than would otherwise be possible.

Take some *chrome alum*, and dissolve it in water; to a teaspoonful of the purplish-green solution, add drop by drop *caustic soda* solution; a green precipitate will first form, which on addition of more soda (with stirring) will redissolve to an emerald-green fluid; on cautiously boiling this in a test-tube, the liquid will soon become turbid, and on standing a short time will allow a green precipitate to subside, whilst the supernatant liquid is colourless. This arises in virtue of a chemical change taking place in the nature of the dissolved compound giving the green colour, solely through the action of heat.

Another instance of every day occurrence is the effect of heat upon "hard" natural water; a "crust" or precipitate of calcareous matter is produced on boiling such water, forming the "furr" inside a tea-kettle, and the "boiler crust" or "scale" in an ordinary steam boiler; this arises from the chemical action taking place under the influence of heat on the substances dissolved in the "hard" water (*vide* Expt. 155).

Expt. 123. To test for Grape Sugar.—Prepare some solution of *sulphate of copper*, add to it some powdered *tartaric acid*, and then add enough solution of *caustic soda* to convert the whole into a clear pretty deep blue fluid; if not clear at first, add more tartaric acid and then more soda, and shake up or stir well. To a little of this blue solution in a test-tube add a few drops of a weak solution of *grape sugar*, or of *treacle*, *honey*, or *golden syrup*, largely diluted with water. On heating the mixture a precipitate will be formed in virtue of chemical action; according to the strength of the solution and other circumstances, the colour of the precipitate will be an orange-yellow, brownish deep red, or dark reddish-brown, in all cases very different from the original blue fluid. By this test you may prove the presence of grape sugar in the juice of a sweet grape, or in the liquid got by cutting up a fig or a date or a raisin in shreds and soaking these in water, the liquid being preferably filtered (Expt. 56) before adding it to the blue copper test liquor. In a certain kind of disease, known as *diabetes*, sugar is contained in the urine; and this test is medically employed for the purpose of examining the progress of the patient towards recovery or otherwise, by noticing whether much or little sugar is present from day to day, and regulating the treatment as this and other symptoms render requisite.

Expt. 124. Formation of a Precipitate, and Re-solution thereof by means of the same Fluid.—It is not unfrequently observed that when a precipitate is formed by the careful addition of a certain quantity of a given solution or test liquor to another fluid, a further addition of the same test liquor will cause the precipitate to dissolve again and form a clear fluid; in such cases the precipitate is said to be "soluble in excess" of the test liquor. Several such cases have already been noticed, *e.g.*, in Expt. 121, where a scarlet precipitate, formed by adding iodide of potassium to perchloride of mercury (otherwise termed corrosive sublimate), was redissolved on adding excess of the first-named solution; and where the white precipitate formed by adding ammonia to zinc sulphate was redissolved on adding excess of the former; and in Expt. 123, where the green precipitate formed on adding caustic soda to chrome alum was soluble in excess of caustic soda.

Take a few drops of solution of *sulphate of copper* in a test-tube, and slowly add, drop by drop, solution of *ammonia*. A blue precipitate will first be formed, which on further addition of ammonia will be redissolved, forming a deep azure blue solution very much deeper in tint than the original sulphate of copper solution. This deepening in tint serves as a test for copper. Thus fill a test-tube nearly full of water, and add to it one small drop of sulphate of copper solution, and shake up. On looking downward into the test-tube held vertically, with a piece of white paper underneath, the water will appear little if anything different in colour from that seen in another test-tube full of pure water similarly viewed; but on adding a few drops of ammonia solution to the first test-tube and shaking up, a very distinct blue coloration will now be visible on looking down the tube against the white paper, and comparing the colour observed with that in the pure water tube held by its side.

Expt. 125. To detect minute Quantities of Lead or Copper in Drinking Water.—Sometimes natural water contains small quantities of lead or copper, derived from minerals containing these metals in the mountains, &c., where the water is collected from the natural rainfall; more frequently pure water supplied for domestic use gets contaminated with lead by passing through leaden pipes, or being stored in leaden cisterns. Such water is very objectionable, being often highly injurious to health; the presence of any serious contamination of this kind may be detected by a *comparison colour test*, worked somewhat after the fashion of the last experiment. Procure two test-tubes as long as possible, better still, two glass cylinders a foot high, standing on flat bottoms (fig. 40); fill each with the water to be tested, and add two or three drops of dilute pure hydrochloric acid to each. To one of the cylinders now add a teaspoonful of freshly prepared clear solution of *sulphuretted hydrogen*, and shake up the water to promote proper intermixture. If there is no lead or copper in the water, both cylinders will appear alike when you look downwards through the long columns of fluid (with a piece of white paper underneath); but if there is any poisonous metal present, a more or less marked difference in colour will be noticed, a distinctly yellower or browner shade being communicated to the contaminated water by the sulphuretted hydrogen than it originally possessed.* If the test is to be made an excessively delicate one, a large bulk of water should be treated with a few drops of hydrochloric acid, and evaporated in a clean glass or porcelain vessel on a water-bath (Expt. 89), until reduced in volume to such an extent as only just to fill the two cylinders; obviously, if the water has been reduced to one-tenth of its original bulk by evaporation, anything present in the water will be concentrated tenfold, and the test accordingly made ten times as delicate.

Expt. 126. To detect Copper and Lead in Pickles, Preserved Provisions, &c.—Copper compounds are sometimes employed to give a bright green tint to preserved vegetables, such as tinned peas, &c., and certain kinds of pickles. In the latter case the presence of copper can generally be detected by filtering some of the vinegar clear into a test-tube and adding solution of sulphuretted hydrogen, when a considerable darkening or formation of black precipitate will take place. To find the copper in the solid vegetables, these should be burnt to ashes over a spirit lamp or Bunsen lamp in a little tray of platinum made by bending up the sides of a slip of platinum foil, the tray being supported by a stand or tripod, preferably resting on a triangle made by threading an iron wire through three bits of clay tobacco pipe stem (fig. 65). The ashes are treated with a few drops of diluted nitric acid, and the liquid filtered through a small filter; a few drops of the

* Good water may naturally have a slightly yellowish shade when thus viewed, owing to traces of clayey matter, peat, &c., being present in quantity too small to affect the quality of the water materially, but large enough to be just visible to the eye when thus examined.

filtered fluid will turn black with sulphuretted hydrogen solution if copper be present; whilst other tests for copper are described in Expts. 95 and 124.

When copper or brass cooking utensils are employed in the kitchen, it often happens that these impregnate the food cooked therein with copper, more especially when the vessels are not scrupulously cleansed before use. Acid fluids and substances containing sugar (also oily matters, such as butter, &c.), if left in contact with copper and air for some time, will often cause a little copper to become oxidised and dissolved, much as ammonia does under similar conditions (Expt. 114). Hence such vessels ought to be *well tinned inside*, so as to prevent the copper or brass from coming into direct contact with the food. In case it is desired to test food to see if copper has been thus taken up, the method employed is usually that above described, viz., drying and burning to ashes (or **incinerating**) in a large platinum dish, and then testing the ashes.

Fig. 65. Pipe-Clay Triangle.

A similar process (with appropriate modifications as to the final tests) is used for the detection of *lead*. One of the best tests in this case is to evaporate a little of the nitric acid solution of the ashes to dryness over a steam bath (Expt. 89), so as to get rid of any excess of nitric acid; the residue is dissolved in a few drops of water, and a drop of solution of *potassium chromate* added, when a yellow precipitate is formed of *chromate of lead*,* if lead is present to any considerable extent; whilst more minute quantities still can be found by means of sulphuretted hydrogen water. Lead is often contained in tinned vegetables (especially where sugar, acids, or sweet juice is largely present, as with canned peaches and apricots, pine apples in slices, &c.), being derived from impurities contained in the tin, and from the solderings of the cans.

Expt. 127. To pour out apparently Blue Ink, Brown Sherry, or Milk from the same Water Bottle into different Wine Glasses. —A curious effect may be produced by taking advantage of the property possessed by ferrocyanide of potassium of forming differently coloured precipitates with different metallic compounds. Rinse out three wine-glasses, one with a solution of *perchloride of iron*, the second with a solution of *sulphate of copper*, and the third with a solution of *sulphate of zinc*, so that a few drops of each

* "Chrome yellow," used by artists, is essentially chromate of lead prepared substantially by this process, but with various modifications for the purpose of obtaining different shades.

liquid may remain in the glasses. Dissolve a few grains of *ferrocyanide of potassium* in a pint of water, and put the solution into a white glass water bottle or decanter. If the strengths of the respective solutions are properly proportioned (which you can easily adjust by making one or two preliminary trials) you will find that on pouring out what appears to be plain water from the decanter into the three glasses respectively, the first will apparently be filled with blue ink, the second with a liquid somewhat resembling brown sherry, and the third with milk. Be careful not to allow any one to drink your fictitious beverages, as the taste by no means equals the appearance, whilst one may easily be made sick or ill with the metallic compounds, which are all more or less injurious to swallow.

A similar kind of effect may be produced in a variety of other ways. Thus fill the decanter with a weak solution of *perchloride of iron* to which a small quantity of *sulphuric acid* has been added, and rinse out five wineglasses, one with solution of *ferrocyanide of potassium*, another with a moderately strong solution of *sulphocyanide of potassium* (a different compound); a third with a very much weaker solution of the same salt; a fourth with a strong solution of *carbonate of potassium* (or instead sprinkle inside a few grains of *carbonate of sodium*); and the fifth with a solution of *nitrate of barium*. On filling up the five glasses from the decanter the first will appear to be filled with blue ink, the second with blood * or port wine (according to the strength of the solution) the third with sherry (if the solutions are not too strong), the fourth with champagne or ginger ale (on account of the effervescence produced by the action of the sulphuric acid on the carbonate, as in Expt. 99), and the fifth with milk, the white precipitate being now due to the formation, not of ferrocyanide of zinc, as in the former case, but of *sulphate of barium*. This kind of experiment may be varied in a large number of ways, according to the chemicals used; but in all cases it is unwise to taste the coloured fluids.

Expt. 128. To produce a Solid Mass by Mixing two Clear Fluids.—Make a strong solution of pearlashes (*carbonate of potassium*) in one glass, and another of *chloride of calcium* in a second; mix the two together, when a chemical change of double decomposition will ensue, causing the precipitation of a solid substance essentially identical with chalk, termed *carbonate of calcium*,

* Conjurors often use two sponges, soaked in solutions of perchloride of iron and sulphocyanide of potassium respectively, for the purpose of apparently producing blood in some of their illusions, the sponges being squeezed so as to express the liquids, and cause them to mix together.

whilst *chloride of potassium* remains in solution. If the solutions are sufficiently strong, the precipitate will be so bulky relatively to the amount of fluid present, that the whole mass will apparently set solid, so that the vessel containing it may be turned upside down without any liquid running out.

Several other pairs of fluids will produce a like result; thus a solution of *silicate of soda* may be used along with diluted *hydrochloric acid:* if the strengths of the two fluids are properly adjusted, on mixing the two and stirring them together with a stick or glass rod, the whole will speedily set to a firm jelly: this effect is due to the formation of *silicic acid*, which when liberated in this way does not precipitate as a powder, but gelatinises like glue or jam.

Expt. 129. To make Invisible Writing Visible.—One interesting application of the property possessed by certain substances themselves colourless, or nearly so, to form strongly-coloured precipitates on intermixture, is the use of one of a pair of such fluids to write with (using a clean quill pen) on white paper. If care is taken not to scratch the paper, no writing will be visible when the fluid is dry; but on brushing over the paper lightly with the other fluid, or on dipping the paper into a basin full of the second fluid, the writing becomes visible, owing to the formation of a coloured precipitate in the pores of the paper or adhering to its surface. Thus, if you write with a solution of *ferrocyanide of potassium*, the pale yellow tint will be so faint as not to be noticeable on most kinds of writing paper, especially if some kind of writing in ordinary black ink be made on the paper, and the invisible writing traced between the lines; if, after drying, the letters be brushed over with a weak solution of *perchloride of iron*, the writing will become visible as blue letters; or if you dip the letter into a basin containing a weak solution of *sulphate of copper*, the letters will be "developed" in a reddish-chocolate colour.

Another way of proceeding is to write with a colourless solution of *acetate of lead*, and then expose the writing to the nasty-smelling gas *sulphuretted hydrogen*, produced in Expt. 13. The letters will appear in brownish-black.

Scarlet letters may be developed by writing with a solution of *iodide of potassium*, and developing with solution of "corrosive sublimate" (or *perchloride of mercury*—a strong poison); or the two solutions may be used conversely, the corrosive sublimate to write with and the iodide of potassium as developer; if used in this way the developing solution must not be too strong nor applied in too large a quantity, otherwise the scarlet precipitate at first formed will be again dissolved, and the writing will wholly disappear (Expt. 121). If the writing be done with the corrosive sublimate solution and weak *caustic soda* be used as developer, yellow characters will appear; and the same result will be obtained by writing with *acetate of lead* and developing with *iodide of potassium*, or better, with a moderately weak solution of *yellow chromate of potassium*.

In all these instances (and many other similar ones which will occur to you as you become more acquainted with the properties of substances and their chemical nature) the essential action going on in rendering the characters legible is the formation of a precipitate of some kind or other; but there are other kinds of "sympathetic inks," which act on a different principle. For example, writing done with a weak solution of *nitrate of cobalt* (a pink fluid) will be all but invisible when dry, but on heating the paper before the fire the characters show forth in *blue*, gradually fading away on removal from the fire and exposure to air (especially if damp) and reappearing on warming again. This is due to a different kind of chemical action brought about by the heat, and essentially consisting in the expulsion of water contained in the pink cobalt compound, whereby a much more deeply-tinted blue substance is formed. On exposure to cool moist air this blue compound gradually absorbs moisture again, and becomes reconverted into the lighter pink compound, so that the writing fades, and becomes almost invisible.

Expt. 130. Summer and Winter Pictures.—A pretty application of the above principle is the "Changeable Landscape" sometimes sold, which in ordinary weather represents an outdoor scene of some kind with a winter aspect. On carefully warming the picture before the fire, the grass becomes green and the trees clothed with verdure instead of snow, and the whole scene changes from winter to summer. On leaving the picture to itself in a cool place, the colour will fade away, and the old winter view will be restored. This effect is brought about by carefully brushing over the parts of the picture to be treated with solutions of compounds of cobalt mixed with other substances, capable of modifying the colour of the more highly-tinted products formed on heating and driving off moisture, the strength and nature of the liquid regulating the tint produced on warming. Copper chloride added to cobalt chloride solution gives a greenish or yellowish hue according to the quantity added; nickel chloride and perchloride of iron also give green shades when in the proper proportion, whilst zinc salts develop a red tone. Overheating must be carefully avoided, otherwise the metallic compounds will be so acted upon as to produce brown stains on the paper, which will not fade away on standing in a cool place.

Expt. 131. Chemical Weather-Prognosticator.—Strips of paper impregnated with similar fluids are sometimes used as **weather-prognosticators.** If the air be comparatively free from moisture, and a long way from being saturated with aqueous vapour (Expt. 45), it tends to exert a drying action on the cobalt salts, causing a lilac or blue shade to become visible; whilst if the air be nearly saturated with moisture, the blue disappears, and a pale pink only is seen; so that violet and blue colours indicate that fine weather is probable, at any rate for awhile, and pink tints indicate that the atmosphere is more or less waterlogged, a condition of things probably tending towards rain. A strip of *sea-weed* dried in the sun is impregnated with the solid constituents of sea-water, and if suitably chosen can be similarly used as a weather-indicator; in damp air the saline matter present takes up moisture, and makes the weed limp and soft; whilst in fine weather it dries up again, and becomes more or less hard and rigid. *Catgut*, *hairs*, and other fibres of animal origin are somewhat similarly affected by moisture. A toy weather-indicator often met with consists of a strip of wood, &c., balanced nearly horizontally on a pivot but not quite, so that the heavier end has a tendency to move forward; at the two ends of the lever are placed figures of a man and woman respectively, and a catgut thread is attached to the lever in such a way that when the thread shrinks it draws in one side of the lever which is placed in the entrance of a miniature doll's house, so that one figure comes out and the other goes in; the arrangement being usually such that the man comes out in wet weather and the lady, gaily attired, in fine.

Expt. 132. To precipitate one Solid whilst dissolving another.—Put into a wine glass a little solution of *chloride of copper*, and then dip into the liquid a clean iron key or bright knife-blade. In a few seconds you will find that a film of red metallic copper has formed over that part of the metal immersed in the solution. This is due to the circumstance that the chemical change called **replacement or displacement** (or "single decomposition") takes place (Expt. 9). The original fluid was a solution of a substance containing two constituents, copper and chlorine respectively; when the iron key comes in contact with this, the iron "decomposes" the compound of copper and chlorine in such a fashion that copper is set free in visible form, whilst the iron combines with the chlorine, forming a new compound, chloride of iron. This latter substance is invisible to the eye, as it dissolves in the water as fast as formed; but its presence may be shown by chemical tests, thus—Let the key stop in the solution for an hour or two; by this time a pretty thick deposit of powdery copper will be formed, and all the copper will be removed from solution; when this is the case take out the key, and divide the fluid into two equal parts; to one half add a dessertspoonful of strong solution of *ammonia;* a muddy brownish light-green copious precipitate will be formed, but no blue solution; whilst the original solution of chloride of copper similarly treated will give a beautiful deep blue fluid. To the other half add some solution of *ferricyanide of potassium* (red prussiate of potash), when a deep blue precipitate of "Turnbull's blue" (Expt. 83) will form; whilst the original copper chloride solution will give a yellowish-green coloured precipitate with the same test. *N.B.*—Neither of these two tests for iron will work satisfactorily unless the key has remained in the liquid long enough to precipitate all the copper from solution.

Expt. 133. To make a Lead Tree or "Arbor Saturni."—Get a good-sized wide-mouthed glass jar or small fish-globe, and fill it full of a solution of *sugar of lead* made by dissolving that salt in hot water in the proportion of about an ounce of the salt to between 1 and 2 pints of water; let the solution stand twenty-four hours to settle before use, pouring off only the clear fluid free from sediment into the jar. Now get some lumps of *zinc* (not too thin) weighing altogether two or three times as much as the sugar of lead used; tie them together with string or copper wire, or fix them to a wooden or wire frame here and there over its surface. Suspend the bunch by means of a string from a stick laid horizontally over the mouth of the jar (like the bell-dialyser in fig. 54), so that the lumps of zinc are all immersed in the sugar of lead solution, which should nearly fill the jar. After a few hours the zinc will have "displaced" the lead from the solution, just as the iron did the copper in the previous experiment; as the lead deposits it forms bright glistening feathery plates and films hanging down from the zinc, and altogether forming, if rightly managed, a beautiful metallic tree, which will remain permanent for a long time if kept immersed in the liquid, and not roughly handled or shaken so as to dislodge the delicate crystals of lead from their support. If, however, you attempt to remove the tree from the fluid, you will probably spoil it entirely.

Expt. 134. To make a Silver Tree or "Arbor Dianæ."—Many other instances of this kind of chemical action are known, especially amongst metals. A solution of *nitrate of silver* into which a teaspoonful of *mercury* is dropped, will on standing form a "silver tree" or "arbor Dianæ," * consisting of crystals of silver

* Diana and Luna were two alchemical names for silver, as was also Saturn for lead, Jupiter for tin, Mars for iron, and Venus for copper, and so on with the other planets

formed by the action of the mercury which becomes dissolved as the silver is removed from solution. If, after all the silver has thus been precipitated, the liquid (now containing solution of *nitrate of mercury*) be allowed to stand with strips of clean copper immersed, the mercury will be again thrown out of solution, forming a white film on the surface of the red copper, from which, if thick enough, small drops of fluid mercury may be wiped off by pressure between the fingers; by and by all the mercury is thus reprecipitated, whilst copper dissolves during the action. Finally, if the resulting solution of "nitrate of copper" be treated with metallic iron, the copper is in turn precipitated again, whilst the iron dissolves, as in Expt. 132.

If a bundle of thin copper wires be placed in a solution of silver nitrate, metallic silver will be deposited in crystals all over the surface of the copper, forming a sort of silver tree resembling the "arbor Saturni" of the last experiment, some of the copper becoming dissolved during the process.

In order to precipitate a metal from a solution of one of its compounds, it is not always necessary to use another metal as in the last experiments; thus, for example, gold may be precipitated by means of fluids.

Expt. 135. To precipitate Metallic Gold from Solution by means of a Fluid.—Obtain a small quantity of solution of *chloride of gold* (a rather costly substance), and add it to a much larger bulk of water. Prepare a solution of *sulphurous acid* (Expt. 117), or one of *sulphate of iron* (sometimes called "green vitriol") and, add a little of it to the diluted solution of gold; a very finely divided precipitate will be formed, which at first will remain suspended in the water, giving it a peculiar greenish hue when looked through, but will by and by subside to the bottom of the vessel as a brownish-yellow powder; this powder is pure gold in a "spongy" condition; if collected by pouring off the fluid carefully, stirring up with water, allowing to subside so as to wash away the adhering solution, and finally drying with a gentle heat, it looks more like iron rust than anything more valuable; but if "burnished" or pressed with a hard smooth surface it becomes bright and yellow, and like ordinary gold in appearance. Many other metals besides gold may be precipitated from solution in similar fashion, by using appropriate chemicals; glass objects are often coated with silver, so as to produce mirrors by processes of this kind (Expt. 198).

Expt. 136. To obtain precipitated Gold apparently in the form of a blood-red Fluid.—Make an extremely weak solution of chloride of gold, and put it in a white glass bottle. Get a small fragment of *phosphorus* (a very inflammable solid substance, kept under water to prevent its taking fire, which must be handled very carefully with wet fingers) about the size of a large pin's head, and drop it into a small bottle or test-tube, and add a few drops of *bisulphide of carbon*, an evil-smelling volatile liquid, which has the power of dissolving phosphorus (Expt. 237). Shake up, and when the phosphorus is dissolved, pour the solution into the bottle containing the chloride of gold; and then cork the bottle, shake well, and allow the whole to stand some hours. The phosphorus will gradually precipitate the gold, but in finer particles than in the last experiment, in consequence of which the fluid has a blood-red tint when looked through against a bright light. Within certain limits, the weaker the gold solution the better is the result; a few drops of an ordinary yellow solution of chloride of gold to a pint of water suffices, the best result being obtained when about half a grain of solid chloride of gold is dissolved in a pint of water. Sometimes the colour is more inclined to purple or amethyst than blood-red; this depends on the size of the particles. When blood-red, they are very fine, and will remain suspended for weeks and months before they subside; if, however, a little salt be added to the water, its colour changes to purple and the particles subside much more quickly. The possibility of obtaining gold in this form was first discovered by Faraday, whence the term "Faraday's Ruby gold," frequently applied to the blood-red fluid.

Expt. 137. To precipitate Charcoal from Sugar.—In both of the last experiments, the reason why the gold precipitates is that the chemical action taking place between the solution of chloride of gold and the sulphate of iron or phosphorus causes the gold and the chlorine in the chloride of gold to become separated from one another, the gold being set free as a solid powder, and the chlorine converted into other compounds which remain dissolved in the water, just as with the chloride of copper in Expt. 132. In the same sort of way sugar may be regarded as a compound of *carbon* (charcoal or lampblack) with water, so that by adding to sugar some substance which has a strong tendency to take up water, the sugar becomes "decomposed" or separated into solid black carbon and water. Such a substance is *sulphuric acid* or oil of vitriol. If this fluid be cautiously poured into its own bulk of strong syrup, much heat is evolved, and steam is thrown off whilst the whole swells up, and becomes black through the separation of impure carbon.

This experiment should be made in a sink or on the ground outside the house, so that if the vessel containing the syrup breaks with the heat, the corrosive acid will not be spilt where it can do damage. The mixture must be made cautiously and gradually (see Expt. 98) in a stoneware vessel, such as a white jam-pot. The black carbon thus formed is prepared on a large scale for making ordinary blacking, treacle being used instead of sugar, and the product being treated by certain processes to remove the acid, which otherwise would destroy the boots to which it is applied.

CHAPTER XI.

Solution of Gases in Liquids and Separation of Gases from Liquids.

Just as in the case of solution of solids by processes involving chemical change, so with gases may the change produced be one invisible to the eye, or one producing a visible alteration, either in colour, or owing to the formation of a precipitate, &c. Of invisible changes, the following experiments furnish illustrations.

Expt. 138. To produce a neutral Solid by the Action of Ammonia Gas on an Acid Fluid.—In Expt. 57 an illustration is given of the action of acids on certain kinds of colouring matter, viz., that derived from the litmus lichen; if a little tincture of litmus (alcoholic solution of the colouring matter) be added to some distilled water, and one drop of diluted hydrochloric acid be then poured in and the whole stirred up, the liquid will turn bright red, owing to the action of the acid on the colouring matter.

This is a general property of all acids; but there exists another class of bodies, of which lime-water (Expt. 152), ammonia solution, and common washing soda are examples, which possess exactly the opposite property; *i.e.*, whilst acids turn the litmus-colouring matter from blue to red, these other substances, or **antacids**, will destroy this effect, and restore the original blue colour on addition to the reddened fluid; this may be easily verified by dividing the red liquid obtained as above into three portions, and adding some lime-water to one, some ammonia solution to the second, and a little solution of washing soda to the third, when in each case, the liquid will turn blue, if too much acid have not been added in the first instance. Lime-water, ammonia, washing soda, and other bodies that produce this effect, are said to exert an **alkaline** reaction on colouring matters, whilst hydrochloric acid and similar bodies exert an **acid** reaction. Many substances, however, are known which have no action of either kind; such substances are said to be **neutral.** Thus, if some pure common salt, or some "Epsom salts" (sulphate of magnesia), be dissolved in water, the resulting liquid will neither redden water made blue with tincture of litmus, nor turn blue such water reddened by addition of a drop of weak acid.

Fill a good-sized dry bottle with ammonia gas, as in Expt. 75; set it on the table mouth upwards, and quickly pour in two or three teaspoonfuls of distilled water to which a few drops of diluted hydrochloric acid have been added; and which, consequently, possesses (to begin with) a strong acid reaction. Cork the bottle, and shake up; the ammonia gas will be dissolved, and will chemically react on the hydrochloric acid, without, however, producing any visible change; owing to the absorption, there will be a partial vacuum formed in the bottle, as in Expt. 77, so that the cork will be pretty firmly pressed inwards by the atmospheric pressure, and the air will rush into the bottle when the cork is extracted. Pour the liquid out of the bottle into a little evaporating basin, and test it with a blue litmus paper; no reddening will now ensue unless too much acid have been added to the water. Should this be the case, fill another bottle with ammonia gas, and shake up the liquid a second time with this gas also. Finally, evaporate the fluid either over a small flame, or preferably on the water-bath (Expt. 89); a white solid residue will be left, which (if pure) will be neutral in character as regards its reaction on litmus colouring matter. This white solid is called "salammoniac," or "*chloride of ammonium*," and is the product of the chemical action taking place between the ammonia and the hydrochloric acid.

If a few drops of diluted sulphuric acid be used instead of hydrochloric acid in this experiment, precisely the same results as regards appearance will be produced; the neutral solid ultimately obtained in this case, however, will be different in chemical character, being the salt called *sulphate of ammonium.*

Expt. 139. To restore the Colour of Clothes spotted by Acid.— The colouring-matter of litmus is by no means the only one affected by acids and alkalies, many other substances being known which behave in a similar fashion, amongst which may be mentioned ordinary ink, and most of the somewhat analogous black dye-stuffs used for dyeing clothes. In experimenting with acids, even with the greatest of care, you are pretty sure sooner or later to get some spilt about or splashed on to your clothes, with the result of burning holes in them if the acids are strong mineral ones, like oil of vitriol; or of making red spots, if weaker. In this latter case the original colour of the cloth can often be made to reappear, and the danger of destroying the fabric so as to produce holes, may be diminished, by dropping solution of ammonia (not too strong) on the stain, so that the spot disappears. If the acid spilt be nitric acid, and the spot have turned yellow, this treatment will usually not suffice to bring back the original colour, but it will often prevent a hole being produced, if the ammonia be applied soon enough.

When clothing becomes faded in colour, the original tint may often be more or less restored by sponging with water to which a little ammonia solution has been added; the fading being due to the acid vapours in the air, perspiration, &c.

Expt. 140. To remove Ink Marks from Paper.—Black ink of the older kind is made from solutions containing iron mixed with materials of the nature of the substances used in tanning leather, such as gall-nuts, &c. (Expt. 294), which bodies give rise to an intensely black compound, especially after the writing has been exposed to the air so as to cause a certain amount of oxidation to take place, which deepens the tint; some of the modern inks, however, are made of different materials. Writing done with the former kind of ink can be removed from paper, &c., by brushing dilute hydrochloric acid over the marks, and by and by washing them well with water, using a soft camel's-hair brush, and well pressing with blotting-paper to remove all acid which otherwise might rot the paper, and make it brittle and fragile. The acid decomposes the black colouring matter and dissolves the iron, so that water can wash out both the tanning substance and the iron compound, and thus entirely remove the marks.

Expt. 141. To remove Iron Mould from Linen, &c.—If ink be spilt on linen (tablecloths, &c.), the black marks will often become yellowish-brown after washing, thus producing stains of "iron-mould;" contact with rusty iron, &c., may also stain linen similarly. Such stains may be removed by the same means as those above mentioned with respect to ink marks, *i.e.*, cautiously treating with dilute hydrochloric acid, and then well washing to remove the acid.

Instead of hydrochloric acid, *oxalic acid* or *citric acid* may be used, or a sour substance known as *acid oxalate of potassium* (sometimes called "salts of sorrel"). *Lemon juice* will often answer the purpose, this fluid containing citric acid. All these preparations have the advantage that they do not rot or injure the fibres as readily as hydrochloric acid might possibly do if used too strong, or allowed to remain on too long.

Expt. 142. Changes of Colour produced by Acids and Alkalies.—Many dye-stuffs now in use are affected by acids and alkalies so as to give one colour with an acid liquid and another with an alkaline one. The colour obtained by mashing up a *red cabbage*, treating the pulp with alcohol, and squeezing out the fluid through a calico bag, becomes green with alkalies, red with acids, and purplish when the liquid is neutral. Violets and purple dahlias contain similar colouring matters. The liquid obtained by similarly treating *cochineal* (a kind of insect) with alcohol furnishes when diluted with water a colour which is orange if the solution is ever so faintly acid, but of a beautiful purple if alkaline. A substance termed *phenol-phthalein*, when dissolved in spirits of wine and used to test an alkaline watery solution, develops a fine red or brilliant pink colour, which becomes absolutely colourless on adding enough acid to neutralise the alkali. *Turmeric test-paper* is prepared by digesting powdered turmeric root with alcohol, straining off the fluid, and soaking fine white blotting-paper in the dark orange-yellow solution. Alkalies turn this paper brown, whilst acids restore the original yellow tint.

As a general rule, a test-paper soaked in a strong solution of an alterable colouring matter like those above mentioned, can be used as a test for acids by previously adding to the coloured fluid a minute quantity of alkali before soaking the paper; or, conversely, as a test for alkali by adding a little acid to the liquor. *Red* and *blue litmus papers* are thus prepared,—the former serving as a test for alkalies, becoming blue when dipped therein; the latter as a test for acids, becoming red when brought in contact therewith (Expt. 57); and similarly with other colouring matters.

If a test-paper suitable for distinguishing alkalies (like red litmus paper) be moistened with water and exposed to air containing only a very minute quantity of ammonia gas, the alkaline colour will soon appear, as the ammonia will dissolve in the water, moistening the paper and neutralising the trace of acid present. In the same kind of way acid vapours and gases in the air will speedily redden a moistened blue litmus paper; *e.g.*, when such a paper is held over hot vinegar, or over the mouth of a bottle of fuming hydrochloric acid. Putrefying urine liberates ammonia, as also does decaying horse-dung, wherefore a red litmus paper hung up in a urinal or in a stable will speedily become blue. Decaying fish produces an alkaline vapour closely resembling ammonia, termed *methylamine*, which will similarly turn a delicate test-paper blue. In all such cases the chemical action taking place is precisely parallel with that occurring in Expt. 138, viz., that the gaseous or vaporous alkaline matter in the air is dissolved by the moisture present in the paper, and reacts on the acid also present, producing a neutral compound; and *vice versâ* in the case of reddening a blue paper by acid vapours.

Expt. 143. To dissolve Carbonic Acid Gas in Caustic Soda.—The substance called *caustic soda* belongs to the class of antacids above mentioned, being one of the best known of such bodies; when dissolved in a large quantity of water it forms a solution of peculiar **alkaline** taste, and when in a smaller quantity, a strongly corrosive and biting fluid, whence the term **caustic.** If the caustic soda is pure, no effervescence will be produced by adding some diluted hydrochloric acid to a solution of it; chemical change, however, will take place, for the acid properties of the hydrochloric acid will disappear, and the alkaline ones of the caustic soda; nothing but pure common salt being formed when the acid and soda are mixed in just the right quantity to "neutralise" one another.

Place some caustic soda solution in a glass, and pass a current of carbonic acid gas (produced by the generator described above, Expt. 100) into it for some time (fig. 66); the carbonic acid gas will dissolve, but at the same time will act chemically, although invisibly, upon the caustic soda, converting it into a different substance termed *carbonate of soda.* If to the resulting carbonate of soda solution some hydrochloric acid be now added, a vigorous effervescence will ensue, thus showing one great difference between caustic soda and carbonate of soda, the latter being produced from the former by making it combine chemically with carbonic acid.

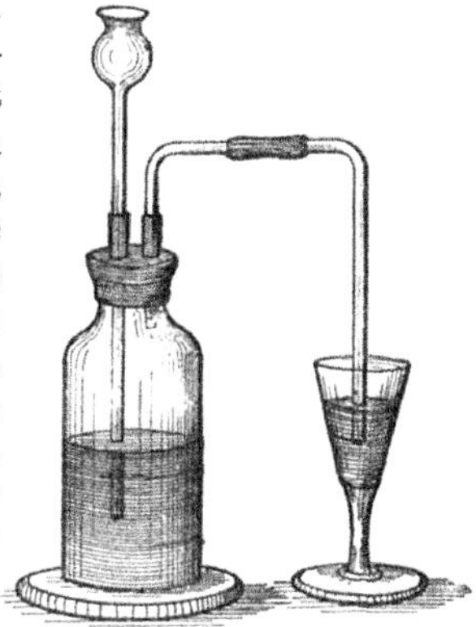

Fig. 66. Carbonic Acid passed through a solution of soda.

In this experiment carbonate of soda has been produced from caustic soda by making carbon dioxide act chemically upon the latter. In practical manufacture this operation is usually reversed; carbonate of soda is first obtained (originally as a natural mineral product, more recently as "barilla" from the ashes of sea-weeds, and subsequently by submitting common salt to various chemical processes), and is converted into caustic soda by taking away carbonic acid therefrom; the effect of this is greatly to intensify the alkaline characters of the soda, and thus to render it more chemically active and better fitted for preparing soap and for other analogous uses. The alchemists applied the term **mild alkali** to natural carbonate of soda (also known for that reason as **mineral alkali**), and to the analogous substance *carbonate of potash* (obtained from the ashes of vegetable matter, and thence distinguished as **vegetable alkali**); after removal of carbonic acid these substances were distinguished as **caustic** alkali, on account of their increased caustic or corrosive qualities. This operation of removing carbonic acid, or "causticising," is performed by treating solutions of the mild alkalies with quicklime, boiling the whole together until the action is complete.

Expt. 144. To causticise Soda Solution.—Take an ounce of washing soda crystals and dissolve it in half a pint of soft water. Add a tablespoonful of well-burnt quicklime, previously reduced to a sort of thin paste or cream by rubbing with water by means of a pestle and mortar. Boil the whole in a glass flask for a quarter of an hour; remove the flame, and allow the sediment to settle; pour off a teaspoonful of the nearly clear fluid into a test-tube, and add a little dilute hydrochloric acid. If the action of the quicklime is complete, little or no effervescence will be produced; but if much gas is evolved, the boiling must be continued until no more escape of carbonic acid gas is brought about on adding acid to a sample of the liquid, an additional quantity of quicklime being added if the first amount used does not suffice to bring about the requisite degree of causticising after an hour's boiling. When the action is complete, allow the contents of the flask to stand till cool, and carefully pour off the clear caustic soda solution into another bottle for use; a bottle provided with a cork that has been dipped in melted paraffin-wax answers best, ordinary untreated corks being easily attacked and corroded by the caustic soda, whilst bottles with glass stoppers are apt to become useless through the stopper becoming tightly fixed, and not easily extracted without breaking the glass.

In order to show that the quicklime has actually taken away the carbonic acid from the carbonate of soda, pour into a test-tube some of the thick sediment left after all the clear liquor has been decanted, and place in a second test-tube about as much of the original lime-cream. Add a teaspoonful of hydrochloric acid to each; if the quicklime was of good quality, there will be but little effervescence produced in the second test-tube, the lime dissolving comparatively quietly in the acid; whilst a copious effervescence will be produced in the first test-tube, owing to the escape of the carbonic acid taken up by the lime from the soda.

If *carbonate of potash* * (pearl-ashes) be used instead of washing

* Crude potashes can easily be prepared by burning almost any kind of wood to ashes, boiling these with water, filtering the liquid, and evaporating

soda crystals, a parallel action will take place; the solution of carbonate of potash (effervescing vigorously with acids) will become converted by boiling with lime into "liquor potassæ," or solution of *caustic potash*, which will not effervesce with acids, but will be far more corrosive and biting to the taste than the original solution of mild alkali.

Expt. 145. To dissolve Sulphuretted Hydrogen Gas in Water and in Ammonia.—Place in the gas generator used in Expt. 143 some lumps of sulphide of iron, and then, after inserting the cork and tubes, pour hydrochloric acid down the thistle funnel; the chemical action taking place (Expt. 13) causes sulphuretted hydrogen gas to escape. Pass the gas through distilled water, just as the carbonic acid gas was passed into the caustic soda solution in the last experiment; the water will dissolve some of the gas, thereby furnishing the solution of sulphuretted hydrogen referred to in some of the preceding experiments (Nos. 12, 95, 121, 125, 126).

Pass the sulphuretted hydrogen gas into solution of ammonia instead of water; the sulphuretted hydrogen will dissolve and will act chemically on the ammonia at the same time, producing a compound called *sulphydrate of ammonium;* this being soluble in water, is formed without any visible action taking place; but its formation is easily rendered evident by the great alteration produced in the properties of the fluid when the gas has been passed through for some time. Thus, into each of two test-tubes pour a few drops of solution of *copper sulphate*, and to the one add some of the original ammonia solution, and to the other some of the sulphydrate of ammonium solution; the first will act as in Expt. 124, producing a blue precipitate, dissolving to a deep blue clear fluid on adding more ammonia; the second will produce a black precipitate, soluble neither in ammonia solution nor in excess of sulphydrate of ammonium. If a little solution of *perchloride of iron* (ferric chloride) be used instead of sulphate of copper, the ammonia solution will produce a red-brown precipitate, and the sulphydrate of ammonium solution a black one; whilst if *chloride of cadmium* solution be employed, the former reagent will throw down a white precipitate, and the latter a yellow one.

Caution.—Sulphuretted hydrogen is a poisonous gas, though not so dangerous to breathe in small quantities as some others; moreover, on account of its chemical action on many metallic compounds, and especially on lead paint, it is apt to stain and discolour many ordinary articles of furniture and painted goods.

down the filtered liquor to dryness; a strongly alkaline mass results. Pearl-ash is this crude substance somewhat purified and refined.

It should therefore be prepared in the open air, or in a stable or outhouse, and not inside a dwelling-house, where, in addition to causing annoyance on account of its frightful odour, it may do other damage to paint and furniture. In laboratories where it is extensively used, the apparatus is usually placed inside a "fume chamber" or "stink cupboard," *i.e.*, a kind of closet furnished with a flue and a glass door in front (fig. 67); by means of a gas jet or otherwise a current of air is kept ascending the flue, so that any evil-smelling substances or dangerous gases and vapours can be manipulated inside the cupboard, the door being closed when the operator does not require to touch the apparatus, so as to avoid leakage of fumes and smells into the apartment.

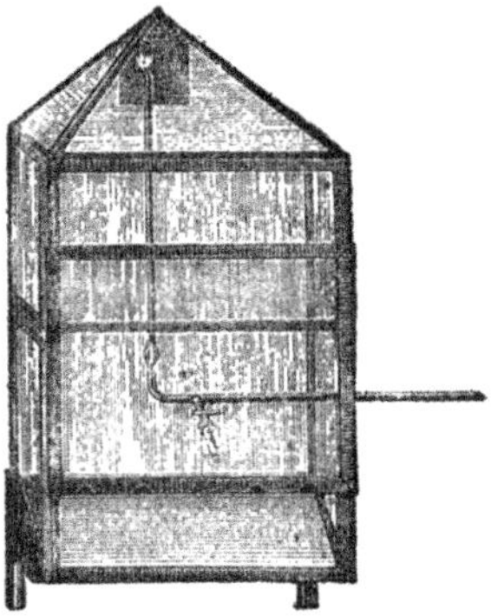
Fig. 67. Stink-Closet.

The unpleasant smell of rotten eggs is due to the fact that sulphuretted hydrogen is produced during their putrefaction. The gas is also produced by certain natural chemical actions taking place underground, so that the water of springs in localities where this occurs dissolves the gas, and acquires its disagreeable odour. Harrogate, in Yorkshire, is one well-known place where "sulphur springs" of this kind occur. In consequence of the property possessed by this gas to form black compounds on coming in contact with certain metallic compounds (Expt. 14), it discolours ordinary paint, because white lead is an ingredient therein, and is a compound acted upon in much the same fashion as silver or copper nitrate. Certain kinds of powder used by ladies as cosmetics contain lead, and others bismuth; in consequence, complexions that have been improved (?) by the use of such cosmetics will not stand the atmosphere of a sulphur spring, but turn grey, brown, or even black on contact with air impregnated with sulphuretted hydrogen.

Expt. 146. To illustrate the Action of Sulphuretted Hydrogen on Paint.—One white metallic compound used for paint, *zinc white*, is not affected by sulphuretted hydrogen. Procure a little of this mixed with linseed oil ready for use, and with it paint one side of the head of a plaster cast, or some similar object. Paint the other side in the same way with ordinary white paint made from oil and *white lead*. Now expose the painted cast to air containing sulphuretted hydrogen, by putting it in a chip hat-box with a small hole cut in the side through which is introduced the delivery pipe of a sulphuretted hydrogen generator. In a very short time the side painted with lead will be turned black, whilst the other side painted with zinc white will be unchanged, provided the zinc white be pure (*vide* also Expts. 14, 129).

Separation of Gases by Solvent Chemical Action.

A very frequent operation, both in laboratory experiments and in chemical manufactures, is to treat a mixture of gases with some solution that will act chemically upon one or more of the gases present, absorbing them in consequence, and leaving the others undissolved and unacted upon. Thus, in the purification of coal-gas for ordinary lighting purposes, methods of this description are largely used. The following experiments illustrate this.

Expt. 147. To separate a Mixture of Carbon Dioxide and Air. —Place a jar mouth upwards on a table, and let the delivery pipe of a generator of carbon dioxide (Expt. 100) pass into it, so that some of the gas may pass into the jar; or otherwise, pour into the jar some of the heavy carbon dioxide gas evolved from an aerated water siphon on running some of the fluid into a tumbler, &c., as directed in Expt. 71; in either case you will have a mixture of carbon dioxide and air in the jar. The presence of the former gas may be proved by the special test described more fully in Expt. 152, consisting of pouring out a little of the gas from the jar (as in Expt. 101, fig. 58) into another one containing some *lime-water*, closing this second jar with a cork, and shaking up vigorously, when the lime-water will be rendered milky if carbon dioxide be present. Now pour into the first jar a wine-glassful of caustic soda solution, quickly cork up with a greased cork, and shake thoroughly; turn the jar upside down, and remove the cork with the mouth under water, so that the atmospheric pressure may force water into the jar (as in Expt. 77), to supply the place of the absorbed gas. On re-corking the jar under water and removing it, it will be found, firstly, that on applying the lime-water test as above described no milkiness is produced, showing that the admixed carbon dioxide has been removed by the action of the caustic soda solution; and, secondly, that a lighted taper introduced into the jar at the end of a wire will burn freely, showing that the air present in the mixture of gases has *not* been similarly absorbed and removed.

Production of Colours by Chemical Changes due to Solution of Gas.

As a general rule, no marked alteration in colour attends the absorption of a gas by a fluid, as long as no precipitation is brought about; but some exceptions occur. Thus, in Expt. 114, the action on copper of ammonia gave rise to solution of copper, and formation of a blue liquid when air was also allowed access, but not in the absence of air.

Although we are accustomed to regard the atmosphere as a typical example of a gaseous substance, yet really more than one gas is present, in even the purest open-country air; whilst the more or less vitiated air of a town, or close unventilated room, really contains a number of gases. In all cases, however, two gases greatly predominate, called *oxygen* and *nitrogen* respectively. The latter roughly constitutes about $\frac{4}{5}$ of the air and the former $\frac{1}{5}$; but although oxygen is not present in the larger quantity, it is by far the more important of the two, as it is the oxygen and not the nitrogen which is necessary to keep up animal life by breathing.

Expt. 148. To procure a dark-coloured Fluid by Absorption of Oxygen from the Air.—In Expt. 114 what happens is that the oxygen present in the air dissolves in the liquid and exerts a chemical action on the copper, converting it into a compound (*oxide of copper*), which dissolves in the ammonia-water, forming a blue fluid; so that ultimately the oxygen becomes removed from the air to a greater or lesser extent, thus affording the means of preparing nitrogen gas (*vide* Expt. 184). A precisely analogous action takes place if a little *pyrogallic acid* (as much as will lie on a shilling) be put into an eight-ounce wide-mouthed bottle, and then a tablespoonful of caustic soda solution poured in; quickly insert a well-fitting cork (previously greased to make it air-tight) and shake up. The pyrogallic acid dissolves in the caustic soda solution, forming a fluid which immediately absorbs oxygen if in contact therewith, but not nitrogen, forming a dark reddish-brown fluid in so doing. After a few minutes shaking the whole of the oxygen present in the air originally contained in the bottle will be dissolved and chemically acted upon; if now the bottle be turned mouth downwards in a basin of water, and the cork cautiously removed, still under water, as in the last experiment (147), it will be noticed that the water will to some extent rush into the bottle, being driven in by atmospheric pressure, somewhat as in Expt. 77, but not so violently; on comparing the quantity of water thus driven in (again corking the bottle under water before removing it for inspection) with the quantity of air originally present, it will be seen that about $\frac{1}{5}$ of the gaseous contents of the bottle have been absorbed, but that the odd $\frac{4}{5}$ have remained unabsorbed, being the nitrogen.

Expt. 149. To show that Nitrogen extinguishes a Candle.—The reason why combustible bodies, such as candles, burn in the air is simply that chemical changes go on between the burning material and the oxygen of the atmosphere; if the oxygen is removed, the residual nitrogen will no longer keep up the chemical change, and the flame is consequently extinguished. Lower a lighted taper at the end of a wire, or a spill of wood or paper, into the bottle containing nitrogen obtained in the last experiment, and you will see that the flame is extinguished, just as it was by carbon dioxide in Expt. 71; and for the same reason in each case, viz., that there is no oxygen present to keep the flame alight.

For other methods of preparing nitrogen, see Expts. 183 and 184.

Expt. 150. To show that a fresh Supply of Air is requisite to keep up the Combustion of a Candle.—Fix a lighted candle at the end of a wire, and dip it into a tall narrow cylinder (fig. 40), and cover the mouth of the cylinder with a plate. In a short time the candle will go out, the flame becoming extinguished, because all the available oxygen in the cylinder has become used up by chemical action.

Now withdraw the candle, relight it, and place it in a second similar jar, and slip into the jar a piece of wet cardboard previously cut to size, so as to form a vertical partition in the middle of the upper part of the jar. If properly placed in position, the candle will now continue to burn, because a ventilating current is set up in the jar, the hot products of combustion rising up on one side of the pasteboard partition, and fresh air descending down the other side, so as to keep the candle supplied with air. That the two currents are set up in opposite directions is easily shown by lighting a piece of brown paper, and blowing out the flame, so that the paper smoulders and gives off a thick smoke; bring the smoking end of the paper torch to one side of the partition, and the smoke will be carried downwards towards the candle by the descending current of air; whilst on the other side the smoke will be carried upwards by the ascending current of hot gases and vapours from the flame.

The pasteboard is wetted to prevent its possibly taking fire at the candle flame; a strip of metal (sheet copper, zinc, or tinplate, &c.) may be conveniently used instead.

This system of ventilation by means of a shaft divided by a "brattice" or vertical partition is sometimes used in mines, a large fire being employed to produce the up-current.

Expt. 151. To produce White Indigo and convert it into Blue Indigo by the action of the Air.—Put into a bottle (capable of being tightly closed with a cork) some indigo solution (*sulphindigotic acid*), and dilute it with water, so as to obtain a moderately deep blue-coloured liquid. Add a tablespoonful of *quicklime* and half as much *glucose* (grape sugar). Tightly cork the bottle, and let it stand some hours, with occasional shaking; a chemical action will take place, resulting in the conversion of the blue indigo into "white indigo"), so that the colour of the liquid fades. By and by, when the blue has disappeared and a yellowish liquor only is contained in the bottle above the sediment, the clear liquor may be decanted into a cup; by exposure to air it will gradually take up oxygen (like the pyrogallic acid dissolved in potash in the last experiment), and the blue colour will be restored.

Into the yellow fluid dip a piece of white flannel or a skein of white Berlin wool, and then hang it up wet for some time; as the blue indigo is formed by the action of the air on the white indigo solution, it becomes attracted to and fixed in the substance of the woollen fibres, permanently dyeing them blue. This is substantially the way in which indigo-dyed woollen cloth is usually prepared.

Production of Precipitates by Solution of Gases in Liquids.

This class of chemical action is of very frequent occurrence; thus the test employed in Expt. 147, for the presence of carbon dioxide, when mixed with other gases, is based upon a reaction of this kind.

Expt. 152. To make a Clear Fluid, apparently Water, resemble Milk by dissolving a Gas therein.—Get a little quicklime (an ounce or less will suffice), and pour on it a few drops of water; if freshly burnt, it will become hot, and by and by fall to powder. This is called "slaking" the lime, and is due to a chemical action taking place between the quicklime and the water (Expt. 240); with large quantities of lime a very high degree of heat can be thus produced, so that wood has been known to be charred and similar damage done, as though by fire, when water has accidentally come in contact with large stores of quicklime. Shake up the slaked lime several times in a clear wine bottle nearly filled with soft water, or preferably distilled water, cork the bottle, and let it stand some hours, shaking it up occasionally. Next day pour off from the sediment the clean *lime-water*, or watery solution of lime, produced, and keep it for use in a well-closed bottle. Fill a small jar with carbon dioxide by displacement (Expt. 101); set it on the table mouth upwards, and pour in a wine-glassful of lime-water; cork the jar and shake up, when the lime-water will become milky, as in Expt. 147; the action being that, firstly, the carbon dioxide gas dissolves in the water present; and, secondly, the solution so formed instantaneously acts on the lime present in the liquid, so as to form an insoluble solid compound, or precipitate of *carbonate of lime, chemically identical with chalk or limestone:* at first the particles of this solid remain suspended, giving the appearance of *milk*; but by and by they subside, and *fall* to the bottom as a powder.

Expt. 153. To turn Water apparently into Milk by the Breath.—The action going on in the lungs of air-breathing animals always results in the formation of carbonic acid gas, which is contained in the exhaled breath; consequently, if you blow a stream of air from the lungs by means of a piece of glass tube through a little lime-water in a goblet, the lime-water will quickly turn milky, precisely as in the last experiment, and for the same reason. In country fairs and such like places, travelling quack doctors and medicine-sellers sometimes "humbug" the unsophisticated yokels by pretending that they have an unfailing means of detecting whether any one is consumptive or not; they say that if any one blows through their marvellous water by means of a tube, if the lungs are affected the water will become milky; but not if the lungs are sound. A stout healthy-looking confederate or "bonnet" is given a glass of water, and blows through it with no result, this particular glass being simply ordinary spring water; he is then pronounced sound, and is perhaps stated to have been cured a few weeks ago of serious lung-disease by taking the wonderful medicine for sale. Some one of the audience then steps up, and is given a glass of lime-water, which of course turns milky as soon as he blows through it; the quack then tells him that he is sure to die (which no doubt is true, though not in the sense meant), but that he may derive benefit by purchasing the infallible medicine; which it is needless to say is of no use whatever as a cure for consumption, even if not absolutely harmful for other reasons.

Expt. 154. To turn Lime-water Milky by means of a Burning Candle.—The reason why carbonic acid gas is exhaled from the lungs is, that during the various life actions that go on in the body, fuel (taken into the body in the form of food) is made to undergo chemical changes exactly analogous to those taking place when a candle is burned in the ordinary way, or when coal is consumed in a boiler-fire. In all cases the fuel consumes air (or rather one of the constituents of air called oxygen, Expt. 148), and produces heat, moderate in the case of a living animal, but far more intense in the case of a candle or fire; whilst the products of the chemical changes taking place are largely gaseous, and mainly consist of carbonic acid gas, which escapes from the lungs of the animal on exhalation, and from the chimney of the fireplace. Fix a piece of wax-taper or ordinary candle to a wire (fig. 41), and lower the burning candle into a clean dry glass cylinder; almost immediately you will see that the glass becomes dimmed inside, just as the chimney of a paraffin or moderator oil lamp, or argand gas-burner becomes dimmed and bedewed for a few moments on first putting it over the flame; this is because water vapour is one of the products of the chemical change going on during burning (in addition to carbonic acid gas), and this vapour condenses on the cool glass as dew (Expt. 43). After the candle has burnt inside the cylinder for a few seconds, quickly pull it out, pour in half a wine-glassful of lime-water, cover the mouth of the jar with the hand (or cork it), and shake up well. If too much time have not been lost, so that the hot gases from the flame may not have all escaped, the lime-water will become milky, just as in the two last experiments, and again for the same reason, viz., that the carbonic acid gas has dissolved in the lime-water, and brought about a chemical change with the lime therein present, resulting in the precipitation of artificial chalk.

Expt. 155. To dissolve Artificial Chalk in Water by means of a Gas, and reproduce it by Boiling.—Fill a wine glass half full of clear lime-water, and then place the delivery tube of a carbon dioxide generator in action inside the wine glass, so that the issuing gas bubbles through the lime-water (fig. 66); in a few instants you will see that the lime-water begins to turn milky, owing to the production of artificial chalk (as in the preceding experiments). Continue the stream of gas for a few minutes, and by and by you will see that the milkiness disappears, so that finally you have a clear fluid just as at first. This is owing to the circumstance already mentioned (Expt. 73), that water strongly impregnated with carbonic acid is capable of dissolving chalk and such like substances; the first portions of carbonic acid gas that dissolved in the lime-water acted chemically on the lime and produced "carbonate of lime," or artificial chalk, which, being insoluble in plain water, precipitated as a finely-divided solid substance; but when more carbonic acid became dissolved in the water later on, this solution of carbonic acid redissolved the artificial chalk formed at first, producing a somewhat different chemical substance, sometimes called *bicarbonate of lime*, and differing from "carbonate of lime" in being soluble in water, instead of insoluble. Natural spring-and river-waters generally contain more or less bicarbonate of lime (and sometimes an analogous substance, *bicarbonate of* *magnesia*); the result of which is that they possess to a greater or less extent the same peculiar property as the solution of bicarbonate of lime prepared as above described; viz., that, on boiling, this compound is chemically affected by the heat in such a fashion that ordinary "carbonate" of lime or magnesia is formed, which being now insoluble in water precipitates in the solid form. Transfer the solution just prepared to a flask, and heat it carefully to boiling over a lamp; solid particles will rapidly be deposited, and in a short time almost the whole of the dissolved lime will be thrown out of solution as carbonate of lime. When ordinary spring water is boiled, exactly the same thing takes place; the "furr" inside a kettle is principally the carbonate of lime thus formed; in steam boilers the production of an internal solid incrustation in this way is constantly taking place, and is a source of great inconvenience in many ways, more especially in that it prevents the heat reaching the water so readily, and thus leads to waste of fuel, as well as damage to the boiler by the outer portion becoming unduly heated.

Water containing lime or magnesia dissolved is said to be "hard;" when present as bicarbonate of lime or magnesia, the hardness is mostly removed by boiling, the lime being thus rendered insoluble and precipitating; lime so removed constitutes what is termed **temporary hardness** for that reason. Most water, however, also contains more or less lime and other mineral matters in forms not removed by simple boiling (chiefly as *sulphate of lime* or dissolved gypsum); and the hardness due to this cause is consequently termed **permanent hardness.** Hardness in the water supply of a town is unobjectionable up to a certain limit; but beyond that it is a serious evil, as very hard water wastes much soap when used for washing purposes, and is much less well adapted for cooking, making tea and coffee, and such like dietetic purposes, than softer water. Some of the softest water supplied to large towns and cities in Britain is that of Glasgow, derived from Loch Katrine. Ordinarily, temporary hardness in water is removed and the water softened by simply boiling the water and allowing it to stand.

Expt. 156. Other Ways of softening Hard Water.—Water rendered temporarily hard, by the presence of much bicarbonate of lime therein, is sometimes softened on a large scale by various methods before distributing it for town use: for the most part these methods are substantially modifications of "Clark's process," which consists in adding to the water just so much lime as will suffice to convert all the bicarbonate of lime (and magnesia, &c.) present, and all dissolved carbonic acid gas into ordinary carbonate of lime, which being insoluble in plain water precipitates, and is removed by standing and running off the clear fluid from the top. Any suspended clayey or other solid matter is thus removed, being carried down by the subsiding carbonate of lime; accordingly the water thus purified is usually of a beautiful blue hue when viewed in large tanks. Sometimes mechanical filtering or straining arrangements are employed instead of simple standing at rest so as to save time.

Permanently hard water is not materially improved by such treatment, except in so far as bicarbonate of lime (temporary hardness) is thereby removed. If, however, carbonate of soda is added to the water, the sulphate of lime present is acted upon by the carbonate of soda, and carbonate of lime formed, which being insoluble precipitates. "Washing soda," "soda crystals," "Scotch soda," "soda ash," &c., are different names for more or less pure carbonate of soda used by laundresses and others for softening hard water and aiding the action of soap.

Prepare some artificial chalk solution (solution of bicarbonate of lime) as in the last experiment (155), and add to it some clear lime-water; a precipitate will be formed, and if the right quantities of each fluid, are used the whole of the lime will be removed from solution; this, however, is not easy to effect properly without making analyses of each solution, so as to calculate in what proportions they must be mixed. In any case, the formation of a precipitate illustrates Clark's purifying process.

Expt. 157. Softening permanently Hard Water.—Get some plaster of Paris and put it in a bottle nearly filled with distilled water; shake up at intervals, and then allow the liquid to subside till next day; pour off some of the clear fluid, and you will have an artificially-prepared excessively hard water, such as is intentionally produced for brewing certain kinds of beer, the process of hardening in this way being termed "Burtonising" the water. Add to the hardened water (solution of sulphate of lime) some clear solution of washing-soda crystals and you will at once see that a precipitate forms, showing that some of the lime has been thrown out of solution. One part of the use of soda in washing consists in its power of thus precipitating lime; besides which, it acts like soap as a "detergent" or cleansing agent.

Expt. 158. To produce Coloured Precipitates by means of a Gas.—Set up a sulphuretted hydrogen generator (Expt. 145), and into different test-glasses pour a little of each of the following solutions:—(1) *sulphate of copper*, (2) *acetate of lead*, (3) *tartar* *emetic* (tartrate of antimony and potassium), (4) *chloride of cadmium*. Pass a few bubbles of gas from the generator through each of these fluids in succession (fig. 66), and you will see that in the first two glasses a black precipitate is formed, in the third an orange one, and in the fourth a bright yellow one. In each case the same kind of action takes place as that occurring in Expt. 121, when solution of sulphuretted hydrogen was used instead of the gas itself; *i.e.*, a metallic sulphide (sulphide of copper, sulphide of lead, sulphide of antimony, sulphide of cadmium) is formed, which precipitates, being insoluble.

Expt. 159. To separate Metals from one another by means of a Gas.—For analytical purposes sulphuretted hydrogen is one of the most convenient substances known to the chemist, in spite of its repulsive odour; the above mentioned experiments show that when solutions containing copper, cadmium, antimony, and such like metals are treated with sulphuretted hydrogen, the metal is so chemically acted upon that it is removed from solution by being converted into an insoluble compound which precipitates. Not all metals, however, are thus acted upon; and in consequence sulphuretted hydrogen serves as a means of separating those that are acted upon from those that are not.

Dissolve a few grains of green vitriol (*sulphate of iron*) in an ounce of water, and add to the solution two drops of a solution of *sulphate of copper*. The liquid will now contain two metals, copper and iron. Pass a current of sulphuretted hydrogen gas through the liquid for a few minutes, until the fluid smells strongly of the gas after well stirring up. If too much copper solution have not been added, this treatment will suffice to render all the copper insoluble as a black precipitate, so that on filtering the fluid and passing a little more gas through the clear filtrate, no more black precipitate is formed (should this not be the case at first, owing to having added too much copper, it will by and by be effected by passing a sufficient quantity of gas). That the black precipitate contains copper may be shown by transferring the filter paper with the black precipitate to a small evaporating basin, pouring on a little diluted nitric acid and warming, when a solution of copper will be formed, which will give a deep blue liquid on adding some strong ammonia solution to it (Expt. 124); whilst a test for iron applied to the clear filtered liquid will show that that metal is not removed from solution. For this purpose a few drops of *ferricyanide of potassium* (*red* prussiate of potash) may be employed, which will give a deep blue precipitate (Expt. 132).

Evolution of Gases from Liquids in consequence of Chemical Action.

In the preceding experiments we have had several illustrations of the production of gases of different kinds during the solution of solids; thus *hydrogen* gas is set free when zinc dissolves in hydrochloric acid (Expts. 10, 104, &c); *sulphuretted hydrogen*, when sulphide of iron is acted upon by the same acid (Expts. 13, 145, &c.); *nitric oxide*, when copper is dissolved in diluted nitric acid (Expt. 116); *sulphur dioxide*, when copper is heated with strong sulphuric acid (Expt. 117); *carbon dioxide*, when chalk is treated with diluted hydrochloric acid (Expt. 99). In similar fashion gases are frequently evolved by chemical actions taking place between liquids of various kinds; sometimes the action takes place in the cold, sometimes the application of more or less heat is requisite. An example of this class of action is afforded by Expt. 143, where the action of hydrochloric acid upon solution of carbonate of soda gives rise to the evolution of carbonic acid gas; the production of effervescence on dissolving "lemon kali," &c., in water (Expt. 73), is really due to exactly the same kind of action; the water dissolves the carbonate of soda present in the powder, and the tartaric acid is also dissolved by the water, so that the acid solution thus formed acts on the solution of the carbonate of soda just as the hydrochloric acid does.

Expt. 160. To generate Ammonia Vapours from Solutions.—Place in a test-tube a teaspoonful of a strong solution of *chloride of ammonium* or *sulphate of ammonium*, add as much solution of caustic soda, and heat the whole; when tolerably hot you will find that a strong smell of ammonia is perceptible at the mouth of the test-tube. Hold over the mouth of the tube a piece of litmus test-paper (Expt. 142) that has been reddened by dipping in a tumblerful of water to which one drop of hydrochloric acid has been added; the escaping ammoniacal vapours will turn the red paper blue.

Expt. 161. To generate Sulphur Dioxide by mixing two Liquids.—Add a teaspoonful of diluted hydrochloric acid to as much of a strong solution of *sodium sulphite;* a strong smell of burning brimstone will then be noticed, owing to the evolution of sulphur dioxide on account of the chemical action taking place. If *sodium hyposulphite* be used instead of sodium sulphite, the liquid will gradually become milky, and develop the smell of burning brimstone, the milkiness being due to the precipitation of particles of solid sulphur.

Expt. 162. To produce Chlorine Gas by mixing two Liquids.

—Put an ounce of *bleaching powder* (chloride of lime) in a bottle with half a tumblerful of water, and shake up well; now filter off (Expt. 56) a portion of the solution formed; a quantity of lime and other substances will usually be left undissolved, whilst a clear solution will be obtained, containing amongst other things a compound called *hypochlorite of calcium.* Add to a little of this solution some diluted hydrochloric acid; a chemical action will take place, the result of which is that a peculiar gas termed *chlorine* is evolved.

Caution.—This gas is extremely injurious to breathe, producing great irritation of the lungs; all experiments made with it are best performed in a draught closet (Expt. 145, fig. 67), or failing this, in the open air, or in a room permitting of free ventilation by opening doors and windows, and great care must be taken not to inhale the gas in any quantity. Should small quantities be accidentally inhaled, and distressing coughing and irritation of the throat be consequently produced, this may be a good deal relieved by pouring a few drops of ether in the palm of the hand and breathing in the vapour. The chlorine gas evolved in this experiment has a greenish-yellow tint (whence the name from the Greek χρολως = pale green); if the solutions used are weak, little or no gas will be evolved as such, almost all being retained in solution in the water present, which will take up a yellow colour.

Expt. 163. To illustrate the Bleaching Action of Chlorine Solution.—To a wine-glassful of water add a few drops of *litmus tincture*, so as to tint the water purple; now add a little of the chlorine solution prepared as in the last experiment, by adding hydrochloric acid to bleaching powder solution; the colour will be discharged or bleached by the action of the chlorine.

Instead of litmus tincture, a little solution of *indigo* may be used with the same result; or some claret, or a little of the red liquid obtained by pickling red cabbage or beetroot. Almost all vegetable colours, in fact, are destroyed by chlorine, and to this circumstance is due the use of bleaching powder for whitening paper, calico, &c., in the arts. Roses and other coloured flowers may be bleached by immersing them for some time in chlorine water or in a jar of the gas.

Dip a piece of pink blotting-paper into the solution obtained by dissolving bleaching powder in water, and then dip it into water to which a little hydrochloric acid has been added; the chlorine thus set free in contact with the colouring matter in the paper will rapidly discharge the pink colour. Instead of pink blotting-paper, a piece of calico soaked in claret and then dried so as to be reddened, may be employed with the same result.

Calico as prepared by spinning and weaving from cotton is not perfectly white, but is rendered so by bleaching it in just the same way; a bath of "bleach liquor," or solution of bleaching powder, is prepared and the calico dipped into it, being unwound from a large roll, and gradually passed through the liquid. After passing through the bath, the calico is then similarly passed through a second vat containing water with a little sulphuric acid dissolved therein; the goods thus "soured" become whitened by the action of the chlorine set free; after washing with water to remove acid, they are far less brownish tinted than at first. In the same kind of way fruit or wine stains may be removed from a tablecloth by bleaching them out with a solution of bleaching powder applied to the stain, a little vinegar or other weak acid solution being subsequently also applied, and the cloth finally rinsed in plain water.

Expt. 164. To turn a colourless Liquid Blue by means of Chlorine Solution.—Get as much starch as will lie on a sixpence, and mash it up with a dessert-spoonful of water to a cream in a cup by means of a spoon. Boil half a tumblerful of water in an evaporating basin, and whilst boiling stir in the cream of starch. The solid white particles of starch will become a thin jelly with the boiling water, forming the material used by laundresses to stiffen collars and shirts, &c. When the starch-jelly is cold pour a little into a tumblerful of water, and add a few drops of solution of iodide of potassium; finally drop in a little chlorine solution, and stir up. A beautiful blue colour will be produced, owing to a chemical action taking place; the end result is the formation of "iodide of starch," a blue compound produced by the combination of starch with the iodine contained in the iodide of potassium, and set free therefrom by the action of the chlorine. If too much chlorine be used, the blue liquid becomes browned.

The blue liquid thus formed possesses a very curious property; if a little of it be boiled in a test-tube, the blue colour will disappear as the liquid gets hot; but on cooling again the blue tint reappears. When the test-tube is heated to such a degree that the blue colour has vanished, plunge the hot tube into a tumbler of cold water so as to cool it rapidly: the blue will first appear at the bottom of the tube, as this cools quickest (Chapter XX.), but will speedily spread through the whole of the fluid. This experiment will not succeed well if too much chlorine-water have been used so as to brown the liquid.

The preceding two experiments may be combined together so as to produce a curious effect. Prepare two glasses of water, one containing a small quantity of indigo solution so as to tint it blue, and the other a little starch paste with a drop or two of iodide of potassium solution. Chlorine water dropped into the first, and stirred up with the liquid, will discharge the blue colour, but will cause a very similar tint to appear in the colourless fluid in the second glass, thus apparently causing the liquids to change places.

The blue colour produced in the second glass can be again discharged by dropping in a little solution of *sulphurous acid*, or of the salt called *sodium thiosulphate*, sometimes termed *sodium hyposulphite*, or *hyposulphite of soda*.

CHAPTER XII.

Intermixture of Liquids accompanied by Chemical Action.

The preceding experiments have shown that when two liquids (solutions) are intermixed, visible change of state may occur, through the formation of a precipitate of solid matter formed by the chemical action and insoluble in the fluids present; or through the formation and evolution of a gas or vapour. But besides actions of these kinds, liquids may act chemically upon one another in such fashion that two dissimilar liquids unite to produce one single resultant fluid, or may form by double decomposition two different liquid products. The following experiments illustrate these kinds of action.

Expt. 165. To neutralise Sulphuric Acid.—We have already seen, in Expt. 138, that when a solution of hydrochloric acid or of sulphuric acid is shaken up with air containing ammonia gas, the ammonia is absorbed and neutralises the acid, forming a saline combination, chloride or sulphate of ammonium. In just the same kind of way sulphuric acid, dissolved in a moderately large quantity of water, can be neutralised or converted into a neutral saline combination by cautiously adding to it a solution of caustic soda or potash. Place some diluted sulphuric acid in an evaporating basin, add to it a few drops of tincture of litmus so as to redden the fluid, and then drop in with continual stirring a solution of caustic potash until the red colour turns purple; if too much potash be added the fluid will become quite blue, in which case a few drops more of dilute acid should be added. Evaporate the liquid, and set by to crystallise; potassium sulphate will deposit in small crystals as the liquid cools. These may be collected on a paper filter and drained from the mother-liquor (Expt. 56), and will then be found to be neutral to litmus test-papers.

Expt. 166. To dissolve Quicksilver.—Introduce a small quantity of mercury into a test-tube (about as much in volume as a small drop of water), and pour over it nitric acid diluted with two or three times its bulk of water, and then warm gently over the lamp. The acid will act on the quicksilver precisely as it did upon copper in Expt. 116, the gas *nitric oxide* being evolved, and the mercury being gradually dissolved, forming a salt called *nitrate of mercury*, corresponding with nitrate of copper.

If mercury be boiled with strong sulphuric acid, much the same kind of change takes place as when copper is similarly treated (Expt. 117); sulphur dioxide is evolved, and the mercury becomes converted into *sulphate of mercury*, a salt much less soluble in water than sulphate of copper, but still not absolutely insoluble.

Expt. 167. To dissolve Bromine in Caustic Soda.—Bromine is a brownish-red liquid, emitting most suffocating fumes; in water it dissolves only to a limited extent, but in caustic soda solution it dissolves readily, forming a solution possessing the same bleaching properties as chloride of lime, and containing an analogous compound termed *hypobromite of sodium.* A few drops of this solution added to water tinted with indigo or litmus tincture, will bleach the colour readily like chlorine-water (Expt. 163).

Expt. 168. To produce Hydrogen by means of Sodium Amalgam.—Certain metals will dissolve in hot quicksilver, forming solutions termed "amalgams," which often become solid at the ordinary temperature. If a piece of the metal sodium as large as a pea be held in a pair of tongs and plunged underneath half an ounce of warm mercury, the mercury will dissolve the sodium (if its surface is cleaned by scraping previously), generally producing a good deal of heat and some amount of burning of the sodium or "deflagration"; if now the solution of sodium in mercury, or fluid *sodium amalgam*, be poured into a cup full of water, a chemical action will be set up, resulting in the evolution of bubbles of combustible gas, hydrogen (Expt. 10), and the formation of an alkaline solution (Expt. 142) containing caustic soda, whilst the mercury remains in its original condition.

Expt. 169. To produce vigorous Combustion by mixing two hot Liquids.—In one crucible melt half an ounce of zinc over a good lamp, and in another fuse as much saltpetre (nitrate of potassium). Firmly grip the crucible containing the melted saltpetre with a long pair of two-handed crucible tongs (fig. 68), and pour the liquid saltpetre into the melted zinc; a vigorous chemical action will take place, and the zinc will burn brilliantly, evolving a large quantity of flakes of "zinc white" or oxide of zinc.

Caution.—In this experiment, as in all others where there is apt to be a sudden development of heat through chemical action, there is always a possibility of hot particles being violently ejected from, and splashed out of, the vessel containing the active substances, whereby a severe burn may be easily occasioned or damage done to the eyesight. The greatest care must, consequently, be taken in performing such experiments; the hands may be protected by means of a pair of gloves, and the eyes by a large pair of spectacles, or, better still, by holding a pane of glass in the left hand in such a way that the face is shielded thereby from splashes, &c., whilst what goes on can be readily seen through the glass. In carrying out experiments where *sodium* is required, or *phosphorus* (which are both substances liable under certain circumstances to set up intensely-vigorous, almost explosive, chemical action, leading to the possible projection of hot burning splashes) these precautions are extremely desirable, as severe injury may easily be occasioned by the incautious employment of dangerous chemicals.

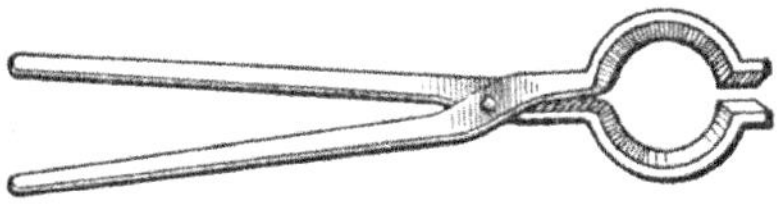

Fig. 68. Tongs.

Another point to be strictly attended to is, that whenever highly active substances are used *they should never be employed in any large quantity at a time;* a few grains of such materials may generally be manipulated with tolerable safety, if proper care be taken, whereas much larger quantities would be highly dangerous.

Separation of Liquids from Solutions by Processes involving Chemical Actions.

Just as solid substances may be precipitated from fluids (Expt. 121), and gases evolved from liquid substances (Expts. 160–162) by chemical actions, so under suitable conditions may liquids be similarly set free by virtue of analogous changes.

Expt. 170. To set free Liquid Fatty Matter from Hot Soap Solution.—Cut up a piece of ordinary household soap weighing about an ounce in small pieces, and put them into a flask containing about half a pint of distilled water; heat the whole on the water bath (Expt. 89), with occasional shaking; the soap will gradually dissolve, forming soap-water such as would be used for blowing soap-bubbles (Chapter XVIII.) Pour into the hot solution some hydrochloric acid; this will act on the soap, and liberate therefrom the "fatty acids" which, together with soda, constitute the soap. The exact nature of these fatty acids depends on the materials used for preparing the soap, being somewhat different according as tallow, olive oil, palm oil, &c., are employed; but in any case they will not dissolve in the watery fluid, and consequently gradually separate as an oily layer on the surface. Enough hydrochloric acid should be added to make the liquid bright red when a little tincture of litmus is added, and the whole well shaken up; if this hot fluid be poured out into a saucer or basin, and left to cool, the fatty acids will by and by set to a solid or pasty mass, according to their nature.

Expt. 171. To precipitate Fluid Quicksilver from Solution.—Boil half an ounce of *granulated tin* (Expt. 16) with strong hydrochloric acid for half an hour; a liquid will be formed containing a compound termed *stannous chloride* or *chloride of tin.* Now dissolve in water a few grains of *corrosive sublimate* (or *mercuric chloride*), and add to the solution some of the stannous chloride solution; the first effect will be that a white precipitate of a compound called *calomel* will be produced; but on adding more stannous chloride solution, and boiling, this will turn grey, and ultimately will become converted into minute globules or drops of fluid quicksilver. In this experiment the action is obviously closely akin to that whereby gold is separated from solution of chloride of gold (Expt. 135).

Expt. 172. Another way of reproducing Mercury from Solution.—Take a penny-piece, or a fragment of brass or copper, and rub it over with a solution of *nitrate of mercury* (Expt. 166), using an old tooth-brush or a piece of rag dipped in the fluid as a rubber; the copper present in the metal employed will act on the nitrate of mercury, forming *nitrate of copper* (Expt. 134), and precipitating *metallic mercury*, which will adhere to the metal and whiten its surface, making it resemble silver; the white coating, however, is not as permanent as actual silvering, but by and by disappears, partly because the mercury gets rubbed off, and partly because it gradually sinks into the interior of the solid metal.

Caution.—Mercury has a strong tendency to adhere to gold and silver articles, and also to penetrate into them if it come in contact with them externally, rendering them brittle in so doing (compare Expt. 69). Whenever any experiments are made involving the use of quicksilver, care should be taken that the quicksilver should not be splashed over rings, watch chains, &c., otherwise these may be seriously injured, and rendered so brittle as to break readily. Should a little mercury get splashed on to a gold ornament, it may often be removed by cautiously heating the article in the flame of a Bunsen or spirit-lamp, so as to volatilise the mercury, after which a little burnishing will restore the lustre, so that little or no mark is visible. Overheating must be avoided, especially in the case of rings in which stones are set. Instead of a lamp, a hot piece of iron may be used.

Expt. 173. To prepare White Vinegar.—The purest forms of vinegar consist of the substance termed acetic acid largely diluted with water, and sometimes coloured up with burnt sugar. Such a liquid may be prepared as follows:—Dissolve an ounce of *acetate of soda* in a tumblerful of hot water; in another vessel dilute a teaspoonful of oil of vitriol (sulphuric acid) with two or three times its bulk of water. If the two fluids are now mixed, acetic acid will be formed, but will not separate from the watery fluid, because it is easily soluble in water. Place the whole in a retort or other distilling arrangement, such as the flask with bent tube and Liebig's condenser described in Expt. 36 (fig. 23), and boil off two-thirds of the fluid, condensing the vapours; the distilled liquid thus obtained will be sour to the taste, owing to the volatilisation of the acetic acid along with the steam, and will in fact be a dilute solution of acetic acid capable of use as vinegar.

Expt. 174. To prepare Dilute Hydrochloric Acid.—Repeat the last experiment, substituting common salt for acetate of soda. Hydrochloric acid will now be formed instead of acetic acid, so that the distilled liquid will be a weak solution of hydrochloric acid. A stronger solution may be obtained by using cold water saturated with salt, and adding cautiously to the brine about one-fifth of its bulk of oil of vitriol little by little, so as to avoid the danger of spurting acid about by the generation of steam, owing to the development of heat on diluting the oil of vitriol (Expt. 98): finally, the liquid is distilled as before, the distillation being carried on until little if any more fluid drops from the condenser. The solution thus obtained is not a saturated solution of hydrochloric acid, but is strong enough for almost all experiments requiring that acid.

Expt. 175. To prepare Hydrochloric Acid Gas.—Many other volatile acids can be obtained in solution in similar fashion; the less water is used in proportion to the other substances, the stronger is the solution obtained. In many instances the water may be omitted altogether, and thus the acid itself obtained, and not merely a watery solution of it. Thus, in Expt. 270, concentrated nitric acid is obtained by heating together saltpetre and oil of vitriol. If common salt and sulphuric acid be thus employed, a gas (*hydrochloric acid gas*) is evolved, which fumes copiously in the air; if this be collected in a stout globular flask or bolthead (the collecting bottle being mouth upwards), and if a cork and glass tube be then placed in the mouth of the bottle, and the whole treated as the similar ammonia flask in Expt. 77, the same result will be brought about, viz., the formation of a *miniature fountain*, because of the pressure of the air forcing water into the flask, owing to the diminution in bulk of the gaseous contents thereof, when water comes in contact with them so as to begin to dissolve them. The liquid resulting is a weak aqueous solution of hydrochloric acid; by employing arrangements in which comparatively small quantities only of water are brought in contact with the gas, stronger solutions are obtained; when the water is nearly saturated with the gas (to produce which result it has to absorb several hundred times its volume of the gas), the liquid fumes more or less in the air. Such solutions of hydrochloric acid are those employed (sometimes after diluting to weaken them somewhat) in various of the experiments above described, such as Nos. 10, 13, 99, 118, &c.

---

§ 5. Chemical Actions producing Change of State without the Employment of Solvents.

CHAPTER XIII.

Chemical Actions of Decomposition or Breaking-up.

We have already seen (Chapter I.) that most chemical actions are referable, in the first instance, to one or other of the three classes—simple **decomposition** or **breaking-up** (*e.g.*, where water is decomposed by electricity into oxygen and hydrogen gases); **combination** or **synthesis** (*e.g.*, where a mixture of oxygen and hydrogen is fired, so as to cause the two gases to unite together and form steam); and **reciprocal decomposition**, where two substances so react upon each other as to produce two new products, each different from the original materials (*e.g.*, where zinc and hydrochloric acid give hydrogen and chloride of zinc, or where iron sulphide and hydrochloric acid give sulphuretted hydrogen and iron chloride; these two changes being reactions of **single displacement** and **double displacement** respectively).

In many instances these actions take place when some or other of the substances involved are employed in the form of solutions, the solvent taking no part in the chemical change, but simply facilitating it by enabling the particles of reacting matter to come together more freely, in virtue of the mobility peculiar to the liquid state; but the use of a solvent is by no means absolutely necessary for the occurrence of chemical changes.

In some instances the application of heat to a solid causes it to melt, and the further application of heat causes the melted substance to decompose, the breaking up being facilitated by the fluid condition assumed on melting; but this state of fluidity through heat is not essential, as solid bodies may often be decomposed without any previous melting. Thus the operation of burning limestone to quicklime is a case in point. When lumps of limestone are strongly heated by fire (practically effected by mixing the limestone and the fuel together and burning the mixture in a large grate or "kiln"), they become decomposed and lose largely in weight; about $\frac{2}{5}$ of the material is converted into gas (*carbon dioxide*, identical with that produced by acting on chalk with hydrochloric acid, Expt. 99), which escapes, whilst the remaining quicklime differs in very many respects from the original limestone; more especially in possessing the property of "slaking" with water, *i.e.*, evolving heat when wetted with water, owing to combination taking place between the quicklime and the water, producing "slaked lime;" and in being somewhat soluble in water, yielding an alkaline solution, or "limewater" (Expts. 152 and 240).

On the other hand, when metals are heated in contact with air, and more especially when heated in an atmosphere of pure oxygen, they usually (though not invariably) combine with the oxygen present, producing "oxides" (Expts. 15, 18). Sometimes the combination is attended with the development of great heat and vivid light, *i.e.*, vigorous "combustion" is brought about, just as when coke or coal burns in the air, owing to the combination of the oxygen with the oxidisable or combustible matter present. Here, obviously, no solvents are employed.

Yet again, if certain oxides, *e.g.*, the oxide of lead thus produced, be mixed with powdered charcoal, and strongly heated in a crucible so that the external air does not get access to the mass, the charcoal will act upon the oxide of lead in exactly the same kind of way that zinc acts upon hydrochloric acid (Expt. 10); the charcoal will take away the oxygen from the oxide of lead, forming metallic lead, and producing oxide of carbon, just as the zinc takes the chlorine away from the hydrogen, producing free hydrogen and chloride of zinc. In this case reciprocal decomposition takes place, again without the aid of a solvent. It is by actions more or less of this kind that most metals are actually extracted from their natural sources or "ores."

Expt. 176. To produce Oxygen Gas by the Decomposition by Heat of a Melted Solid.—The substance termed *potassium chlorate* is a crystallised solid substance at the ordinary temperature, but on heating it melts and becomes a liquid. Place in a dry test-tube as much of the salt as will lie on a sixpence, and gently heat it in the Bunsen lamp; at first it will quietly melt to a transparent colourless liquid; but as it gets hotter it will begin to effervesce vigorously, owing to the escape of bubbles of the gas oxygen set free in consequence of the chemical decomposition taking place in the salt through the effect of heat. Light a wooden match or spill, and blow out the flame, so that a spark still remains alight; introduce this smouldering end into the mouth of the test-tube; the oxygen will so accelerate the slow combustion of the spark as to cause the wood to burst into flame. A wax taper or candle with smouldering wick will generally behave in the same way.

Expt. 177. To collect a Jar of Oxygen Gas.—In the last experiment the potassium chlorate did not begin to decompose and evolve oxygen until it was heated to a temperature considerably above its melting point. It is a curious fact that when certain solid substances are added to powdered potassium chlorate, these facilitate the decomposition to such an extent that oxygen is evolved before the mass is heated sufficiently high to melt, although the solid substances themselves are not affected by the action; accordingly, for the sake of convenience, it is usual to prepare oxygen by heating gently a mixture of powdered potassium chlorate with about a quarter of its weight of *black oxide of manganese* (otherwise termed *manganese dioxide*). Place a tablespoonful of the mixture in a small dry flask provided with a tightly fitting cork, through which passes a piece of quill glass tubing bent as indicated in fig. 69, so that the further end of the tube can be made to dip into a basin half full of water. As with all other gases that do not dissolve very freely in water, a method of collecting oxygen is conveniently applicable, by means of the **pneumatic trough**, which substantially

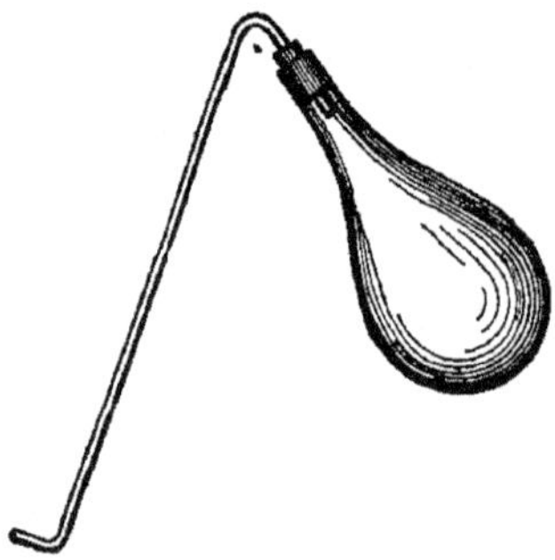

Fig. 69. Flask and Bent Tube.

consists in first filling the bottle or jar (to be ultimately filled with gas) with water by immersing it therein in a tub or basin, and then raising it bottom upwards partially out of the water, but so that the mouth of the jar is still under water; or otherwise, by filling the jar with water and corking it, or covering the mouth with the palm of the hand, or with a saucer, and rapidly inverting it, and lowering it till the mouth is under the water in the basin, the cork, hand, or saucer being then removed; if this is dexterously done, the jar remains filled with water, not more than a comparatively insignificant bubble or two of air entering during the operation.

The jar thus filled with water, and partly raised out of the basin, remains full of water in consequence of the atmospheric pressure; but if the delivery tube of a gas generator producing an appropriate kind of gas (such as hydrogen or oxygen), be dipped into the basin and brought under the mouth of the jar, the bubbles of gas passing out at the end of the delivery tube will rise up into and be caught in the jar, which will thus be gradually filled with gas, which takes the place of the water. When the jar is full, a gas tray or saucer (fig. 70) is slipped under its mouth whilst still immersed, and the saucer and jar are then lifted out of the basin together, and set by for use afterwards; whilst another jar is made to take the place of the first one, and is similarly filled, escape of gas or entrance of air being prevented by the water in the saucer which "seals" the jar with a "water lute" or "hydraulic valve," a kind of arrangement for closing temporarily the mouths of vessels containing gas extensively used in another form in coal gas works and chemical factories, where gases of various kinds are dealt with. As soon as one jar is thus filled, another may be made to take its place, and thus several may be filled in succession.

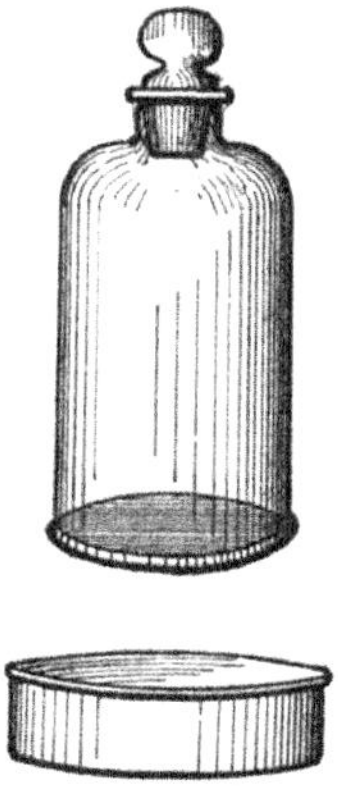

Fig. 70. Bell jar and Gas Tray.

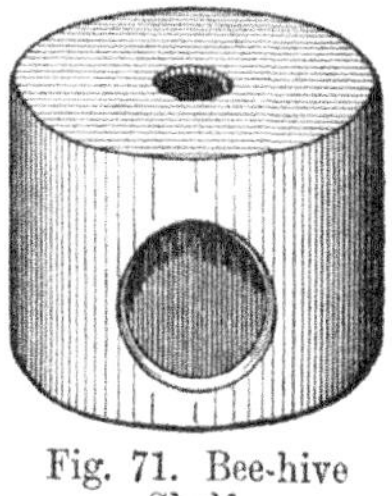

Fig. 71. Bee-hive Shelf.

To save the trouble of holding the jar whilst filling with gas, a couple of half bricks may be arranged side by side in the basin half an inch or so apart; the delivery tube is made to pass into the vertical slit between them, whilst the jar stands partly on one and partly on the other brick in such a position that the bubbles rise up into it. At the instrument-dealer's a pottery "bee-hive shelf" (fig. 71) may be purchased, answering the same purpose; this consists of a sort of earthenware cylinder or basin with a wide flat base perforated by a hole, and provided with a slit or circular orifice in one side; this stands in the basin of water or "pneumatic trough" upside down, the delivery tube of the gas generator passing through the slit in the side so that the gas bubbles into the interior of the "bee-hive" or inverted basin-shaped cavity, and thence passes through the orifice into a jar standing on the flat base or "shelf." Fig. 72 represents the mode of use of the bee-hive shelf with a trough constructed of glass plates.

Any gas not too readily soluble in cold water can be thus collected

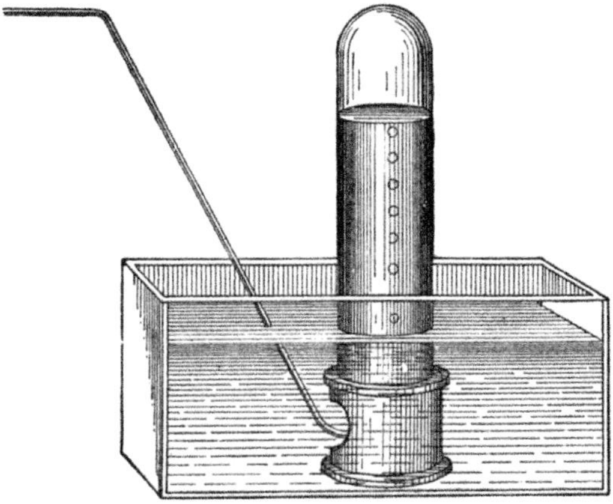

Fig. 72. Collection of Gas at the Pneumatic Trough.

by means of the pneumatic trough in jars or such like receptacles. When gases easily soluble in cold water are to be examined, hot water may sometimes be employed, as, for example, in the case of nitrous oxide or "laughing gas" (Expt. 187). In other cases a "mercurial trough" is employed (fig. 73), *i.e.*, a basin containing mercury instead of water, but manipulated in much the same way as that above described. Large vessels cannot well be employed in this way on account of their great weight when filled with mercury. When the gas is considerably lighter or heavier than air, it may be collected by displacement; thus, in Expt. 75, a jar full of ammonia gas was thus collected, and in Expt. 104, one of hydrogen, both gases lighter than air, the jars being held mouth downwards, and the delivery pipe being so turned that the escaping gas passes upwards; on the other hand, carbonic acid gas (carbon dioxide) being much heavier than air (as illustrated by Expt. 100), may be collected by displacement by simply passing the end of the delivery tube of an appropriate generator down to the bottom of a jar standing on a table mouth upwards. In the case of the ammonia, the lighter gas accumulates in the inverted jar, and thus gradually displaces the air originally contained in it, the ammonia being unable to rise through the solid glass; in the case of the carbon dioxide, the heavier gas accumulates in the erect jar in exactly the same way as water would do if run in gradually by a pipe.

If the position of the collecting jars be reversed in such experiments, no gas at all will be retained; thus if you attempt to collect ammonia gas in a jar mouth upwards, all the ammonia will rise out of the jar and escape into the air as rapidly as it passes in

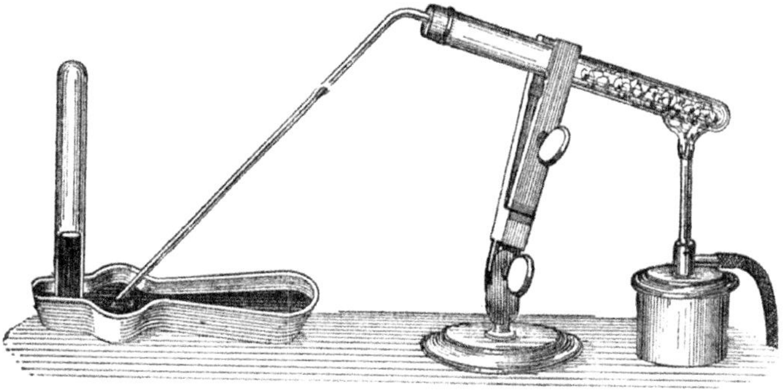

Fig. 73. Mercurial Trough.

from the generator; and, conversely, if you try to collect carbon dioxide in a jar held mouth downwards, the jar will never become properly filled, because the heavy gas will fall out of the jar as quickly as it passes in, just as water would do under similar conditions. In the same kind of way a jar of hydrogen (lighter than air) held mouth upwards for a few seconds, allows all the gas to float up, heavier air taking its place, so that no hydrogen is left in the jar (Expt. 105); whilst a jar of carbon dioxide held mouth downwards for a few seconds, similarly becomes emptied of that gas, the relatively lighter air taking its place (Expt. 101).

For storing gases not readily soluble in water, gasholders are used such as that shown in fig. 74. A copper drum, surmounted by a reservoir for water, is connected with the latter by means of two pipes, one of which passes down to the base of the drum, and the other only communicates with the top. If the taps in each pipe are both turned on, water will run down the first of these pipes, and force out the gas through the other, causing it to bubble up through the water in the reservoir, and thus enabling jars, &c., to be filled just as when working with the pneumatic trough. If the first tap be turned on, and the other turned off, the pressure of the water can be made to force gas out through a side tube on turning on the tap connected therewith; so that by connecting this tap with a bladder, as represented in the diagram, this may be filled with gas; or if connected with any required apparatus by

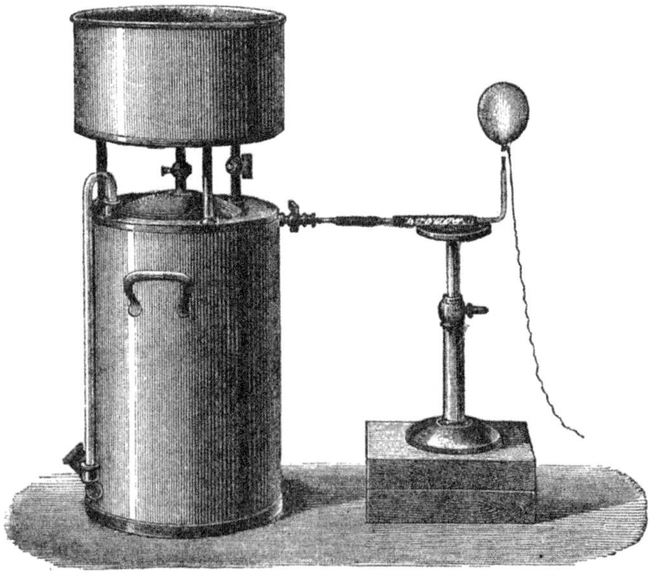

Fig. 74. Gasholder.

means of an india-rubber tube, a stream of gas can be passed continuously through the apparatus, as long as there is any in the gasholder to force out.

In order to fill the gasholder with gas in the first instance, all three taps are turned on, and water poured into the reservoir until the drum is quite full of water, and all air originally in it is displaced by the water. The three taps are then turned off, and the plug at the bottom unscrewed; the short bit of slanting piping fixed into the side of the drum, and ordinarily closed by the plug, acts like the lower part of the "bird fountain," or vessel for supplying water to birds in cages, the bulk of the water being prevented from escaping by the pressure of the atmosphere, whilst some can escape now and then when required, bubbles of air entering to supply its place. The delivery tube of a gas generator evolving the gas to be stored in the holder is then passed into the drum through this short slanting tube; the bubbles of gas rise into the drum, and water flows out until the drum is filled with gas and emptied of water (which is known by means of the gauge), when the delivery pipe is withdrawn, and the plug inserted again.

Fig. 75 represents a model of the kind of gasholder used in gas works and chemical factories. A large reservoir is constructed of sheets of iron bolted together, and is suspended by chains passing over pulleys and counterbalanced by weights, so that it can rise and fall as required. Gas is led into the reservoir by the pipe *a* leading from the generators, and passes out through the pipe *b*. If *b* be closed by turning off the valve connected therewith, the gas entering in by the pipe *a* accumulates in the reservoir, and by and by raises it up out of the cylindrical basin of water in which it rides up and down; thus the reservoir is filled. In order to force the gas out, as for example when a town is to be supplied with coal gas, the counterpoising weights *c*, *c* are partly removed, so that the weight of the reservoir partly presses on the contained gas, and forces it out through the tube *b*, when the valve therewith connected is opened.

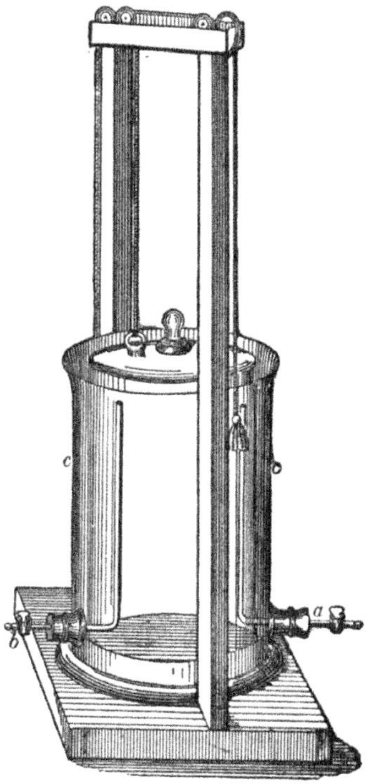

Fig. 75. Model Gasholder.

For laboratory experiments and for producing the "limelight" in theatres, &c. (Expt. 211), gases are often stored in india-rubber bags, filled by directly connecting them with the delivery tubes of gas generators, and emptied by placing a board with weights on it on the bags (fig. 76), so that the pressure of the weights forces the gas out in a current through an attached pipe leading to wherever the gas is required. For experiments on the small scale, an ordinary

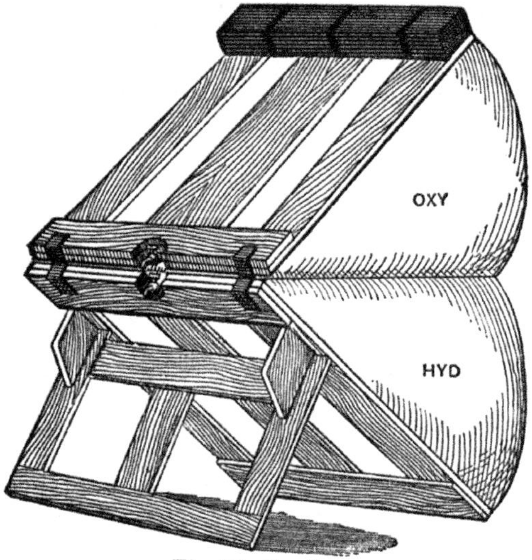

Fig. 76. Gas Bags.

ox bladder used in the same kind of way answers very well. To fill the bladder, it should be softened and made pliable by dipping in water, and when soft should be blown full of air, and then well rubbed with glycerine; the air should then be squeezed out and the bladder connected with the gas generator, preferably by fixing in the mouth of the bladder a piece of brass tubing with a tap attached, and tying the bladder firmly round the tube, so that when the tap is turned off no gas can escape, whilst when turned on it can pass in and out through the tube.

Instead of bags or gasholders, gases are often stored in strong iron cylinders, into which the gas is pumped under great pressure, so that by slightly opening a valve the gas issues forth spontaneously; owing to the great diminution in the bulk of the gas thus occasioned such cylinders are far more conveniently portable than large india-rubber bags (fig. 77).

Fig. 77. Cylinder of Compressed Gas.

Expt. 178. To prepare Oxygen and Quicksilver by heating a Solid.—In Expt. 176, where oxygen is prepared by heating potassium chlorate, the chemical action is one of simple decomposition. Potassium chlorate is really a compound of oxygen with another substance termed *potassium chloride*, and when heated breaks up into these two constituents, the former escaping as a gas, whilst the latter remains as a fused mass in the test-tube. Many other compounds containing oxygen behave in similar fashion when heated; *i.e.*, they break up, evolving oxygen and some other substance which together with oxygen form the constituents of the compounds.

Place in a test-tube as much *red oxide of mercury* as will lie on a sixpence, and heat it pretty strongly in the Bunsen lamp; the red powder will first turn almost black, and will then become decomposed by the heat; oxygen gas will be liberated, as may be shown by introducing into the mouth of the tube a glowing spill of wood (Expt. 176), which will burst into flame; simultaneously small globules of *quicksilver* will condense on the side of the tube. Let the tube cool, and then introduce a pencil or glass rod, and rub together the film of condensed mercury therewith, when you will see that it actually does consist of globules of quicksilver.

For experiments illustrating the chemical properties of oxygen, *vide* Chapter XIV.

Expt. 179. To prepare Chlorine Gas by heating a Solid.—Into a test-tube introduce one or two grains of solid *chloride of gold*, and apply heat; decomposition will ensue, and *chlorine gas* will be evolved, whilst *metallic gold* remains in the tube in a "spongy" state, *i.e.*, in somewhat loose particles. To test the production of chlorine, the fumes evolved at the mouth of the tube may be cautiously wafted towards the nose, when the peculiar smell of diluted chlorine will be smelt; care should be taken not to take a strong sniff directly at the mouth of the tube, on account of the choking and irritating properties of chlorine when breathed in any quantity. (Compare Expt. 162.)

Another test for chlorine is its property of turning blue a mixture of potassium iodide and starch (Expt. 164). Prepare a mixture of starch paste and potassium iodide, dip a piece of white paper therein, and place the paper over the mouth of the test-tube; the chlorine passing off will turn the paper blue where it comes in contact with it.

Expt. 180. To prepare Chlorine in larger Quantity.—The ordinary method of preparing chlorine gas it to heat *strong hydrochloric acid* solution with powdered *manganese dioxide*, using the arrangement indicated by fig. 55. The gas should be collected by displacement in jars held mouth upwards, as the gas is heavier than air; unless a properly constructed draught cupboard (Expt.

145) be used, or the experiments made in the open air, it is very difficult to avoid inhaling more or less of the gas, and so producing much distressing coughing and irritation; consequently only moderately small quantities of materials should be employed, say 1 ounce of manganese dioxide and 4 ounces of hydrochloric acid.

Although chlorine dissolves somewhat freely in cold water, so that it cannot well be collected at the pneumatic trough in the ordinary way, it is much less soluble in strong brine, especially when warm, so that it may be collected over this fluid without much difficulty. Instead of using hydrochloric acid and manganese dioxide, the following materials may be employed. Mix 4 parts of common salt and 3 of manganese dioxide well together in a mortar; transfer the mixture to a generating flask, and pour over it a mixture of 8 parts of oil of vitriol and 3 of water, previously mixed together and allowed to cool. Chlorine will be evolved, especially if the materials are slightly warmed from time to time.

Expt. 181. Another Way of preparing Chlorine.—Into a gas generator (fig. 56) put two or three ounces of good *bleaching powder* (chloride of lime), and then pour diluted hydrochloric acid down the funnel; chlorine gas will be evolved, as there will not be sufficient water present to keep all the gas generated in solution. For experiments illustrating the chemical properties of chlorine gas and solution thereof in water, *vide* Nos. 163, 215, 216, 217, 228, &c. (Compare Expt. 162, *Caution.*)

Expt. 182. To prepare Cyanogen Gas.—Heat in a dry test-tube a few grains of powdered *cyanide of mercury;* the substance will behave like oxide of mercury (Expt. 178), *i.e.*, it will split up into two constituents, *mercury and cyanogen*, the former of which will condense in globules on the sides of the tube in the cooler part, whilst the latter will pass off as a gas of peculiar peach-blossom-like odour and inflammable nature, so that on applying a light to the mouth of the tube the evolved gas will burn with a violet-purple flame. The gas can be collected over water at the pneumatic trough, if the cyanide of mercury be heated in a small test-tube generator like that shown in fig. 73.

Caution.—Cyanogen gas is very poisonous, therefore great care must be taken lest it be inhaled in any quantity.

Expt. 183. To prepare Nitrogen Gas.—Put into a flask about equal weights of salammoniac and sodium nitrite with a little water; fix in the mouth of the flask a delivery tube, as in Expt. 177 (fig. 69); gently heat the mixture in the flask, and nitrogen gas will be evolved. The action taking place is that the two compounds, ammonium chloride (or salammoniac) and sodium nitrite, react upon one another by double decomposition, forming sodium chloride and ammonium nitrite, which latter substance is a ternary compound of nitrogen, hydrogen, and oxygen; being unstable, it is immediately decomposed by the heat, forming a mixture of water vapour and nitrogen gas. The latter may be collected over water at the pneumatic trough, like oxygen.

Expt. 184. Another Method.—Ordinary atmospheric air is shown by Expt. 148 to be a mixture of oxygen and nitrogen, from which the former can be removed by the action of pyrogallic acid dissolved in caustic soda. Many other substances may be employed in the same way; for instance, phosphorus.

Float on water a large bung, and on it put a bit of phosphorus the size of a pea, carefully dried in blotting paper. Touch the phosphorus with a hot wire, and it will begin to burn; as soon as this occurs hold over the bung and burning phosphorus a jar or bell-shaped vessel (fig. 78), so that the bottom of the jar just touches the water. At first the heat of the burning phosphorus expands the air, and drives a little out, so that a few bubbles escape under the edge of the jar; the jar should then be lowered so that its mouth is an inch or two under water. As the phosphorus burns it absorbs the oxygen, and the compound formed produces a white cloud of solid oxide of phosphorus particles; so that the bulk of gas in the jar diminishes from the removal of the oxygen. Soon the flame of the phosphorus goes out, all the oxygen being consumed; let the jar remain at rest for half an hour; by that time the fumes will have all subsided, and the gas in the jar will be clear; it will consist of nearly pure nitrogen. (Compare Expt. 169. *Caution.*)

Fig. 78. Preparation of Nitrogen.

Expt. 185. To show that Nitrogen extinguishes Flame.—Prepare nitrogen in a bell jar with a stopper in the top; when the gas is clear remove the stopper, taking care that the jar is sunk so deep in the basin that the water stands at the same level inside and out. Introduce a lighted candle by a wire; the candle will be extinguished, just as in Expts. 71 and 149.

Expt. 186. To reproduce Air from Nitrogen and Oxygen.—Prepare some nitrogen, and transfer it to a jar so as to fill it about four-fifths full; this is done by depressing the jar containing the nitrogen under water in a bucket or large basin, and tilting it so that the gas gradually escapes, holding over it the second jar filled with water, so as to collect the escaping gas. Now add sufficient oxygen from another jar containing that gas to fill up the remaining fifth of the jar; the two gases will rapidly intermix by diffusion (Chapter IX.), and ordinary atmospheric air will result, as may be tested by introducing a lighted candle, which will burn just the same in the jar as in another similar one filled with ordinary air.

Expt. 187. To prepare Laughing Gas (Nitrous Oxide).—When nitric acid and ammonia in solution are brought together they combine, forming a neutral salt, *nitrate of ammonium*, just as sulphuric acid and ammonia form sulphate of ammonium (Expt. 138). If this solution be evaporated, and the residual salt heated till all water is driven off, the pure nitrate of ammonium is left as a fused liquid which sets to a crystalline solid mass on cooling. If, however, the heat be continued, the substance decomposes, somewhat as nitrite of ammonium splits up (Expt. 183); but with this difference, that steam and the gas *nitrous oxide* are formed instead of steam and nitrogen. The nitrate of ammonium is, like nitrite of ammonium, a ternary compound of nitrogen, hydrogen, and oxygen, but contains more oxygen; the result of which is that, instead of nitrogen being produced, a compound of nitrogen with the surplus oxygen present is formed.

Nitrous oxide dissolves somewhat readily in cold water, consequently it should be collected at the pneumatic trough over water as hot as the hands can possibly bear.

Expt. 188. Nitrous Oxide supports Combustion.—Into a jar of nitrous oxide introduce a lighted candle; the candle will burn freely, even better than in air. Similarly a bit of sulphur will burn vigorously in the gas if introduced by means of a "deflagrating spoon"(fig. 79); the reason for this is that the heat of the burning substance introduced splits up the nitrous oxide into nitrogen and oxygen, and the latter keeps the material alight, just as air would do, and even better. A glowing spill of wood will generally burst into flame when placed in a jar of nitrous oxide, just as it would in oxygen (Expt. 176).

Fig. 79. Burning Phosphorus in Nitrous Oxide.

When breathed, however, nitrous oxide behaves very differently from air or oxygen; great excitement and exhilaration is first brought about, which often produces laughing (whence the name "laughing gas"), and subsequently much the same kind of intoxication as alcohol, but not so long in duration. A remarkable degree of insensibility to pain generally accompanies this latter stage, whence the use of laughing gas as an "anæsthetic" (producer of insensibility to pain) for dental and other surgical purposes. On account of the danger of fatal results, should too much of the gas be inhaled by persons suffering from weak heart and other complaints, anæsthetics (laughing gas, vapour of chloroform, or ether, &c.) *should never be breathed* except under medical direction, or in the presence of some one acquainted with the proper methods of use and precautions against danger requisite in the employment of such materials. Dentists generally keep the gas stored in strong iron bottles, into which it is pumped under great pressure (fig. 77), and employ a specially constructed piece of apparatus for the purpose of enabling the gas to be safely inhaled.

Expt. 189. Decomposition of Sugar into Carbon Dioxide and Alcohol.—Several varieties of sugar are known to chemists; thus the ordinary table sugar is derived mostly from the beetroot or sugar-cane, and is generally distinguished as *cane sugar*; grape juice and many other kinds of fruits (also honey) contain a different sugar, known as *grape sugar;* cows' milk contains another variety much less sweet to the taste, termed *milk sugar* (Expts. 89 and 120); and manna (an exudation from a species of ash tree) contains a still different variety, called *mannite.* All these and several other kinds known are capable of crystallising; but besides these, uncrystallisable varieties of sugar exist in treacle, golden syrup, molasses, and such like substances.

Many of these sugars, but not all, possess the remarkable property of undergoing *fermentation* under suitable conditions; that is, if the seeds or germs of certain minute vegetable growths are introduced into solutions containing these sugars, the vegetable matter grows and increases, and in so doing causes the sugar to become split up or decomposed into *carbon dioxide* and *alcohol.* In ordinary brewing the fermentable saccharine matter is mostly derived from *malt*, or barley which has begun to germinate and sprout by damping and keeping in a warm place, and has then been killed by heating to such an extent as to destroy the vitality of the seed without otherwise decomposing it; if the heat be high, the sugar present in the sprouting grain becomes more or less browned or "burnt" (like the caramel or browned sugar used by cooks for browning gravy, &c., made by heating sugar in a hot oven), and gives a dark-coloured liquor on "mashing" or treating with water to dissolve soluble matters, suitable for making stout and porter; whilst the light-coloured malts made by not heating so highly are employed for beer and pale ale. The vegetable germs are introduced into the "wort" or liquor produced by treating the malt with water in the form of *yeast*. As the action of fermentation proceeds, the "yeast plant" increases in quantity, and either sinks to the bottom of the vat as a sediment (when the fermentation is carried out at a low temperature, producing the German "lager" beers), or rises to the top mixed with froth and scum (when the temperature is higher, as in the English system of brewing). Large quantities of carbon dioxide gas are evolved from the fermenting liquid, whilst a corresponding quantity of alcohol is simultaneously formed; for bottled beer, the liquid is bottled before the fermentation is quite finished, so that a further development of carbon dioxide goes on in the bottle, causing the beer to froth greatly on uncorking (Expt. 71).

In the manufacture of wine, the general character of the action is much the same; for "sparkling" wines (champagne, &c.), the wine is bottled, like the beer, before the fermentation is finished; whilst "still" wines (claret, sherry, port, &c.), are carefully treated to prevent any such action going on after bottling. In hot climates, there are usually sufficiently large quantities of yeast plant germs floating about in the air to start fermentation in a saccharine fluid when exposed to the air for a short time so as to take them up; thus freshly expressed grape juice will begin to ferment apparently spontaneously in such a climate if kept in an open vessel. If, however, the juice be pressed out into clean bottles with as little exposure to air as possible, the action of fermentation is greatly delayed, as the requisite quantity of germs has then not been taken up; whilst by taking extreme precautions to avoid contact with air containing germs, the grape juice may be sealed up in glass vessels, and kept for an indefinitely long time without fermenting at all.

Dissolve a quarter of a pound of moist sugar (cane sugar, containing a little uncrystallisable sugar and colouring matter) or of honey in a quart of water, and add to this solution some fresh brewer's yeast, or a little German yeast rubbed up with water to a cream (German yeast is the frothy yeast from brewing collected on a cloth filter, drained, and pressed as dry as possible). Place the whole in a flask with a cork and delivery tube attached, and keep the whole in a moderately warm room. After the lapse of a few hours the liquid will begin to froth, and if you then place the end of the delivery tube in a wine glass containing lime water (fig. 66), you will see that gas is being evolved by the fermentation now beginning; and that the gas is carbon dioxide will be shown by the milkiness or precipitate produced in the lime water as it passes through (Expt. 152). This action will go on for some days; when it slackens, the fluid may be distilled (Expt. 38) until about one third has passed over; this distillate should be again distilled until almost one-third has passed over. This twice "rectified" spirit will probably be strong enough to burn if a little be poured on some cotton wick, or warmed in a spoon, and then lighted; but if too weak to burn, a pretty strong spirit may be extracted from it by putting an ounce or two of pearlashes (potassium carbonate) into a bottle, pouring in the weak spirit, shaking up, and allowing to stand, when watery solution of potassium carbonate subsides and strong alcohol floats up to the top (Expt. 39).

Expt. 190. To make Ginger Beer and Nettle Beer.—Dissolve loaf sugar in boiling water in the proportion of from $\frac{3}{4}$ to 1 lb. of sugar to the gallon of water; add to this mixture the rind of a lemon chopped very fine and 1 oz. of powdered ginger; stir up well, cover, and allow to stand six hours; then add 1 oz. of cream of tartar and the juice of a lemon, with a tablespoonful of good yeast, and mix thoroughly; allow to stand in a sufficiently warm room to enable the fermentation to proceed; after some hours the mixture may be poured off from the sediment and bottled, the corks being well tied down; if kept in a warm room the fermentation will be sufficiently advanced after twenty-four hours more to make the ginger beer fit for use, but a somewhat longer time is usually preferable.

In some country districts analogous drinks are prepared by gathering certain herbs (in particular, young nettles), and pouring boiling water over the green shoots or leaves, &c.; sugar is dissolved in the decoction (sometimes treacle is used instead), and the whole allowed to stand some hours so as to settle, and the clear liquid poured off; yeast is then added (with lemon juice or other flavouring matters according to taste), and the whole kept in a warm room until fermentation is well developed, when the liquid is bottled and stored awhile, so that the fermentation may become fully developed whilst in bottle.

In these and all other beverages, a certain proportion of alcohol is necessarily present, owing to the nature of the changes produced in the saccharine matter by the action of fermentation; a somewhat similar kind of non-alcoholic "ginger beer," however, is prepared by aerated water manufacturers by processes essentially the same in principle as that described in Expt. 72, but carried out on a much larger scale, the carbon dioxide gas employed being generated from chalk and sulphuric acid, and compressed and dissolved in the fluid to be aerated by means of a powerful pump.

Expt. 191. Alcohol from Starch.—During the "malting" of grain, and in general during the sprouting of most kinds of seeds, the life action of the growing vegetable brings about a change of the insoluble starch stored up in the grain or seed into soluble sugar which is contained in the sap. This chemical transformation can be imitated artificially, and by fermenting the sugar thus produced alcohol can be readily formed from starch. In this way rice starch, potato starch, and many other starchy materials are employed for producing spirituous liquors, in most cases of far inferior quality, viewed as beverages, than those produced from natural sugar, more especially that of the grape.

Grind up some starch with water to a thin cream, pour this into boiling water, add a few drops of sulphuric acid (previously diluted with water) to the fluid, and boil the liquor for some hours, adding water from time to time to replace that lost by evaporation. Finally add some chalk, whereby a more or less marked effervescence will be produced, carbon dioxide being evolved, and *sulphate of calcium* formed; the sulphuric acid will thus be neutralised (compare Expt. 165), and as the sulphate of calcium formed is not very soluble in water, most of it will be precipitated as a solid. Strain off the liquor and filter it; the presence of grape sugar in the clear solution may be readily shown by the copper test described in Expt. 123; if the solution be evaporated on the water bath (Expt. 89), a sweet solid or semi-solid residue will be left, consisting of impure grape sugar. If there is enough of it, this may be dissolved in water and fermented with yeast, and the production of carbon dioxide and alcohol ascertained as in the last experiment.

Large quantities of converted starch are thus prepared and sent into the market under various names, being chiefly used as sources of alcohol in the production of fermented liquors of various kinds. Some of the purer sorts of sugar thus produced are nearly white and solid, and are often termed *glucose;* others contain so much uncrystallisable sugar that they can only be obtained as semisolid soft masses, or a thick syrupy fluid. The coarser kinds are largely used, under the name of *malt substitutes*, in brewing various kinds of beers, being far cheaper than genuine malt; but the advantage to the consumer of the use of such materials is at best doubtful, whilst in many instances the beer thus made is distinctly of inferior quality, and sometimes decidedly objectionable; the better kinds of starch sugar are used in the preparation of wines (mixed with grape juice of inferior richness as regards sugar, so as to enable the mixture to produce a wine of sufficient alcoholic strength), and for many other purposes, the light-coloured uncrystallised varieties being sometimes sold under the name of honey. A great deal of the starch sugar made is directly fermented and distilled in order to prepare spirits of wine for manufacturing purposes and as a beverage; some kinds of rectified spirit thus produced are eminently unwholesome, on account of the presence of injurious impurities, more especially *fusel oil* (Expt. 85). In converting starch on a manufacturing scale the materials are usually not boiled in open pans, but heated together in pressure boilers, which facilitates the action.

Expt. 192. Alcohol from Cotton and Linen or Paper.—The woody matter of most vegetables mainly consists of a substance called *cellulose*, which is obtained in different conditions from different plants; thus the fibres of hemp and flax stalks, the down or filaments of the cotton seed, and the analogous products derived from jute, Esparto grass, and numerous other sources of vegetable textile fabrics and paper pulp, all consist essentially of cellulose in some form or other. Chemically, cellulose is closely related to starch and sugar, being formed from and transformed into bodies of these natures during the growth and life of plants and trees, and being also capable of transformation into sugar by chemical means. Thus the action of certain mineral acids, more especially sulphuric acid, will convert cotton and linen into sugar, which is one reason why splashes of acid are apt to rot and burn holes in clothing (Expt. 139).

Cut up a piece of calico or linen (part of a shirt will do) into small shreds, and stuff them into a large flask; pour on some sulphuric acid, and shake it about so as to moisten the shreds thoroughly with the acid; allow the whole to stand an hour or two, then add some water (about half a pint to every ounce of acid used), and boil gently for some hours, adding a little water from time to time to replace that lost by evaporation. Pour the whole out into a basin, and gradually add powdered chalk little by little with continual stirring; copious effervescence will ensue, owing to the production of sulphate of calcium and carbon dioxide, as in the last experiment; when all the acid is neutralised, strain off the clear liquor (adding a little more water if necessary), and press the sulphate of calcium mud in a calico filter or bag, so as to squeeze out as much fluid as possible. Filter this liquid clear through paper, and evaporate the solution on the water bath. Finally a tolerably clear solid sugar will be obtained, which may be used to sweeten a cup of tea, or fermented with yeast, as in Expt. 189.

Paper and even sawdust can be similarly treated with much the same result, excepting that the more compact nature of the ligneous matter of certain kinds of sawdust renders the chemical action less rapid. Alcohol has thus been made commercially from sawdust, but of somewhat inferior quality; that from pine saw-dust is said to have a natural resinous flavour, so as to form a sort of gin. As with starch, the converting action of the sulphuric acid is a good deal quickened by heating the materials in pressure boilers rather than in open vessels.

In the ordinary process of baking household bread, the flour is mixed with water and some yeast to a thick paste or "dough," which is then set by in a somewhat warm place until the dough begins to "rise." After the "working" has gone on for some time the dough has greatly increased in bulk, and has become honeycombed all over with bubbles. This arises from the action of the yeast upon the starchy matter present, first converting it into sugar, and then splitting up the sugar into alcohol and carbon dioxide; this latter swells up the dough and makes it porous, so that when baked, bread of the usual vesicular structure is produced, instead of a solid mass like a suet dumpling. One result of this is that all bread prepared with "leaven" is to some extent impregnated with alcohol, so that by taking a portion of the crumb from the interior of a newly baked loaf, soaking it in water, and distilling the fluid, minute quantities of alcohol can be extracted, sufficient to recognise by chemical tests. Most of the alcohol formed during the rising of the dough, how-ever, is driven off by the heat of the oven, so that only very small quantities remain in the baked bread.

In the preparation of bread on the large scale the baker's men sometimes knead the dough by treading; in any case a good deal of handling is required. As this is by no means an appetising operation, machines have been constructed to take the place of this mixing and kneading by hand or foot. What is termed "aerated" bread is prepared in a different way. Flour is mixed in a specially constructed machine with water impregnated with carbon dioxide under considerable pressure, this pressure being maintained during the mixing; when the dough is thoroughly mixed the pressure is relaxed, and consequently a good deal of gas is liberated, just as when a bottle of soda water is opened; this swells up the dough and makes it porous, just as though leaven had been used. Aerated bread, consequently, contains no alcohol, and from the absence of leaven in it is often considered more whole-some for persons suffering from weak digestion and other ailments, besides being made in a far more cleanly way than ordinary leavened bread. Owing to the different mode of preparation, it often has a slightly different taste; this is partly due to the fact that the action of the yeast in the ordinary process of bread making converts a little of the starch into sugar, which escapes decomposition into alcohol and carbon dioxide, and remains in the bread, rendering it somewhat sweeter to the taste than aerated bread, unless a little sugar is purposely added to the latter during manufacture, as is sometimes done.

Expt. 193. To prepare Vinegar.—It is worth noticing, that whereas the formation of alcohol from sugar by splitting up requires the presence of no free oxygen for its occurrence, and can therefore go on in a corked bottle or closed cask, the subsequent process of **souring** wine and beer so as to produce vinegar (from the French *vin aigre* = sour wine), will not take place under such conditions, access of air being indispensable as well as the growth of a peculiar organism or *vinegar ferment*, analogous to yeast, but not identical therewith. Half-empty casks of ale and bottles of light wines are peculiarly apt to undergo the "acetic fermentation," and turn sour, because the remaining liquid is necessarily in contact with the air, which supplies the oxygen requisite to enable the change to take place; strong wines containing a large proportion of alcohol are not so easily affected as lighter ones. Sometimes vinegar is prepared from injured batches of wine, damaged cider, and such like materials, but preferably from properly fermented malt worts, by filling a wooden cylinder with shavings and allowing air to pass freely into it, whilst the liquor to be soured into vinegar is allowed to trickle slowly over the shavings, and thus to be exposed in thin layers to the air. The shavings soon become coated with the vinegar plant, so that the souring action goes on rapidly. Another class of vinegar, preferred by some, is made by simply diluting pure acetic acid (Expt. 173) to taste, and sometimes colouring with a little burnt sugar.

Opinions differ widely as to the relative goodness of vinegar prepared in these two ways; some prefer that made by a fermentation process, as it has a somewhat different flavour; thus old-fashioned housewives sometimes prepare their own vinegar by fermenting to sourness a mixture of sugar, &c., and certain flowers and plants, *e.g.*, primroses. Others consider that pure acetic acid diluted with 25 or 30 or even more times its weight of water, is preferable to all fermented products.

Expt. 194. To detect Sulphuric Acid in Vinegar.—Sulphuric acid is sometimes used to adulterate vinegar, more especially the coloured varieties; formerly this addition was permitted by law to the extent of 1 part in 1000, but now it is forbidden in this country. Add to some of the clear vinegar a solution of *nitrate of barium* or *chloride of barium;* if sulphuric acid be present, a cloudiness will be produced, and after standing a perceptible precipitate (of *sulphate of barium*) will become visible.

Expt. 195. To show the Decomposition of Carbonic Acid by Growing Plants.—The direct splitting up of carbon dioxide into carbon and oxygen is not easy to effect, the very highest temperatures attainable by the aid of the electric current being requisite; but the action may be indirectly effected by the aid of growing vegetation. We have seen, in Expt. 153, that the exhaled breath contains carbon dioxide, the chemical action taking place in the body of an animal during living and breathing being that oxygen is dissolved from the air in the lungs, and is carried by the circulation of the blood all over the body; in its passage this dissolved oxygen oxidises certain of the constituents of the blood and tissues, forming carbon dioxide, which is finally exhaled by the lung. Living vegetation effects exactly the opposite kind of change; *i.e.*, carbonic acid gas is dissolved from the air, &c., in the moisture of the soil, and absorbed by the rootlets of plants, and so conveyed to the leaves by the circulation of the sap; under the influence of sunshine the carbonic acid is decomposed, oxygen being evolved in the free state, whilst the carbon of the carbonic acid, together with the hydrogen and oxygen of the water present, are transformed into solid "cellulose" (the woody part of plants) and similar products. By charring these in closed vessels (Expt. 201), charcoal is obtained, so that indirectly this charcoal and free oxygen are thus procured from carbon dioxide by decomposition.

Fill a jar or tumbler with water by immersion in a basin, raising the jar bottom upwards out of the water, but so that the mouth is still covered; pass into the jar some sprigs of fresh mint, or of some green water plant, and expose the whole to sunlight for some hours. By and by bubbles of gas will form on the submerged leaves, and will gradually collect at the top of the jar; these are nearly pure *oxygen gas*. If the water employed be mixed with a quarter of its bulk of pure aerated water, so as to give a larger supply of carbon dioxide, it will be practicable ultimately to collect a sufficient quantity of oxygen to show that the gas will relight a glowing match, and form an explosive mixture with twice its bulk of hydrogen (Expts. 176 and 208).

In order that vegetation may thus decompose carbonic acid, the action of light is indispensable; some plants (*e.g.*, mushrooms) will grow in the dark, but in such cases the green colouring-matter known as *chlorophyll* (the tinting material in all green leaves, &c.) is never formed; apparently, the presence of this compound is necessary for the evolution of oxygen from carbonic acid.

Expt. 196. Action of Light on Silver Compounds.—Certain

compounds of silver are decomposed by exposure to light, which circumstance is the basis of the majority of the processes employed in photography. Precipitate some *chloride of silver* by adding some nitrate of silver solution to brine (Expt. 120), and wash the precipitate once or twice by decantation. Place the washed silver chloride in sunlight for some time, when it will become blackened through decomposition. The presence of animal or vegetable matter in contact with the silver compound greatly promotes the action and setting free of metallic silver in the form of black particles; thus if a piece of paper be soaked in brine and dried, and then written on with a quill pen dipped in a solution of silver nitrate, it will become charged with particles of silver chloride precipitated in its pores by double decomposition wherever the pen has touched it; on exposing the prepared paper to light it will blacken much more quickly than pure silver chloride alone, speedily rendering the writing visible.

Expt. 197. Marking Ink for Linen, &c.—Write with a clean quill pen dipped in a solution of nitrate of silver on the linen to be marked, having previously soaked the spot in salt water and dried it again. *Chloride of silver* will be formed where the pen passes, as in the paper in the preceding experiment; on exposure to sunlight the chloride of silver will be decomposed, and metallic silver deposited in the pores of the linen so as to mark it permanently.

Ordinary marking inks are usually solutions of *nitrate of silver* or some other silver compound coloured with sap-green, indigo, &c., to make the writing legible at first. In use, the decomposition is generally brought about by means of heat rather than light, a hot flat-iron being placed over the writing, which produces the same result more rapidly. India-rubber stamps and types made of metal that will not act chemically on silver compounds (unlike copper, Expt. 134) are often conveniently used instead of quill pens; if the metal be acted upon, the silver will be more or less precipitated on the type by the action, and the liquid will not communicate the decomposable silver compound to the linen, so that only a comparatively faint mark will be produced, which will not bear frequent washing.

Expt. 198. To Silver a Hollow Glass Vessel internally.—The action of heat in decomposing silver compounds so as to precipitate metallic silver is made use of for the purpose of coating hollow glass vessels internally with silver in such a fashion as to form a bright glistening surface, converting the vessel into a mirror. One way of proceeding is as follows:—Dissolve a small quantity of *tartaric acid* in water and add *ammonia* solution until the liquid smells of ammonia after shaking, and turns a red litmus paper blue. Now add some solution of *chloride of calcium* and warm the whole; a precipitate will be formed which will by and by become granular and sandy; this consists of *tartrate of calcium.* Wash this three or four times by decantation (Expt. 120) and then transfer the wet mass to a flask, add two or three drops of *ammonia solution*, and a crystal of *solid nitrate of silver* the size of a pea, or more if necessary. Now gently warm the flask over a lamp, moving it about so that the pulpy material may flow over the whole of the interior surface in succession; when the temperature rises sufficiently, decomposition will take place rapidly and the interior of the flask will be coated with a film of bright silver. Let the flask cool and carefully wash out the interior with a little cold water; drain mouth downwards, and allow it to stand in a warm room a few hours so as to dry the interior thoroughly. If the silvering has been properly done the flask may be corked and the cork waxed over to exclude air, and the mirror preserved for years; but if air containing sulphuretted hydrogen gets access to it, the silver will soon be attacked and the mirror spoilt. Some little dexterity and practice is requisite to coat a large vessel thus; but a small flask (holding from two to six ounces) can be easily silvered after one or two preliminary trials. To clean the silver off a flask partly silvered, but not satisfactorily, put a teaspoonful of diluted nitric acid into the flask and slightly warm it, bringing the acid successively in contact with every part of the interior by turning the flask round; the film of silver will thus be removed and the flask made fit for another experiment, after well rinsing out with water. In the actual manufacture of mirrors various other chemicals are employed along with silver solutions, the effect in all cases being the "reduction" of metallic silver as a brilliant film.

Expt. 199. To prepare Spongy Platinum.—Metallic platinum can be prepared in the form of a loose powdery substance termed "spongy" platinum by mixing together solutions of *platinum chloride* and *salammoniac* (chloride of ammonium); combination first takes place between the two materials forming a yellow solid substance sparingly soluble in water, which consequently precipitates: the mixture is made in an evaporating basin, and the whole evaporated to dryness over a steam bath (Expt. 89). The dry mass is heated red hot in a porcelain crucible, when decomposition takes place, various gases and vapours being evolved, and nothing but *spongy platinum* being left.

Expt. 200. To prepare Coal Gas.—The ordinary illuminating gas used in our houses is prepared by heating coal in closed vessels, under which circumstances a complicated decomposition takes place, a large variety of substances being formed. *Coke* is left behind in the vessel, being non-volatile, whilst a mixture of volatile products passes off; some of which will condense to liquids on cooling, whilst others will not.

Obtain a clay tobacco pipe and fill the bowl with small pieces of coal. Make a covering over the bowl with good clay so as to enclose the coal in a vessel having as outlet only the shank of the pipe. Dry the prepared and charged pipe in an oven, and plaster up with fresh clay any cracks that may form, repeating the operation until the pipe is covered with an air-tight lid of clay. Now heat the bowl either by placing it in an ordinary fire, or by means of a large Bunsen lamp; in a short time gas will be formed by the action of the heat on the coal, and will issue from the stem of the pipe so as to burn steadily when a light is applied.

In this way the production of "tar" is not perceptible; instead of a tobacco pipe, employ a piece of thin glass tube 8 or 10 inches long and $\frac{1}{4}$ inch diameter internally, drawn out and sealed up at one end in the lamp flame (Expt. 48) like a test-tube. When cold introduce a small teaspoonful of powdered coal into this tube, shaking it down to the sealed-up end, and then apply heat, holding the tube by a clamp-holder (fig. 12) in a nearly horizontal position; the coal will be decomposed as before, and inflammable gas will pass off at the mouth of the tube, where it can be lit; between the heated part of the tube and the mouth a brown evil-smelling thick liquid will condense on the glass, and also a little watery fluid which will not mix with the brown tar.

By using a small glass retort instead of the tube a larger supply of condensed tar and watery fluid may be obtained. On testing the watery fluid with a sensitive test paper, such as a red litmus paper (Expt. 142), you will observe that it has an alkaline reaction, being in fact a very impure solution of ammonia; so that besides coke three other products are evidently formed by the decomposition, viz., the *inflammable gas* (itself a mixture of several constituents); the *watery fluid*, containing dissolved ammonia gas and other substances; and the *tar*. In large gas works these two latter substances are carefully condensed, and certain impurities removed from the gas before it is supplied to our houses for burning purposes; from the watery fluid ammonia and various compounds thereof are obtained; whilst the tar serves as the source of a large number of useful products; amongst which may be mentioned artificial oil of almonds (*nitrobenzol*), used for scenting soap and similar purposes; and many beautiful dyestuffs.

Expt. 201. Destructive Distillation of Wood.—The term "destructive distillation" is applied to the process of heating matter of animal or vegetable origin in closed vessels so arranged that the substance is not burnt by access of air, but is decomposed or broken up under the influence of heat, such products as are liquid at the ordinary temperature being condensed by suitably cooling the gases and vapour evolved; thus in the last experiment, coal was destructively distilled in the retort.

For coal substitute chips of wood; it will now be found that a large quantity of permanent gas is produced, which, like coal gas, will burn with a more or less luminous flame; besides this a variety of tar will condense and an aqueous liquor which, with most kinds of wood, will exhibit an *acid* and not an alkaline reaction, *i.e.*, it will turn blue litmus paper red, and not *vice versâ*. The acid thus produced is in consequence termed *pyroligneous, i.e.*, produced from wood by heat; it mostly consists of *acetic* acid, the acid of vinegar (Expt. 193). Besides this acid, the watery fluid also contains an ill-smelling spirituous substance termed *wood spirit* (*pyroligneous spirit* or *pyroxylic spirit*); this spirituous fluid is extracted and mixed with pure alcohol to render it undrinkable, the "methylated spirit" (Expt. 40) thus produced being still capable of use for burning, varnish-making, and such like purposes.

Expt. 202. Destructive Distillation of Bones.—Substitute bones or gristle for wood in the previous experiment. A very foul-smelling tar will be obtained, and a strongly alkaline watery fluid, much ammonia being produced during the destructive distillation of animal substances. For this reason the alchemists termed ammonia the *volatile alkali* (compare Expt. 143), whilst the name *spirit of hartshorn*, sometimes applied to solution of ammonia, is also derived from the fact that ammonia is produced by heating horn. The coke left in the retort when the distillation of bones is complete is known by the name of *bone charcoal* or *bone black*, and is largely used as a means of decolorising syrups in sugar refining (Expt. 289), on account of its remarkable power of attracting certain substances into its body and thus removing them from solution. Charcoal from dried blood is still more energetic in this way.

Expt. 203. Action of Lime and Caustic Soda on Nitrogenous Matter.—Many substances are known which contain as one of their constituents the element *nitrogen;* most of these possess the property of forming ammonia (a compound of nitrogen and hydrogen) when they are heated with slaked lime, or powdered caustic soda, or with a mixture of the two (*soda-lime*).

Heat in a dry test-tube a few grains of *uric* acid (the chief constituent of the sediment forming in many kinds of urine on standing) previously mixed with two or three times its bulk of powdered soda-lime; the smell of ammonia (mixed with other matters of a tarry nature) will soon become perceptible, and on putting a red litmus paper in the fumes issuing from the mouth of the test-tube it will be rendered blue.

The production of ammonia in this way is employed as a test to see whether substances are "nitrogenous" or not; *i.e.*, whether or not they are compounds containing nitrogen. You may easily prove in this way that nail parings or human hair cut up fine and mixed with soda-lime will evolve ammonia, and are consequently nitrogenous bodies. Similarly, shreds of woollen cloth and silk will produce ammonia when thus treated; but threads of cotton and linen, pure paper, calico, and such like vegetable products will not evolve any ammonia on heating with soda-lime, because they are not nitrogenous compounds.

CHAPTER XIV.

Chemical Actions of Combination.

Development of Heat.

In general, wherever chemical combination takes place a more or less considerable development of heat is brought about; the principal sources of artificial heat are in point of fact due to this action, ordinary *fires* being simply the result of combination of the oxygen of the air with certain bodies of the "combustible" class, *e.g.*, wood, charcoal, coal, oil, tallow, &c. In such cases, the heat produced is not as great as it is when pure oxygen is used instead of air, because the action is somewhat retarded owing to the presence of the inert nitrogen; accordingly, to produce the highest degrees of artificial heat by means of such actions, pure oxygen is employed, one of the hottest known flames being produced by using hydrogen gas as the combustible substance, and supplying the flame with oxygen. Platinum, a metal almost impossible to melt in a furnace heated by any other kind of flame, can be melted in large quantities at a time in a suitable vessel (made of quicklime) heated by the oxyhydrogen flame (Expt. 211), *i.e.*, a flame produced by supplying a jet of hydrogen with the right quantity of oxygen requisite to burn it all to water vapour or steam.

As a general rule, chemical combination will not take place to any considerable extent between substances in the solid condition, principally because their texture is too stiff and rigid to allow of the two different kinds of matter being brought into sufficient proximity except just at the border line where their outsides touch. If, however, one of the substances be a liquid or gas the action is greatly facilitated; hence, with many kinds of solid chemical substances, one or both of the bodies to be operated upon must be converted into a liquid (either by heating it until it melts, or by dissolving it in a suitable solvent, which amounts to much the same thing) before the chemical action can be brought about.

In many cases, also, the chemical change will not take place at the ordinary temperature, but ensues readily if the materials are more or less heated; thus a jet of ordinary coal gas escaping from a burner does not take fire spontaneously, and no chemical action ensues between the coal gas and the air; but if a light be applied (*i.e.*, if a portion of the mixture of gas and air be considerably heated by a flame) the gas and air undergo a chemical action, the result of which is that the oxygen of the air combines with the constituents of the gas; in so doing heat is produced, and thus the next portions of gas passing out from the burner are heated sufficiently to set up the chemical change of burning or oxidation; the heat thus produced propagates the change in the next portions of gas that issue, and so a continuous state of chemical action is kept up, with the result of producing a steady flame. Candles, lamps, torches, &c., burn in just the same kind of way, with this difference that in the case of the coal gas the chemical action takes place directly between the gas and the air; whereas in an oil lamp the first effect of the heat upon the oil is to decompose it or break it up into gaseous products by a sort of destructive distillation (as in Expts. 200, 201, 202), these products then becoming oxidised by the oxygen of the air as before; so that an oil lamp or candle flame is actually a kind of gas factory in miniature.

Expt. 204. To show that Combustible Gases are formed in a Candle or Lamp Flame.—Obtain a tallow candle with a thick cotton wick, or an oil lamp (with a round wick rather than a flat one); light the lamp and spread out the wick so as to produce as large a flame as possible. Hold a piece of quill glass tubing a few inches long with the lower part of the tube in the centre of the lower part of the flame (fig. 80). If the tube be not too big, and the flame be perfectly steady, it will be seen that a current of smoke and gases passes up the tube, forming a small jet of combustible vapours which will burn steadily on applying a light to them.

Expt. 205. To combine Copper and Sulphur together.—Fill a large test-tube with loosely packed thin copper turnings, and then introduce some flowers of sulphur, shaking the tube so that the flowers may be tolerably equally spread about amongst the copper turnings. No change whatever takes place, the copper and sulphur being solids. Now heat the tube in a spirit lamp or Bunsen burner flame, holding it by means of a test-tube holder (fig. 10); the sulphur will soon begin to melt, and as soon as it is liquid it will begin to act on the copper; heat is developed, and by and by, when the sulphur is nearly boiling, the copper will unite with the sulphur so rapidly that the whole mass becomes visibly red hot, the copper actually **burning** in the act of combination with the sulphur. When the action is over and the contents of the tube are cool enough to handle they will be found to consist chiefly of a black brittle substance easily rubbed to powder in a mortar and quite different from either the original sulphur or copper; this substance is the product of the combination of the two, and is termed *copper sulphide.*

Fig. 80. Candle Flame and Tube.

Expt. 206. To combine Iron and Sulphur.—Make a mixture of two parts iron filings and one flowers of sulphur, and heat it in a test-tube as in the last experiment; when the mass is sufficiently hot combination will take place between the iron and sulphur, with the production of heat sufficient to make the whole glow visibly; the product is termed *iron sulphide*, and, as shown in Expt. 13, will produce the evil-smelling gas sulphuretted hydrogen when treated with hydrochloric acid, which the original iron filings would not do, hydrogen only being formed by the action of the acid on them. A solid iron rod heated nearly white hot and touched with a stick of brimstone will rapidly combine therewith, fused sulphide of iron dropping down and the rod being apparently melted away.

In these experiments the chemical action takes place between the solid metal and the sulphur, either in the melted state or in the state of vapour, but not when the sulphur is solid; the product being in each case a substance which is solid at the temperature at which the action takes place, although capable of fusion at a higher temperature. It is a general rule in cases of combination that the product is usually nearer to the solid condition than the materials; thus when two gases combine together to form a compound, that compound is either a solid (Expt. 214), or a liquid which will solidify on further cooling sooner than the original gases (*e.g.*, water produced by the combination of oxygen and hydrogen gases); or a gas which will condense to a liquid and subsequently freeze sooner than the original gases.

Combination of Gases to form Liquid or Solid Products.

Expt. 207. To show that Water is produced by burning Hydrogen.—Place some granulated zinc (Expt. 16) in the gas generator (fig. 56); pour down the thistle funnel some hydrochloric acid solution diluted with two or three times its bulk of water; vigorous effervescence will take place, owing to the dissolving of the zinc by the acid (Expt. 10), and the evolution of hydrogen, in consequence of the chemical action taking place. At first the gas generator contains a quantity of air, so that what will escape by the delivery tube will be a mixture of air and hydrogen; this mixture has the property of exploding violently when a light is applied to it, so that care must be taken at this stage that the issuing gases are not fired, otherwise the generator will probably be shattered, and damage done by the broken glass and spilt acid. After a few minutes, however, practically all the air originally present will have been expelled, and only hydrogen will issue from the delivery tube; and at this stage a light may be safely applied to the issuing gas, which will then burn with a scarcely visible blue flame. Over this flame hold a cold dry glass tumbler; dew will be immediately seen to form on the glass from the condensation of the water vapour formed by the union of the hydrogen with the oxygen of the air, showing that water is a product of the combination. If a large bell jar be used instead of a tumbler, the moisture produced by the burning of the hydrogen will by and by collect to such an extent that distinct drops of water will form and drip down, showing that the dew was really caused by the condensation of water vapour, and not by any other cause. In order that the gas may burn freely, the delivery tube should be bent so as to point upwards; the glass should be drawn out in a flame (Expt. 48), and cut off at the tapering part thus produced, so as to leave an orifice considerably smaller than the bore of the original tube before drawing out. To insure that practically all air is expelled from the generator before lighting the issuing gas, a sample should be collected in a test-tube like the ammonia in Expt. 75 (fig. 59), and a light applied to the mouth of the test-tube, after removal mouth downwards; if on lighting this gas only a very faint pop is heard (owing to the admixture of air and hydrogen unavoidably taking place at the mouth of the tube when uncovered to light it), the gas is fit for collection and use; but if a more vigorous small explosion occurs, the gas is still mixed with air.

Expt. 208. To illustrate the Explosive Nature of Mixed Air and Hydrogen.—Fill a soda-water bottle with water, place it in the pneumatic trough, and partly fill it with hydrogen, so that from one-third to one-fourth of the water is displaced by that gas; now lift it out of the trough altogether, so that the rest of the water runs out, air entering in its place. Fling a towel over the bottle (to avoid being hurt by splinters of glass, should the bottle be burst, which, however, is not likely to occur with a stout soda-water bottle, though it might happen with a thin flat medicine-phial), and apply a light to the mouth by means of a long wax taper or spill of paper. The mixture of air and hydrogen will detonate with a somewhat loud report. Explosions in coal mines are produced in similar fashion. An inflammable gas somewhat resembling hydrogen, termed *fire damp*, is naturally produced from the coal, and when this becomes mixed with air in certain proportions, the mixture explodes with great force when fired, thus leading to disastrous accidents. Ordinary coal gas behaves in the same way, whence the great danger of explosion should a pipe leak in an ordinary house, or should a tap be turned on so as to allow of a large escape of gas.

A still more vigorous explosion may be produced if pure oxygen be substituted for air. Fill the soda-water bottle two-thirds full of hydrogen, and then collect oxygen in it (Expt. 177), so that there shall ultimately be about half as much oxygen as hydrogen present; the explosion on applying a light will now be more energetic, because the explosive gases are not now diluted with the inert nitrogen contained in air.

In all these cases the cause of the explosion is simply that the chemical action generates suddenly a large amount of heat, which greatly expands the gaseous products of combustion (Chapter XIX.), and so causes a violent motion of the air, on account of the pressure produced. In just the same way, when gunpowder is fired, a large quantity of hot gaseous matter is suddenly produced; if the powder be in the barrel of a gun, the sudden expansion projects the bullet with great force, just as a pea is shot out of a peashooter by a sudden expulsion of air from the lungs into the tube behind the pea. If the powder be in a cavity bored in rock or stone, as in blasting for mining and quarrying, &c., the sudden pressure rends and tears the rock asunder, a result which sometimes happens to a gun if the metal is not strong enough to resist the force and prevent its bursting.

Expt. 209. To produce Musical Sounds by means of a Hydrogen Flame.—Arrange a hydrogen generator with an upright jet so as to burn the gas experimented with, as in fig. 81. Hold over the flame and surrounding it like a lamp chimney a glass tube half an inch internal diameter and a foot long; if the size of the flame is properly proportioned, the flame will be thrown into a state of vibration by the current of hot air and steam passing up the tube, so that the hydrogen issues from the jet in a series of pulsations, instead of a perfectly uniform stream. The effect of this is that the column of air in the glass tube is thrown into vibration too, like that in an organ pipe, and a musical sound produced, the tone of which depends on the length and diameter of the tube used, being generally the shriller the shorter the tube. A musical instrument, called the *hydrogen harmonicon*, has been constructed on this principle, furnished with a keyboard like a piano, so arranged that when a key is pressed down a current of hydrogen is turned on to a jet in a suitable pipe, and at the same time is lighted by means of an electric spark, so as to develop sound; a separate jet and properly tuned pipe is of course required for each note. As yet, however, this instrument has not found its way into musical orchestras, the reason for which will doubtless be appreciated by the experimenter whilst trying different sizes of tube, some of which are pretty sure to produce most doleful wails and squeaks, reminding one of the effect of a "cat concert," where each performer takes a cat in his or her arms, eliciting a note by judiciously pinching the tail gently, or otherwise coaxing pussy to emit a yell or a growl as the case may be. Fig. 82 represents a series of four tubes, any of which can be used simultaneously,

Fig. 81. Singing Tube.

hydrogen being supplied to the central tap from a generator or gas bag, and thence to the four burners; the lengths of the tubes are so proportioned as to give the 3d, 5th, and octave of the note yielded by the longest tube.

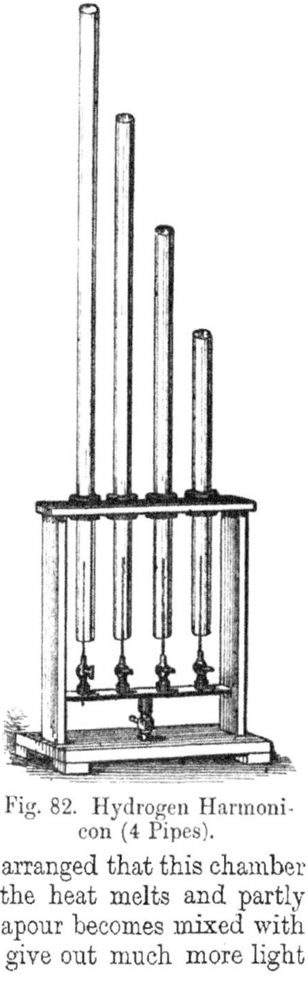

Fig. 82. Hydrogen Harmonicon (4 Pipes).

Expt. 210. To render the Flame of Hydrogen Luminous.—Down the thistle funnel of a gas generator producing hydrogen pour a teaspoonful of *ether*, or, better still, *benzoline;* these substances will be partly volatilised, and their vapours mixing with the hydrogen will yield a gas that will burn with a much more luminous flame than pure hydrogen. If the generator has become somewhat warm with action, *oil of turpentine* will produce the same result.

This method of "carburising" a gas which of itself will not give a very luminous flame, is extensively practised on the large scale. A current of ordinary air passed through a vessel containing a highly volatile suitable fluid (extracted from petroleum) will take up sufficient vapour to burn at a jet like ordinary gas. The "albo-carbon" gas burner acts similarly. The gas supply is connected with a chamber containing the substance known as *naphthalene* (a pretty readily volatile solid obtained from coal tar), so arranged that this chamber becomes heated when the gas burns; the heat melts and partly volatilises the naphthalene, so that its vapour becomes mixed with the coal gas, and enables the latter to give out much more light whilst burning.

Expt. 211. To produce an Oxyhydrogen Flame.—A much greater brightness of flame may be produced by burning hydrogen and supplying the flame with a current of oxygen gas blown into the flame from a narrow orifice, so as to blow the flame out in a sloping nearly horizontal direction (fig. 83), like the "dart" or flame produced by a plumber by means of a lamp and "blowpipe." To produce the current of oxygen a gasholder like that represented in fig. 74 may be used, so that by pouring water into the upper reservoir the gas is forced out at the side tap, and so led by an india-rubber pipe to the jet. Instead of this a jar of oxygen with a cork and tap in the top may be used, the jar being pressed down in a bucket of water so as to force the gas out in a current; or a bladder provided with a tap may be filled with oxygen by first collecting in such a jar and then transferring to the bladder, or by directly connecting the bladder with the oxygen gas generator, or from a gasholder, as in fig. 74. If the oxyhydrogen flame is made to impinge on a lump of chalk or marble, the heat will burn the outer surface of this to quicklime (Chapter XIII.), and will make it glow most brilliantly, so as to give out an intense light.

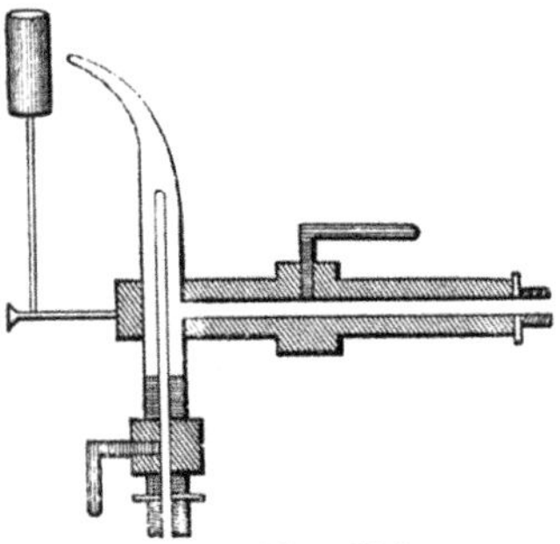

Fig. 83. Lime Light.

Without a suitable holder it is difficult to obtain a steady light in this way; fig. 83 represents one form of compound burner arranged for the purpose, oxygen being supplied by one tap and hydrogen (or coal gas) by the other from a pair of gas bags (fig. 76), or compressed gas cylinders (fig. 77); and a cylinder of quicklime being supported by an iron rod so that the flame shall impinge nearly horizontally upon its surface. Fig. 84 represents another arrangement of the kind, such as is used in theatres to

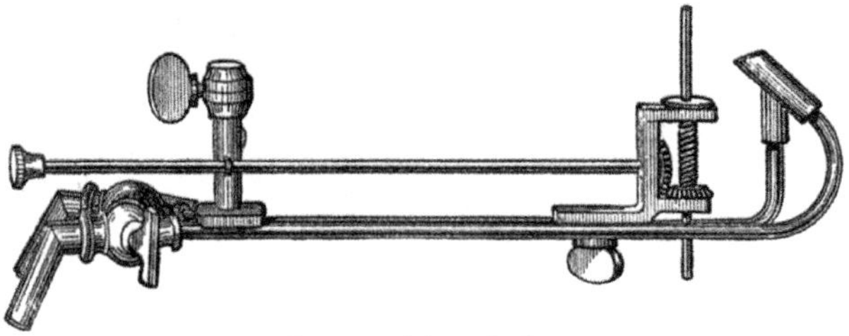

Fig. 84. Lime Light.

produce the "lime light"; by means of a pair of bevel wheels turned by a long rod, the lime cylinder can be turned round from time to time, so that the flame may heat up a fresh smooth part of the surface when the former one becomes roughened and injured by use.

Expt. 212. To make Oxygen burn in Hydrogen.—Arrange a

jet of oxygen in a vertical position; fill a cylinder or narrow-mouthed bottle holding a quart or more with hydrogen at the pneumatic trough, lift it up mouth downwards, and apply a light, and then depress the bottle over the oxygen jet so that the latter is inside and near the middle, like the candle in Expt. 107 (fig. 85). The oxygen will be *lit* as the jet passes inwards (unless the pressure be too great, so that the oxygen issues too violently to burn steadily), and two flames will be seen, one that of oxygen burning in hydrogen inside the bottle, and the other hydrogen burning in contact with the air at the mouth of the bottle. Do not allow the flame to burn longer than a minute or so; otherwise it is possible that the flame at the mouth may go out and air be sucked into the bottle, which mixing with the remaining hydrogen may produce an explosive mixture of gases; withdraw the oxygen jet, and extinguish the hydrogen flame at the mouth by pressing a towel against it.

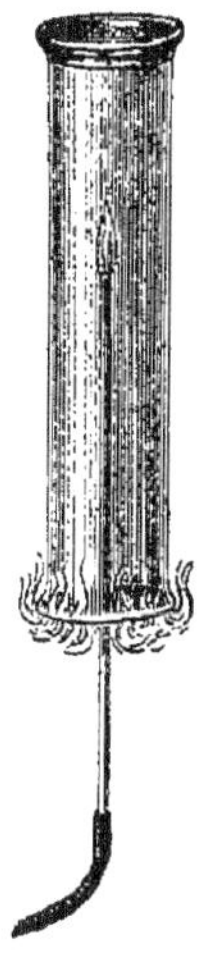

Fig. 85. Oxygen burning in Hydrogen.

Coal gas may be used instead of pure hydrogen in this experiment, the bottle being filled by holding it over an india-rubber tube attached to the gas main, and filling by displacement mouth downwards.

Expt. 213. Air burning in Coal Gas.—Fig. 86 shows a somewhat different way of illustrating the same kind of action, a jet of *air* here burning in an atmosphere of coal gas, whilst the surplus coal gas burns in the air. A lamp chimney is fitted with a doubly perforated cork, through which pass two glass tubes—one, *a*, connected with the coal gas main, the other, *b*, opening into the air. The chimney is removed and the coal gas turned on slightly, so that a small flame burns at the end of *a*; the chimney is now placed in position, when the flame burns steadily on *a* if the tube *b* is wide enough to admit sufficient air. The coal gas supply is now increased gradually by slowly turning on the tap; with a certain rate of supply more coal gas comes in than the air can burn properly. At this stage the flame disappears from *a*, but reappears at the top of *b*, so that a flame of air burning in coal gas is now produced; the surplus coal gas escapes at the top of the chimney, where it may be lit, producing a flame of the ordinary kind, *i.e.*, coal gas burning in air. By properly regulating the gas supply, the flame can be made at will to burn on *a* or *b* as desired.

Expt. 214. Production of a Solid by the Chemical Action of two Gases on one another.—The gas ammonia (collected as in Expt. 75) possesses the remarkable property of directly combining with certain other gases without the aid of any liquid solvent to facilitate the action, and without any heating being requisite to start the chemical change; usually the products of the combination are solid at the ordinary temperature.

Fig. 86. Air burning in Coal Gas.

Prepare two gas generators with the delivery tubes bent downwards. Into one put a tablespoonful of common salt, and pour over it a wineglassful of strong sulphuric acid; the mass will foam up, a choky corrosive acid gas being disengaged, termed *hydrochloric acid* (Expt. 175) (the solution of hydrochloric acid used in so many of the preceding experiments is simply a solution of this gas in water, just as solution of ammonia is ammonia gas dissolved in water). Into the other put some "liquor ammoniæ fort." (strong solution of ammonia), and warm it so as to evolve ammonia gas, in the same way as in Expt. 160. Bring the delivery tubes of both generators into the same bottle or gas jar, and gently heat each generator so as to drive off gas. The two gases meeting together in the central bottle will unite, forming a white solid which will deposit in the bottle and on its sides. This solid is in fact *sal-ammoniac*, which (as already shown) can be formed by the combination of ammonia and hydrochloric acid in solution (Expt. 138); the action of lime on salammoniac, as described in Expt. 263, is really to reverse this action, and cause the salammoniac to be decomposed into ammonia gas which is evolved, and hydrochloric acid which is retained, the lime combining with it.

If a glass rod moistened with strong hydrochloric acid solution be brought near the mouth of a test-tube in which ammonia is being generated (as in Expts. 160 or 203), more or less copious white fumes will be visible, the formation of which serves as a test for ammonia and analogous volatile alkaline vapours when it is desired to know if such bodies are being evolved. Conversely, a rod moistened with strong ammonia will give similar thick smoky fumes if brought near to a vessel from which acid vapours are escaping, such as the mouth of an unstoppered bottle containing strong hydrochloric acid solution.

Moisten two tumblers or wide-mouthed bottles internally, one with concentrated hydrochloric acid solution, so that a film of the fluid adheres to the whole of the inner surface (this is easily effected by nearly filling the tumbler with the liquid, and pouring back as much as possible into the bottle), and the other similarly with strong solution of ammonia. Hold the latter bottom upwards, covering the mouth with a glass plate or card, the former being similarly covered and standing on the table bottom downwards; bring the two tumblers together, one over the other (fig. 87), and withdraw the two cards *a* and *b*; the ammoniacal and acid vapours will intermix and combine, filling both glasses with thick smoky fumes, which will by and by condense in solid flakes on their sides.

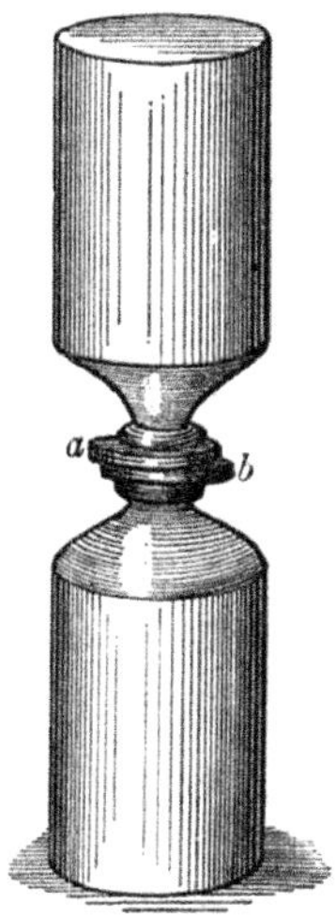

Fig. 87. Ammonia and Hydrochloric Acid.

The air of stables often contains a sufficient amount of ammonia (from the decomposition of dung and other animal matter), to cause a thick fume when a little strong hydrochloric acid is poured out on the ground or into a saucer; similarly, the presence of ammonia is soon recognised by the change of colour of a moistened red litmus paper or other analogous test-paper held in the stable for a short time (Expt. 142).

A number of substances closely akin to ammonia, and either gaseous at the ordinary temperature, or easily converted into vapour, are known to chemists, all of which possess the same property as ammonia itself, viz., that of combining with hydrochloric acid gas to form solid compounds; also certain other acid gases are known, which will behave in the same way as hydrochloric acid gas, *e.g.*, the gases termed *hydrobromic acid* and *hydriodic acid*. Thus the gas termed *phosphine* (phosphoretted hydrogen) will combine with hydriodic acid gas to form solid crystals of the compound termed *phosphonium iodide;* whilst the vapour of the highly volatile compound *methylamine* similarly combines with hydrobromic acid gas to form a solid compound analogous to salammoniac.

Combination of Gases to form Gases.

In actions of this kind there is obviously no change of physical state; but it generally happens that the compound formed by the union of two gases condenses to a liquid on cooling more readily than either of the two original gases. Thus, when hydrogen burns in the air, uniting with oxygen to form water (Expt. 207), what really takes place at first is the combination of gaseous hydrogen and gaseous oxygen to form water vapour or steam, which does not condense to liquid water until it has become greatly cooled down, much heat being produced by the chemical action. By substituting certain other gases for oxygen or for hydrogen in this experiment, it is possible to produce combustion or development of heat in consequence of chemical action, where the products of the action are all gaseous at the ordinary temperature; thus the gas termed *carbon monoxide* will burn with oxygen, producing a third different gas *carbon dioxide* (Expt. 220). Just as water vapour condenses to a liquid more readily than either hydrogen or oxygen, so does carbon dioxide become more readily liquefied by considerable chilling than either oxygen or carbon monoxide. The same result is generally brought about in other analogous cases.

Expt. 215. Combination of Hydrogen and Chlorine Gases.—Collect a small wide-mouthed jar full of chlorine by displacement (Expt. 180), and another similar one full of hydrogen (Expt. 104); bring the two jars together mouth to mouth (fig. 87), and then invert them so that the hydrogen jar is lowest and the chlorine jar uppermost; in a minute or so the heavier chlorine and lighter hydrogen will have intermixed sufficiently to form a mixture of gases capable of exploding with some report on cautiously separating the two jars, and applying a lighted taper to their mouths. The explosive action in this case arises from the same cause as that operating in Expt. 208, where a mixture of oxygen and hydrogen behaves similarly; a chemical action of combination takes place, hydrogen and chlorine combining to form hydrochloric acid, just as hydrogen and oxygen combine to form steam; and so much heat is developed during this action that a considerable expansion of the gases is suddenly brought about at the moment of combination. With pure hydrogen and chlorine already mixed in proper proportions, exposure to light will often bring about combination with explosion, without any flame being requisite.

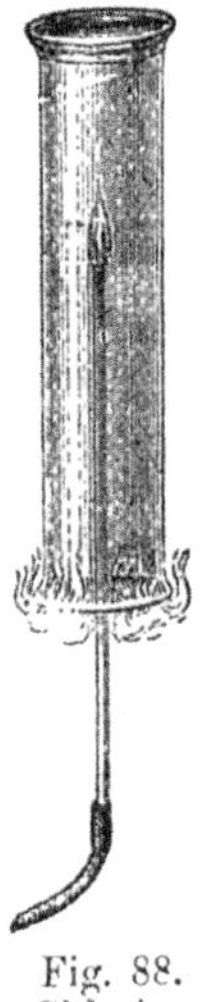

Fig. 88. Chlorine burning in Hydrogen.

Expt. 216. Chlorine burning in an Atmosphere of Hydrogen.—Another way of showing the combination of chlorine and hydrogen is illustrated by fig. 88; a small chlorine generator is set in action, the delivery pipe pointing upwards; a jar of hydrogen is collected, and brought near the chlorine delivery tube mouth downwards. A light is applied to the mouth of the jar, so that the hydrogen takes fire; the jar is then quickly placed over the chlorine delivery tube, so that the tube projects upwards well inside the jar; the jet of chlorine then takes fire and burns inside the jar, combining with the hydrogen, and forming hydrochloric acid gas, the action being very similar to that of oxygen when burning in an atmosphere of hydrogen (Expt. 212).

Fig. 89 shows the converse experiment; a jet of hydrogen from an ordinary hydrogen generator is lighted and introduced into a jar of chlorine, where it continues to burn.

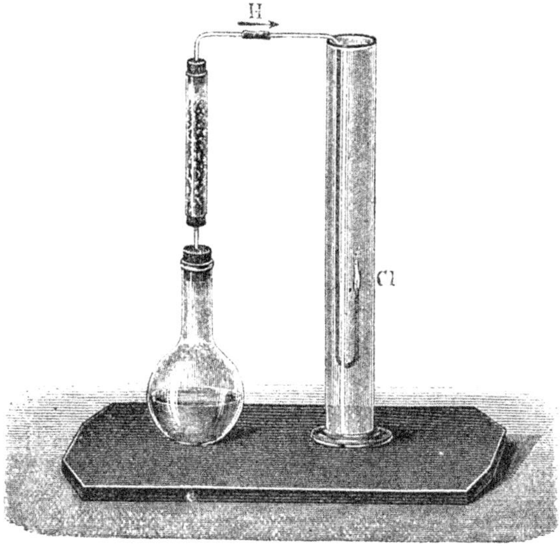

Fig. 89. Hydrogen burning in Chlorine.

Expt. 217. Burning of a Candle in Chlorine Gas.—Although chlorine has a strong tendency to combine with hydrogen, as indicated in the last experiment by the energy with which the combination takes place, the same is not the case as regards all other substances; so that when hydrogen combined with certain other substances is brought into contact with chlorine, it will often happen that the chlorine will combine with the hydrogen present and set the other substance free, thus bringing about an action of displacement (Chapter I.). This is particularly the case with bodies of the ordinary illuminating and combustible class, such as candles, lamps, and coal gas flames. If a lighted candle be tied to a wire and lowered into a jar of chlorine gas, the candle will not go out entirely, but will burn with a dull red lurid flame, emitting clouds of smoke; the carbon contained in the wax or tallow, &c., of the candle does not combine with the chlorine under these circumstances, and is consequently liberated as soot, whilst the hydrogen present does combine with the chlorine, forming hydrochloric acid gas, and evolving sufficient heat in so doing to prevent the flame from becoming completely extinguished.

A jet of coal gas, burning at the end of a narrow glass tube and connected with the gas supply by a flexible tube, will behave like a candle, if lowered into a jar of chlorine.

Expt. 218. Action of Chlorine on Oil of Turpentine.—Soak a piece of thin blotting paper in oil of turpentine, and then introduce it into a jar of chlorine; a vigorous action will at once commence, and clouds of smoke and vapour of hydrochloric acid will be emitted; if the jar be large enough, and the chlorine tolerably pure, the action will become so energetic that the paper will take fire, and a flame will be produced. This arises from the same cause as that operating in the last experiment; turpentine contains carbon and hydrogen combined together (being for that reason termed a *hydrocarbon*); when brought into contact with chlorine the hydrogen combines vigorously with the chlorine, whilst the carbon is set free.

Caution.—In carrying out experiments with chlorine gas, such as the last three, it is indispensable to operate either in the open air, or at least in a freely ventilated room, unless a draught cupboard (Expt. 145) be available to carry off the fumes; if even a very small quantity of chlorine escape into the air and be breathed, most distressing coughing and irritation of the air passages will inevitably be brought about; whilst serious injury may be occasioned by breathing in larger quantities of the gas (compare Expts. 162 and 180).

Expt. 219. To prepare Carbon Monoxide.—Place in a test-tube or small flask half a teaspoonful of solid *sodium formate*, and pour over it a little strong sulphuric acid (oil of vitriol); effervescence will take place (especially if a little heat be applied), owing to the escape of a combustible gas termed *carbon monoxide*, which will burn at the mouth of the test-tube on applying a light. By employing a generator with a delivery pipe turned downwards, the gas may readily be collected over water in the pneumatic trough, as it is not easily soluble in water. The chemical action taking place in this experiment is somewhat complicated; first, the sulphuric acid and sodium formate act on one another by double decomposition in such a way as to form a substance termed "acid sulphate of sodium," together with a volatile acid body closely akin to vinegar, and known as *formic acid;* and then this formic acid becomes decomposed into water vapour or steam and carbon monoxide which comes off as a gas.

Expt. 220. To show that Carbon Monoxide forms Carbon Dioxide on burning.—Collect a small jar of carbon monoxide, remove the saucer, and apply a light (whether the jar is held mouth downwards or upwards is of but little consequence); the gas will burn with a blue flame if pure, not giving out much light. Quickly place the jar on the table, mouth upwards, and pour in some *limewater;* close the jar with the palm of the hand or with a cork, and shake up; the limewater will be turned milky, showing that *carbon dioxide* is now present (Expt. 152). In this case gaseous carbon monoxide and gaseous oxygen have united to form gaseous carbon dioxide, so that no change of state has occurred on the whole.

Expt. 221. Explosive Nature of Mixtures of Carbon Monoxide and Oxygen.—Collect in a soda-water bottle some carbon monoxide (as in Expt. 208), until about two-thirds of the water present in the bottle are displaced; then fill it with oxygen, so that the mixture of gases present will be approximately two volumes of carbon monoxide and one of oxygen. Apply a light to the mouth of the bottle, and a tolerably loud report will ensue. It is as well to wrap a towel round the bottle first, to avoid danger from splintered glass should the bottle break. After the explosion, quickly pour into the bottle some limewater and shake up as in the last experiment; the limewater will be rendered milky by the carbon dioxide formed.

Expt. 222. To prepare Nitric Oxide Gas.—Into a gas generator (fig. 56) introduce some copper turnings, and then pour diluted nitric acid down the funnel; the acid will speedily act on the copper, dissolving it and evolving a gas termed *nitric oxide* (Expt. 116) different from the nitrous oxide or laughing gas obtained in Expt. 187. When the air has been all expelled from the apparatus the gas may be collected over water in the pneumatic trough.

Expt. 223. To produce Red Fumes by combining Nitric Oxide and Oxygen.—Nitric oxide is, when pure, a colourless gas, but when brought into contact with oxygen the two gases combine to form a red coloured gas differing from either, especially in being much more soluble in water. When nitric acid comes in contact with copper (as in Expt. 222), red fumes are at first visible; these arise from the action of the nitric oxide first evolved upon the oxygen of the air in the flask; later on, when the air has been all expelled, no more red colour is visible.

Fill a cylindrical jar half full of nitric oxide at the pneumatic trough, and then lift it up out of the water for a second or two, so that the remaining water may run out and air take its place; dense red fumes will be visible. Lower the mouth of the jar under water, and let the whole stand awhile; the red fumes will dissolve in the water, which will consequently rise in the jar owing to the absorption.

Instead of allowing air to enter, pass some oxygen into the jar; the production of red fumes, and subsequent absorption of them by the water and the consequent diminution in bulk of the gas, will be still more marked.

Combination of Gases with Solids or Liquids to form Gaseous Products.

The burning away of charcoal in the air (or in oxygen, Expt. 224) affords an example of this class of action, the products of combustion being not only gaseous at the temperature at which the combination occurs (that of the burning charcoal), but also remaining gaseous after cooling down to the ordinary temperature. In all such cases it is evident that, although the solid matter has become invisible, yet it has not been actually destroyed, but is still present in a form capable of being rendered evident to the sight by the use of appropriate tests, *e.g.*, by limewater. In the same kind of way when a candle burns, although the wax or tallow disappears, yet it is not destroyed, but only converted into other invisible forms of matter or gases. If a cold dry tumbler or lamp chimney be held for a few moments over a burning candle, it will become visibly bedimmed with dew; this arises from the fact that when chemical change takes place between the materials of the candle and the air *water* is one of the products formed; owing to the heat developed this is evolved in the form of vapour or steam, which condenses as dew upon the comparatively cool glass, just as dew is formed from the air upon a cooled surface in Expts. 43 and 207. On putting a chimney over a moderator or paraffin lamp, or over an argand gas burner, precisely the same thing is seen, *i.e.*, the cool chimney becomes dimmed with dew, which speedily evaporates again as the glass gets hotter.

On the other hand, Expt. 154 has shown that when a candle burns the gas carbon dioxide, capable of making limewater turbid, is also formed; so that, on the whole, although the solid material of the candle has disappeared during burning, yet we have in its place at least two vaporous or gaseous products, water vapour and carbon dioxide gas. If a candle be burnt under such circumstances that the oxygen of the air used in burning can be weighed, and if this weight be added to the weight of the candle which has disappeared, it will be found that the total weight of materials undergoing chemical change is exactly equal to the weight of the products formed by the change, viz., the water and carbon dioxide gas. To carry out this experiment accurately, however, requires much skill and a variety of complex apparatus.

Expt. 224. To burn Charcoal in Oxygen.—Collect a jar of oxygen (Expt. 177); fix a piece of hard charcoal (preferably that made from boxwood) to the end of a wire by coiling the wire round one end of the charcoal (copper bell wire answers well). Light the charcoal by holding it in a flame, so that it is glowing and burning at the far end, and then plunge it into the jar of oxygen; the combustion will be greatly accelerated, and the charcoal will burn and "scintillate" or throw out sparks vigorously. After the combustion is over let the jar stand awhile to cool (so as to avoid cracking it if the sides are hot), with a cork in the mouth or covered over with a saucer, &c., then pour in some limewater and shake up; the limewater will become milky, showing that carbon dioxide has been formed by the combustion, *i.e.*, the solid charcoal has become changed into an invisible gas.

Precisely the same result occurs when charcoal is burnt in air; but in this case the rate of chemical action is much slower, and one result of this difference is that a different gas, called *carbon monoxide* (Expt. 219), is often produced as well as carbon dioxide. As carbon monoxide is poisonous when inhaled, charcoal fires in badly ventilated apartments are somewhat dangerous from this cause.

Expt. 225. To burn Sulphur in Oxygen.—Fill a bell jar with oxygen, the top having an orifice fitted with a stopper or cork. Through this introduce a deflagrating ladle with a little burning sulphur in the bowl (fig. 79); the sulphur will burn much more brilliantly in the oxygen, producing the same suffocating gas (sulphur dioxide) formed when it burns in air. After the combustion is over shake up a little distilled water in the jar; it will dissolve some of the sulphur dioxide, forming a solution of sulphurous acid capable of discharging the blue colour given to starch paste by a drop of solution of iodine (Expt. 164).

Expt. 226. To produce a stream of Blue Fire.—Melt some sulphur in an iron ladle covered over with a lid of metal or earthenware (a bit of tile, the round bottom of a canister, &c.) and continue the heat until the sulphur is hot enough to burn. From an open window above a stone pavement or gravel walk, &c., pour the melted sulphur; a stream of blue fire will be thus formed, the sulphur being so readily combustible that when once hot and lighted the flame will not be extinguished even by falling 15 or 20 feet through the air; a lighted candle will generally go out when dropped from this height, the rapid motion through the air producing the same extinguishing effect, as a heavy puff of wind from the lungs in "blowing it out."

If in Expt. 219 the stream of carbon monoxide issuing from a generator be somewhat rapid, or better, if a small bladder or gas-holder be filled with this gas (Expt. 177) and a rapid stream of gas be then expelled again therefrom, it will often be found that the gas will not burn with a steady continuous flame, as the rapidity of the current of gas produces of itself the effect of blowing out.

Expt. 227. To Bleach Flowers by means of Sulphur Dioxide.—The gas produced when sulphur burns in air or oxygen has the power of acting on certain kinds of colouring matter, converting them into colourless substances; for which reason it is used as a bleaching agent, more especially for straw and some kinds of silk. Burn some sulphur in a jar of oxygen or air, and then place some flowers moistened with water inside (red roses answer well), the colour will in many cases shortly disappear; with a rose, the tint may to some extent be restored by dipping the bleached flower into water to which one or two drops of sulphuric acid have been added.

Combination of Gases with Solids or Liquids to form Non-Gaseous Products.—When a metal rusts in the air, and especially when a melted metal forms a film of dross on the surface (Expt. 15), the action is simply due to combination of the metal with the oxygen of the atmosphere, forming a solid product different from either the metal or the oxygen combined with it. In all such cases there is a greater or less development of heat, according to the intensity of the chemical action and the length of time over which it is spread; when the action takes place rapidly between materials that react vigorously with one another, the development of heat is frequently very intense, quite comparable in extent with the modes of production of heat above discussed where combustible substances are burnt to gaseous products. Thus when certain metals are heated sufficiently in the air they take fire and burn readily, evolving much light and heat in so doing; naturally the action becomes more intense when pure oxygen is substituted for air. Certain metals when brought into contact with chlorine do not require any heating at all to bring about vigorous action with production of light and heat.

Expt. 228. To burn cold Antimony in Chlorine.—If powdered antimony be sprinkled into a jar of cold chlorine (fig. 90) the antimony particles will take fire and glow, combustion being brought about on account of the absorption of the chlorine by the antimony, and the production of vigorous chemical action in consequence; thin leaves of Dutch metal (a kind of brass) will act in the same way.

These experiments, however, like all others involving the use of chlorine gas, are not to be recommended excepting in a room where there are facilities for freely ventilating and so carrying off the chlorine which unavoidably escapes into the air, otherwise distressing coughing and possibly serious injury to the lungs may be brought about, owing to the corrosive action of the gas (Compare Expts. 162 and 180).

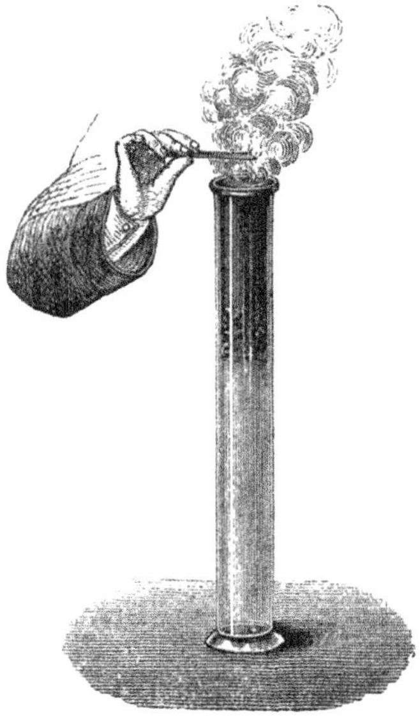

Fig. 90. Antimony burning in Chlorine.

Expt. 229. To Dissolve Gold.—Chlorine has a powerful action upon most metals tending to combine with them producing compounds termed *chlorides*; even gold is affected by this gas, although most chemical agents have no action on this metal. If gold leaf be placed in a jar of chlorine gas it will by and by disappear becoming converted into *chloride of gold,* a compound easily soluble in water, from the solution of which particles of metallic gold can be precipitated by suitable agents (Expt. 135). Solution of chlorine (Expt. 162) behaves in the same way; a few gold leaves placed in a bottle which is then filled with chlorine water, will soon disappear by becoming converted into chloride of gold and so dissolved.

Expt. 230. To distinguish Gold from other Metals.—Nitric acid has no action on gold; thus gold leaf may be placed in a wine glass and nitric acid poured over it without any of the metal being dissolved; whilst a leaf of "Dutch gold" (an imitation made of a variety of brass) would be instantly attacked and dissolved. If, however, some hydrochloric and nitric acids be mixed together, gold will readily dissolve in the mixture, for which reason the compound fluid has long been known as *aqua* *regia*, from its power of dissolving the royal metal. The solution takes place because the two acids react on one another in such a fashion as to develop chlorine, which combines with the gold and is the real agent in forming a soluble substance.

A locket, ring, chain, &c., may be tested to see if it is made of gold or not, by rubbing the article on a tile or stone so as to form a streak of yellow metal like a pencil mark; if this disappears instantly on moistening with a drop of nitric acid the metal is not gold; but if the acid has little or no action on the streak, the metal is gold of tolerably good quality.*

Expt. 231. To obtain a bright light from Magnesium.—The metal magnesium is sold in the form of wires and flattened ribbons for the purpose of burning so as to produce a bright light. Hold a bit of magnesium wire in the flame of a Bunsen lamp for a few moments; the wire will soon take fire and burn with a most brilliant light, a light white substance termed *magnesia* or *magnesium oxide* being produced instead of the solid wire. This arises from the circumstance that the magnesium (when hot enough) absorbs oxygen from the air and a vigorous chemical action is set up, the heat of which keeps the magnesium active as long as air is supplied (just as a candle keeps burning under similar conditions).

Expt. 232. To burn Iron and Zinc in Air.—Into the flame of a Bunsen burner or spirit-lamp sprinkle iron or zinc filings (best from a pepper-castor); brilliant sparkles will be produced owing to the absorption of oxygen from the hot air by the metal and the chemical action (oxidation) set up, just as in the case of magnesium. Iron or steel or thin ribbons or wires can be readily burnt in oxygen gas although not so easily in air; zinc when rolled out into very thin sheets can be burnt in a candle flame almost as easily as paper.

Many other metals besides magnesium, iron, and zinc behave in a similar fashion; when heated sufficiently in contact with air

* The quality of gold is usually expressed by imagining the whole of the material to be divided into twenty-four parts called *carats*, and mentioning the number of these parts which are pure gold, the balance being alloy of various kinds added either to harden and render capable of resisting wear and tear, or to cheapen. The standard for British gold coins or "guinea-gold" is 22 carats; *i.e.*, $\frac{22}{24}=\frac{11}{12}$ of the whole is pure gold, and $\frac{1}{12}$ other metal added to harden; in Australian sovereigns a little of this $\frac{1}{12}$ is silver and the rest copper, whence the yellower shade; in ordinary coins the whole of the $\frac{1}{12}$ is copper. French gold coin contains $\frac{9}{10}$ of its weight of true gold and $\frac{1}{10}$ of alloy, so that English and French coin have not quite the same composition, the former containing $91\frac{2}{3}$ in 100 of gold, and the latter only 90 in 100. 15 carat gold contains $\frac{15}{24}$ or $\frac{5}{8}$ of its weight of precious metal; 9 carat gold only $\frac{9}{24}$ or $\frac{3}{8}$, and so on.

they absorb oxygen, a chemical action being set up sufficiently energetic to cause "combustion." In all cases, the action is far more energetic in pure oxygen than in ordinary air.

Expt. 233. To burn a Steel Watchspring in Oxygen.—Obtain some old watchsprings from a watchmaker, and straighten them by holding them in a flame and then bending with pliers whilst red hot; tie two or three straightened springs together in several places with thread, and warm one end of the bundle and then dip it into flowers of sulphur; a little of the sulphur will stick to the steel, tipping it something like a match; light the sulphur tip at a candle and then plunge the bundle in a jar of oxygen; the heat given out by the burning sulphur will be sufficient to light the steel, which will burn brilliantly throwing out "scintillations" or sparks on all directions. An oxide of iron is thus formed which will drop down in molten globules, which will usually be hot enough to melt their way into the glass if a bottle of oxygen be used, or into the glaze of the saucer if a bell jar and earthenware saucer be employed, even though there be a layer of water half an inch deep at the bottom.

In the preceding experiment the result of the chemical actions taking place between the solid and gas employed was to produce a solid product, viz., the oxides of magnesium, iron, and zinc respectively. If the experiment is made in such a way that the whole of the solid product formed can be collected and weighed, it is always found that the weight of the oxide formed is greater than that of the substance burnt, the excess of weight in the product being exactly equal to the weight of the oxygen fixed by the metal during the chemical change; in other words, the weight of the materials employed is exactly equal to the weight of the product formed, so that on the whole there is neither gain nor loss of weight during a chemical change, but only an alteration in the form and properties of the substances used. This is an invariable law in all chemical changes, whether the products and materials are solid, liquid, or gaseous.

Expt. 234. To burn Phosphorus in Oxygen.—Cut under water a small piece of solid phosphorus about the size of a pea, dry it with blotting paper, and place it in the bowl of a deflagrating ladle (not hot from any previous experiment, and not wet); touch the phosphorus with a hot wire and it will immediately take fire; introduce the ladle and burning phosphorus into a jar of oxygen, and a most brilliant light will be produced.

Caution.—Phosphorus is a highly inflammable waxy solid which requires to be kept under water, and is best cut into pieces also under cold water; it ignites easily on slightly warming, and therefore must not be held long in the fingers, otherwise it may light and produce painful burns which do not heal readily. If lighted whilst moist it is apt to sputter and throw out blazing particles, so that to avoid cracking glass jars when burnt inside them it should be dried by lightly pressing in blotting paper. So little heat is requisite to ignite phosphorus in contact with the air that inflammation can generally be brought about by rubbing a knitting needle or other piece of metal vigorously with a rough cloth so as to warm it by friction, and then making the warm metal touch the phosphorus. Similarly a test-tube half full of water with a few drops of oil of vitriol dropped in (*vide* Expt. 98) will become sufficiently warmed by the heat produced on diluting the acid to fire the phosphorus if made to touch it.

Expt. 235. Luminous Writing.—If a stick of phosphorus be held by a towel or pair of tongs and then used to write with on a wall, pencil fashion, the writing will be found to be luminous in the dark; this arises from the fact that small particles of phosphorus are abraded from the stick and adhere to the surface of the wall, just as blacklead from a pencil; these particles combine with the oxygen of the air and evolve light in so doing, but are too minute to set up actual flame.

Expt. 236. Luminous Faces.—Olive oil can dissolve a small quantity of phosphorus. To prepare the solution, put a little oil into a test-tube and warm it over the lamp; drop in a small fragment of phosphorus previously dipped in oil (so that it shall not remain at the surface and take fire), and cork the test-tube, now and then giving it a shake; the oil will soon take into solution some of the phosphorus, so that if a little of the phosphorised oil be rubbed over the face and hands they will become luminous in the dark, for the same reason as the phosphorus writing in the last experiment, viz., that the phosphorus oxidises, evolving light, but does not take fire owing to the small quantity present.

Expt. 237. Fenian Fire.—When somewhat larger quantities of phosphorus are spread over a considerable surface, the action of oxidation may become sufficiently energetic to bring about actual inflammation. Half fill a small bottle with *carbon disulphide* (a compound of carbon and sulphur) and drop in one or two fragments of phosphorus cut from a stick, each the size of a pea; these will quickly dissolve.* Shake up the liquid and pour a teaspoonful out on to a sheet of blotting paper supported by an

* "Yellow" phosphorus, such as is sold in sticks, requires to be kept under water because of its inflammability; this variety dissolves freely in carbon disulphide; but there is another form of phosphorus known as "red" or "amorphous" phosphorus which is insoluble in that fluid; this form of phosphorus

iron stand so as to wet the paper thoroughly with the liquid. The carbon disulphide, being highly volatile, will soon evaporate leaving a film of phosphorus on the paper; this will soon begin to emit fumes, and by and by will burst into flame. The popular term "Fenian fire" is derived from the reputed use of this liquid by the Fenians for the purpose of setting fire to houses, &c., by throwing in through the window or down a chimney a bottle full of the phosphorus solution, so that the bottle might break and thus cover carpets, curtains, bedclothes, &c., with phosphorus solution which would speedily set them on fire.

Caution.—On account of its tendency to produce spontaneous inflammation when the carbon disulphide evaporates, this liquid is highly dangerous to keep; it should only be prepared in small quantity, and when done with any left over from the experiment should be poured away on the ground in the open air so as to obviate the risk of fire.

Production of Acids by Combustion.

Shortly after the discovery of oxygen it was noticed that when such substances as sulphur and phosphorus are burned therein and the products of combustion dissolved in water, **acid** fluids were obtained (Expts. 225 and 234), neutralising alkalies and reddening certain vegetable colours, such as litmus. Accordingly the term *oxygen* was applied to the gas (from ὀξύς = acid and γεννάω = I produce), it being supposed that it was indispensable to the formation of acids. Later experiments have shown that this was an incorrect idea and that the name is consequently a misnomer; because, firstly, the substances formed when sulphur and phosphorus, &c., are burnt in oxygen *are not themselves acids, but only become so on further combination with water*; secondly, many other substances such as magnesium, iron, and zinc (Expts. 231, 232, 233) when burnt in oxygen do not form acids, even on further addition of water; and thirdly, because acids exist containing no oxygen, *e.g.*, hydrochloric acid, which consists of hydrogen and chlorine only.

Expt. 238. To prepare Sulphurous Acid.—Into a bottle of oxygen introduce a ladle with a little burning sulphur, as in Expt. 225; when the sulphur has burnt away quickly introduce a wineglassful of distilled water, cork the bottle, and shake up well; the water will dissolve the gas (sulphur dioxide) formed by the

is not prone to spontaneous combination with the oxygen of the air and consequently taking fire, so that it may be kept in an ordinary bottle or box freely exposed to the air without inflaming.

union of the sulphur and the oxygen, and form a solution containing *sulphurous acid;* on dipping a blue litmus paper or other test-paper indicating the presence of acids (Expt. 142) into the liquid it will be reddened. Sulphurous acid solution destroys the colour of many vegetable substances, especially when strong, as it exerts a bleaching action upon them.

Expt. 239. To prepare Phosphoric Acid.—Burn phosphorus in a bottle of oxygen as in Expt. 234; cork the bottle and allow it to stand an hour or two; at the end of this time all the white fumes of oxide of phosphorus formed by the combination of phosphorus and oxygen will have subsided and become dissolved in the few drops of water which will have remained adhering to the glass when the bottle was filled with the gas. Rinse out the bottle with a teaspoonful of distilled water, and cautiously taste the liquid; it will be found sour to the taste (intensely so if only very little water be used), and highly acid to blue litmus and other similar test-papers.

By burning phosphorus cautiously in perfectly *dry* oxygen it is possible to prepare "pentoxide of phosphorus" as a white powdery solid; if a little of this is brought into contact with a few drops of water a hissing sound is produced and steam is formed on account of the great heat developed during the combination of the pentoxide of phosphorus with the water to form phosphoric acid, which results as a warm concentrated solution if too much water have not been used.

Expt. 240. To Produce Heat by the Combination of Water with Lime.—On to a lump of good quicklime gradually sprinkle water; the lime will gradually get hot and finally crumble to pieces and give off steam; if too much water have not been added the resulting "slaked" lime is a powder dry to the touch, notwithstanding that it weighs considerably more than the original quicklime (3 parts of quicklime furnish nearly 4 of dry slaked lime).

Pack a tin can holding a pint of water inside a flower pot, filling up the space between with lumps of good quicklime the size of a walnut. Now gradually sprinkle water on the lumps so as to slake the lime; the heat produced will warm up the water in the tin can so that under favourable circumstances it can be made to boil, or at least to become hot enough to scald the finger if dipped in.

In this experiment the chemical action is of the same general character as in the last, viz., the combination of water with something else, producing a new substance and developing great heat in so doing; but the compound formed is of very different nature in the two cases; when phosphorus pentoxide is

hydrated, or combined with water, the product is a powerful acid; whereas slaked lime is an almost equally powerful antacid (Expt. 138).

Expt. 241. Combination of Liquids to form Liquids and Solids.—Chemical actions where two substances both in the liquid state directly combine together (without the intervention of a solvent to produce liquefaction) are comparatively rare, but are not unknown. If melted sulphur be poured into a crucible containing melted lead or zinc the two liquid substances combine together and form metallic sulphides; much as iron and copper combine with sulphur (Expts. 205, 206). If a little mercury be poured into a cup and a few drops of bromine added thereto and the whole stirred together a vigorous action will take place, and the two liquids will unite together forming solid *bromide of mercury*. This experiment is one which should only be performed inside a glassed-in draught place (Expt. 145), not only on account of the danger of inhaling vapour of bromine (Compare Expts. 167 and 162) but also because fumes of mercury and bromide of mercury are apt to be evolved, which also are highly deleterious when breathed.

Actions of Double Combination.

Expt. 242. Combustion of Sulphuretted Hydrogen.—If sulphuretted hydrogen be generated instead of hydrogen in the apparatus represented by fig. 56, using fragments of sulphide of iron instead of granulated zinc, it can be burnt at a jet just as well as hydrogen, the same precautions being requisite as to complete expulsion of air from the generator before applying a light in order to avoid the chance of explosion (Expt. 207). Hold over the flame a cold glass tumbler and you will see that *water* is deposited, just as with hydrogen; but besides this, the gas *sulphur dioxide* is formed, as can be readily shown by its odour, and by collecting some of the hot products of combustion in an inverted jar held over the flame, shaking them up with water, and adding the solution to water rendered blue with starch paste and a drop of weak iodine solution (Expt. 164). In this case the sulphuretted hydrogen may be regarded as first being decomposed into the two constituents sulphur and hydrogen, each of which then burns, uniting with the oxygen of the air to form sulphur dioxide and water vapour respectively; thus affording an illustration of an action of double combination (Chapter I.).

Expt. 243. Combustion of Carbon Disulphide.—Carbon disulphide is a volatile bad-smelling fluid consisting of carbon united with sulphur; it burns readily with a pale flame like that of a spirit lamp, producing as the result of the combustion *carbon dioxide* and *sulphur dioxide* gases. As in the last experiment, the substance may be regarded as being first split up into its constituents, which then burn each to its respective compound with oxygen. The burning of this fluid in a room with closed doors and windows affords a convenient means of producing an atmosphere charged with sulphur dioxide for the purpose of disinfecting clothing, bedding, &c. in cases of contagious diseases and similar illnesses.

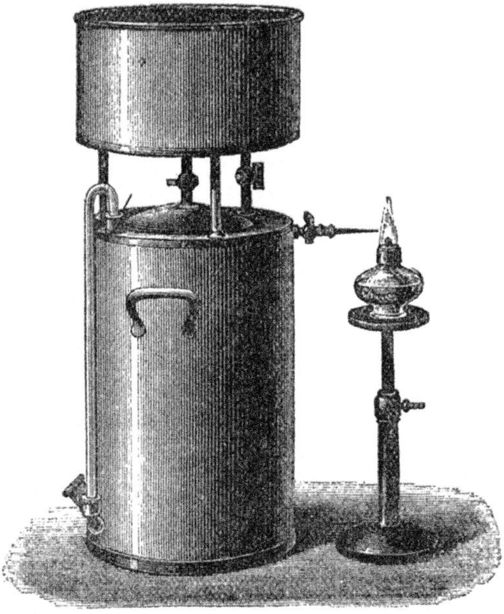
Fig. 91. Oxygen Alcohol Blowpipe Flame.

Expt. 244. Combustion of Alcohol.—Alcohol when burnt in a spirit lamp similarly gives rise to two different products, both gaseous at the temperature of the flame; one, *carbon dioxide*, is still gaseous after cooling; the other, *water vapour*, condenses on cooling. The production of these two substances may be proved in much the same way as that above described for an ordinary candle flame (p. 200); a cold tumbler held over the flame becomes bedewed with condensed water; whilst an inverted jar held over the flame will contain some carbon dioxide, which will render limewater turbid on quickly pouring some into the jar and shaking up before the hot gases have time to escape.

If a spirit lamp flame be blown on one side by means of a jet of oxygen so as to produce an oxyalcohol blowpipe flame (Compare Expt. 211) most of the effects producible by an oxyhydrogen flame can be imitated, as the resulting flame is intensely hot; a piece of chalk or marble placed so that the flame impinges on it will glow intensely and emit a brilliant light; steel springs or thin iron nails held in the flame will be heated sufficiently to burn and scintillate brightly, especially if a good supply of oxygen is afforded so that there may be a surplus to combine with the iron as well as to burn the alcohol. Fig. 91 indicates a convenient way of producing an oxygen alcohol blowpipe flame.

Expt. 245. Coloured Alcohol Flames.—When pure alcohol is burnt the light given out by the flame appears of a blue colour, but has extremely little intensity or illuminating power; by sprinkling a pinch of common salt on the wick of a spirit lamp the flame is altered in colour to yellow, and is rendered a good deal brighter. The yellow light thus emitted causes many objects seen by its aid to appear entirely different in colour from their ordinary appearance by daylight; brightly tinted dresses or other objects seem dull and grey, and the face assumes a most ghastly hue. A soup plate containing a handful of salt on which a wineglassful of methylated spirit has been poured answers well to develop this kind of colour transformation, the room being otherwise dark; or large torches may be made by winding tow round broomsticks, and saturating them with spirit mixed up with powdered salt; or better still, the tow may be first soaked in brine and then dried before winding round the sticks. The effect of illuminating a room full of people with such torches is most remarkable, especially if some one clothed in a white sheet suddenly makes his appearance unexpectedly.

In a similar fashion red flames may be produced by using powdered *nitrate of strontium* instead of salt; and green ones by employing *nitrate of barium* or *boric acid.*

Caution.—Care must be taken not to upset the plate containing the alcohol, otherwise a serious accident may easily happen by objects being fired. The plate should be placed on a tea tray so that nothing will be spilt if the plate break with the heat. In the case of torches, it is desirable to wind some copper wire loosely round the tow, so as to prevent any of the strands being burnt through and fragments of blazing tow being dropped in consequence.

CHAPTER XV.

Chemical Actions of Reciprocal Decomposition.

Actions resulting in the Evolution of Gas.

Expt. 246. To produce Hydrogen from Water.—Certain substances have the power of acting upon water, bringing about replacement (Chapter I.) of the hydrogen by the substance used; various metals possess this property, some of them acting energetically upon liquid water or even solid ice, whilst others require the help of a considerable increase of temperature before the action commences, and consequently will not decompose liquid water, but will act more or less readily on steam.

Wrap up in a bit of wire gauze a fragment of *sodium* the size of a small pea, and hold it by means of tongs under cold water (fig. 92); owing to the interstices in the gauze the water will not be prevented from coming in contact with the sodium; displacement of hydrogen by sodium will rapidly take place, the hydrogen thus evolved bubbling up through the water; so that by placing a small jar filled with water over the wire gauze the hydrogen evolved can be collected. Caustic soda is the complementary product formed by the action; this dissolves in the water, so that on dipping therein a red litmus test paper it will be turned blue owing to the alkaline action of the caustic soda. In Expt. 168 the chemical action taking place was the same as in this case, the mercury in the amalgam used serving to dilute the sodium and render its action less violent.

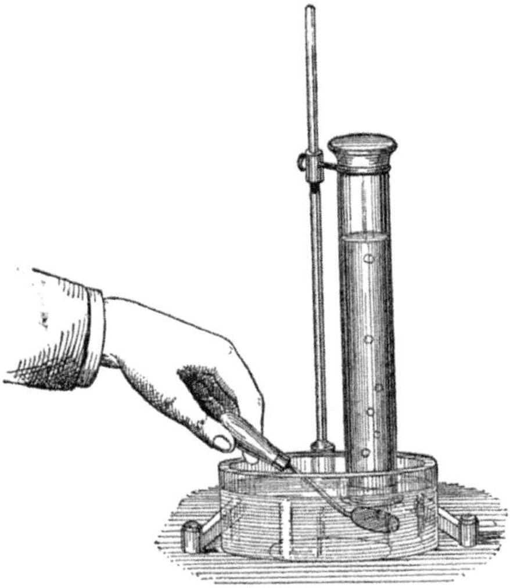

Fig. 92. Hydrogen from Water and Sodium.

Caution.—When making experiments with sodium, be careful how water is brought in contact with the metal in a glass, porcelain, stoneware, or other glazed vessel; if a fragment of sodium be thrown into water in such a vessel without the protecting wrapping of wire gauze, the sodium will float up and become melted by the heat produced by the chemical action; and if the globule of fused metal come in contact with the smooth glazed surface, the action becomes very vigorous and explosive, so that splashes of sodium are thrown about in a burning state, the metal and the hydrogen evolved being kindled through the violence of the action. If the water be warm, the evolved hydrogen is lighted and burns steadily even though no explosive projection of burning particles take place; this is most simply effected by putting a small fragment of sodium on a wooden floor and pouring a little hot water over it. Even in this case, one may easily be burnt by the flying about of burning bits of sodium.

Expt. 247. To set Ice on Fire.—The metal *potassium* is even more energetic in its action on water than sodium, so that if thrown on to cold water, the hydrogen evolved is almost always lighted. The same result takes place with ice; put a small pellet of potassium on the top of a block of ice, and press the two gently together; action will commence, and the heat produced will generally be sufficient to inflame the hydrogen.

This experiment illustrates a somewhat rare kind of action, viz., where two solids directly act on one another. As a general rule in order to bring about chemical change between substances in the solid state, one or other of the substances present must either be melted by heat, or reduced to a quasi-liquid condition by solution in some solvent (*vide* also Expt. 263).

On account of the great tendency of sodium and potassium not only to oxidize in the air, but also to decompose water and water-vapour, these metals require to be preserved either in hermetically sealed vessels, or in bottles filled up with paraffin oil or some similar fluid incapable of acting on the metals. Sometimes the metals are stored by casting them in brick-like ingots, which are then dipped in melted paraffin wax which forms a protecting coating over their surface.

Expt. 248. To set a Pond on Fire.—Into a dry gallipot pour two or three teaspoonfuls of ether, and also put therein a few fragments of potassium; if the ether is pure the potassium will not be acted upon. Now drop the whole into a pond, or into a bucket of water; the ether and the potassium will float up, and the latter will act vigorously on the water evolving hydrogen and setting fire thereto, and to the ether as well.

Expt. 249. To decompose Steam by Red Hot Iron.—Obtain a piece of hard glass tubing ("combustion tubing," made of a particular kind of glass capable of standing a pretty high temperature without melting) about a foot long, and fill it with fragments of iron turnings; at each end fix tightly a cork or india-rubber bung with a piece of glass tube passing through. Connect one end with a flask of boiling water, after heating the central part of the combustion tube as hot as possible by means of one (or preferably more) Bunsen burners, or better still, a series of burners playing into a furnace made of tiles (fig. 93). If the current of steam be not so rapid as to chill the hot iron borings too much, the iron will decompose part of the steam, evolving hydrogen gas and producing a film of oxide of iron on the surface of the borings, the action being closely akin to that of sodium on water, excepting that the resulting oxide of iron is not soluble in water, and that a pretty high temperature is requisite to bring about the reaction. The hydrogen thus produced may be collected and examined by passing the issuing mixture of steam and hydrogen through a Liebig's Condenser (Expt. 36) to condense the steam, and collecting the hydrogen by displacement (Expt. 104). The pneumatic trough is hardly safe to use, because should anything occur to interrupt the current of steam, water might be forced back by the condensation of steam into the hot tube, producing an explosion, or at least breaking the tube. Fig. 93 represents an arrangement by means of which this danger is obviated by the simple device of connecting a funnel tube with the steam generator; this serves as a sort of safety valve, inasmuch as *air is sucked in through the funnel* instead

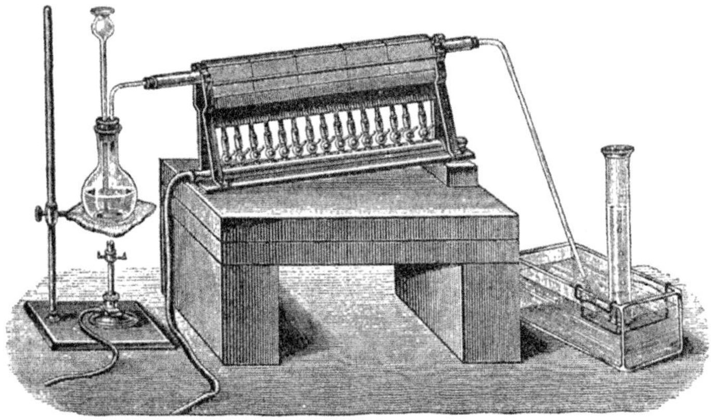

Fig. 93. Hydrogen from Steam and Red Hot Iron.

of water through the delivery tube, in the event of the steam supply becoming interrupted by any accident.

Expt. 250. To burn Phosphorus in Nitrous Oxide.—Prepare a jar of nitrous oxide (Expt. 187) and introduce into it a bit of phosphorus in a deflagrating spoon just lit by touching with a hot wire as in Expt. 234. The phosphorus will burn nearly as brilliantly as in oxygen. The chemical action here taking place is that the nitrous oxide splits up into nitrogen and oxygen, the phosphorus combining with the latter with the evolution of great heat and light, whilst the nitrogen is set free.

Expt. 251. To burn Phosphorus in Nitric Oxide.—Prepare a jar of nitric oxide as in Expt. 222, and introduce into it a deflagrating spoon with a bit of dry phosphorus which has been touched with a hot wire and is just beginning to burn; if the phosphorus has not begun to burn too vigorously, it will be extinguished. Now withdraw the spoon and relight the phosphorus by holding it in a flame for an instant or two, and when it is well alight introduce it again into the jar; if burning vigorously when introduced it will keep alight, but will not burn as freely as it does in nitrous oxide (Expt. 250). The reason for this is that whilst nitric oxide, like nitrous oxide, is a compound of nitrogen and oxygen, it is not decomposed by heat so readily as the latter, so that unless the phosphorus flame is hot enough to decompose it, no oxygen is set free to feed the flame, which consequently dies out. If, however, the heat be sufficient, the nitric oxide becomes split up into nitrogen and oxygen, and the latter gas keeps the phosphorus burning just as with nitrous oxide.

Expt. 252. To get an Egg inside a Bottle.—The brittle shell of an egg chiefly consists of animal membrane with carbonate of lime forming a thickening and strengthening solid mineral deposit thereon; by soaking an egg in dilute hydrochloric acid, the same action is produced on the carbonate of lime as on the chalk in Expts. 99 and 100; *i.e.*, a double decomposition first takes place between the carbonate of lime and hydrochloric acid, followed by another action, the result of which is ultimately that chloride of calcium is formed together with water, whilst carbon dioxide gas is evolved. The effect of this is that the mineral matter is dissolved away whilst the animal membrane is left soft and pliable; in this condition by careful manipulation the egg can be squeezed through the neck of a bottle far narrower than the original untreated egg; when inside, the bottle may be filled with brine, or weak solution of ammonia, when the egg will gradually return to its original shape, and become somewhat hardened so as to preserve its form instead of being soft and pulpy. By placing the bottle full of brine in a saucepan of cold water and very gradually heating this up until it boils, removing the source of heat after boiling a few minutes, and allowing to cool again slowly, it is possible to *cook* and hard-boil the prepared egg inside the bottle without cracking the glass; so that if the bottle be then broken and the egg extracted, it will be found hard and sound and fit to eat (if the original egg was a fresh one). To any one unacquainted with the way in which the matter is managed it seems an impossibility to get a genuine hard-boiled egg inside a narrow-necked bottle.

Expt. 253. To make Bones pliable.—The bones of animals differ from the outside skeletons or shells of molluscs and crustaceans (snails and oysters, lobsters, and such like creatures) in that the latter consist of animal tissue strengthened up and hardened by mineral matter, chiefly carbonate of lime like eggs; whilst the bones of animals (such as oxen, sheep, and mankind) are in the main built up in the same sort of way, but with a different sort of calcareous matter as stiffening agent, chiefly *phosphate of lime.* Both phosphate and carbonate of lime, however, are dissolved out by diluted hydrochloric acid; so that if the shoulder blade of a sheep, the rib of an ox, the claw of a crab, or any such animal part, be immersed for some days in water to which some hydrochloric acid has been added, the mineral matters are dissolved out, and a flexible gristly soft animal tissue is left; in this way a long thin rib or leg bone may be softened and a knot tied on it.

On the other hand, if a bone be strongly heated in a vessel from which the air is excluded, the animal matter is destroyed, and the mineral matter left as a brittle porous mass retaining the shape of the bone, and black with charcoal deposited throughout its substance by the decomposition of the animal matter (Expt. 202); whilst if the blackened bone be heated again in contact with air, or if air be admitted freely during the first heating, the carbonaceous matter is burnt entirely away, and a nearly white porous fragile mass results, consisting solely of the mineral matter originally present in the bone.

Expt. 254. To set fire to Tinfoil.—Spread out on a plate a piece of thick tinfoil, place on it a teaspoonful of solid *nitrate of copper* reduced to powder in a mortar, sprinkle a little water on the powder, and quickly wrap up the whole in the tinfoil. The moist copper nitrate will act rapidly on the tinfoil producing a large amount of heat, sometimes sufficient to cause the tinfoil to take fire visibly; the action in this case is not unlike that taking place in Expt. 251, but is of a more complicated character.

Expt. 255. To melt a Threepenny piece in a Walnut shell.—During the deflagration of combustible bodies in contact with substances containing a good deal of oxygen like saltpetre, a very intense heat is often produced whilst the action lasts. Mix together two parts of finely powdered loaf sugar and one of potassium chlorate, both substances being powdered *separately* (Expt. 260, *Caution*), and mixed by pouring from one paper to another and back again several times. Fill a walnut shell with the mixture and place in the centre a threepenny piece, and then set fire to the whole with a spill; a vigorous flare-up will result, and so much heat will be produced during the action that the coin will usually be melted down.

Expt. 256. To set fire to a Solid by touching with a Liquid.—The mixture of potassium chlorate and sugar used in the last experiment possesses the curious property of bursting into flame when a drop of *oil of vitriol* comes in contact with it, the reason for which is that the sulphuric acid sets free from the potassium chlorate a substance termed *chloric acid* (just as it sets free nitric acid from potassium nitrate—Expt. 270) which acts in such an energetic fashion upon sugar that enough heat is produced to set fire to the whole mass. One of the earliest forms of match for striking a light without using flint and steel with tinder box * consisted of a splinter of wood tipped with sulphur and with the end tied up in a little bag containing a small quantity of the mixture of potassium chlorate and sugar, and a small hollow glass ball containing a drop of oil of vitriol and hermetically sealed; when a light was required the end of the match was struck with a hammer, &c. so as to break the glass ball; the acid coming in contact with the mixture set fire to it, and consequently lit the sulphur match. The numerous accidents occasioned by this kind of matches prevented their ever being extensively used.

Conjurors often use this action as an illusion, fire being produced by touching objects with a magic rod; the rod is tipped with glass with a drop of oil of vitriol at the end, and the object has a little of the potassium chlorate and sugar mixture applied to it.

Caution.—Be extremely careful lest the oil of vitriol should drop off the glass rod or be otherwise spilt, as it is excessively corrosive.

Expt. 257. Potassium Chlorate burning in Hydrogen or Coal gas.—Just as a candle will burn brightly in a jar of oxygen because the wax is so far heated as to give off vaporous matter analogous to coal gas, which unites with the oxygen, producing heat, thus keeping up the formation of vaporous matter and maintaining the action; so can substances capable of giving off oxygen take fire and burn in an atmosphere of hydrogen or coal gas. Fig. 94 represents an arrangement whereby this can be demonstrated. A piece of wide glass tubing (a lamp chimney will answer well) is fitted at the lower end with a perforated cork through which a rapid supply of hydrogen is admitted from a large generator, or of coal gas from the gas main; the upper part is loosely covered with a piece of tinplate forming a sort of lid,

* Before the invention of matches, when a light was required a quantity of "tinder" had to be prepared, consisting of linen rags fired in a pan partially covered up, so as to smother the combustion, all air being carefully excluded when the flame ceased, leaving a carbonised mass. A spark being struck by means of flint and steel, and received on tinder thus made and contained in a small box, would cause the tinder to take fire; by blowing carefully the glowing portion of the tinder would become larger and hot enough to light a sulphur-tipped splinter of wood; when this was done the tinder box lid was closely shut so as to extinguish the glow.

with a perforation in the centre through which a deflagrating spoon can be admitted. Some potassium chlorate is placed in the spoon and heated in a flame until the salt is melted and begins to decompose rapidly evolving oxygen (Expt. 176); the spoon is then quickly introduced into the cylinder of gas through the opening in the lid, when the fused chlorate burns freely, any surplus gas escaping and also burning at the orifice.

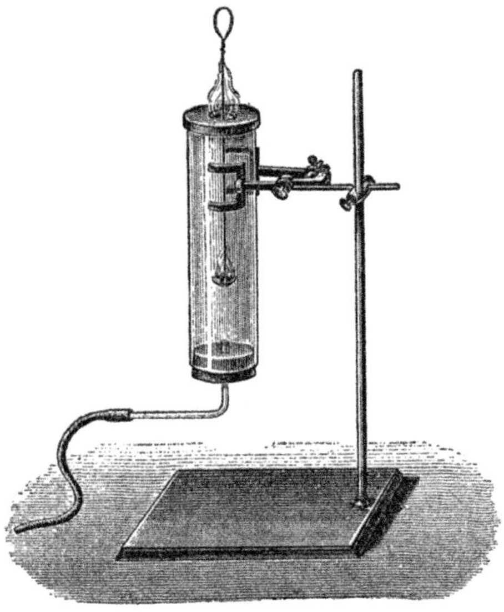

Fig. 94. Oxygen from Potassium chlorate burning in Coal Gas.

Here the action (with hydrogen) is one of simple displacement. Potassium chlorate is a compound of potassium chloride and oxygen (Expt. 178), and when hydrogen acts thereon the oxygen separates from the potassium chloride and unites with the hydrogen, so that potassium chloride and steam are the products. With coal gas the action is similar but of a somewhat more complex character owing to the more complicated composition of coal gas.

Expt. 258. To deflagrate Charcoal with Nitre.—Fuse some *saltpetre* in a crucible and then throw in a piece of *charcoal* previously heated in the lamp flame so as to make it glow and commence to burn; a brilliant combustion or deflagration will be produced, the chemical action taking place being a somewhat complex one; the saltpetre contains potassium, nitrogen, and oxygen, and in contact with the carbon of the charcoal forms a mixture of carbon dioxide and nitrogen gases which escape, whilst *potassium carbonate* (a compound of potassium, carbon, and oxygen) remains. On the whole, therefore, the charcoal is partly burnt by oxygen derived from the nitre, which accounts for the large amount of light and heat developed. Sulphur, zinc filings, and many other combustible solids will similarly "deflagrate" when thrown on to melted saltpetre.

Throw a handful of saltpetre into a good cinder fire; a most brilliant deflagration will result, the action being precisely the same as with the charcoal and fused saltpetre above described.

Paper soaked in a solution of saltpetre and dried will undergo a peculiar kind of rapid smouldering combustion when lit by touching with a red-hot cinder or otherwise; *touch paper* is thus prepared to form the portion of squibs, rockets, and suchlike fireworks intended to be lit. Tobacco and cigars largely owe their power of keeping alight whilst not being actually smoked in the mouth to the presence of small quantities of saltpetre naturally contained in the tobacco leaf, or added during manufacture for the purpose. *Pastilles* are somewhat similar in their action, being generally composed of powdered charcoal and saltpetre with various aromatic gums, &c., added so as to develop a pleasant scent whilst smouldering; a kind of tape known as "Ribbon of Bruges" and by other names is sometimes prepared in a similar fashion, for the purpose of producing an incense-like odour when a piece of it is lighted and allowed to smoulder.

Expt. 259. To prepare Gunpowder.—Gunpowder is manufactured by grinding saltpetre to powder and mixing with it finely ground charcoal and sulphur, the proportions used varying with the quality of powder to be prepared, but usually being somewhere about

Saltpetre, ...	75	parts in 100
Charcoal, ...	12½	,,
Sulphur, ...	12½	,,
	100	

The charcoal employed is prepared from special kinds of wood, dogwood and alder being preferred; the pulverised substances are moistened with water to diminish the risk of firing, and then well ground together to form an intimate intermixture; the mass is then strongly compressed, and the "press cake" obtained broken up by a suitable machine, the fragments being sifted, dried by steam heat, and glazed or polished by placing them in revolving barrels, so that the grains rub against one another; for large grains or "pebble powder" where each grain is as large as a hazel nut or walnut, a somewhat different method of granulation is requisite. These mechanical operations of intermixing, compressing, and graining are indispensable in order to obtain a mixture suitable for fire-arms; on the small scale, a pestle and mortar may be employed to intermix the ingredients, previously well damped so as to form a very stiff paste when ground together; this paste may be spread out in a thin layer on a plate to dry in the air, and when nearly dry carefully broken up into small fragments by chopping with a blunt knife and rubbing in a mortar; even when completely dry the mixture will not be as effective as machine-made powder, and when fired in the form of a train will not go off as rapidly, but more like the composition in a squib, which is very much the same as ungrained powder. The chemical action aking place when powder explodes is much the same as when charcoal and sulphur are thrown into melted saltpetre, these combustible bodies burning by the aid of the oxygen derived from the saltpetre and suddenly producing a large volume of gas in so doing (*vide* Expt. 258).

Expt. 260. A Mixture that detonates when struck.—Carefully powder a few crystals of potassium chlorate in a clean dry mortar; turn the powder on to a sheet of paper, add to it about half its bulk of flowers of sulphur and mix the two by pouring the powders from one paper to another and back again several times, *taking special care to avoid all rubbing with heavy substances.* Place a few grains of the mixture in a mortar and rub with the pestle; a smart crack, or succession of cracks will be heard; place a pinch on an anvil and strike it with the flat face of a hammer; a loud report will be produced.

Caution.—Never prepare this mixture by rubbing together crystals of potassium chlorate and sulphur so as to pulverise the former whilst mixing the two ingredients; if this be done the mixture will probably explode and painful cuts be produced or damage done to the eyes by solid pieces of crystallised chlorate being projected about in all directions. Further, never prepare the mixture in larger quantities at once than one or two good-sized pinches of chlorate crystals, say $\frac{1}{4}$ ounce at the most.

Expt. 261. To prepare Coloured Fires.—The materials employed by firework makers for preparing coloured fires are usually more or less of the nature of gunpowder, containing saltpetre and sulphur with other substances giving a colour to the flame produced when the composition is fired. *Blue fire* (for Bengal lights) can be prepared by powdering separately and then well intermixing together 4 parts of *saltpetre*, 2 of *sulphur*, and 1 of *sulphide of antimony*. *Red fire* can be similarly made by mixing 8 parts of well dried *nitrate of strontium*, 2 of *saltpetre*, 2 of *sulphur*, and 1 of *powdered charcoal;* and *Green fire*, by using *nitrate of barium* instead of nitrate of strontium. Makers often use chlorate of potassium as an ingredient, which causes the composition to burn better; but the mixture is unsafe on account of the explosive properties of potassium chlorate mixed with sulphur (Expt. 260).

What are sometimes sold under the name of "Lightning papers" are pieces of tissue paper that have been converted into a highly-inflammable substance resembling guncotton (by dipping them into nitric acid mixed with sulphuric acid, and washing) and subsequently impregnated with certain salts, giving the colour to the flame; *e.g.*, *strontium chlorate* for red, and *barium chlorate* for green. These burn with a bright flash of light when a flame is applied; whilst no particular danger attends the burning of small strips of such prepared papers, they are highly dangerous to keep in any quantity, on account of the ease with which they can be ignited. Many such compositions as red fire, &c., sometimes possess the power of taking fire spontaneously, especially when kept in a warm place; several fires in theatres and similar places of entertainment are believed to have been thus caused.

Expt. 262. To Deodorise Sulphuretted Hydrogen.—Into a jar full of air pass a few bubbles of sulphuretted hydrogen (Expt. 145), whereby a fœtid smell will be communicated to the air. Pass in a few bubbles of chlorine (Expt. 180); the smell of sulphuretted hydrogen will disappear, and the air will no longer blacken a strip of paper soaked in nitrate of lead solution (Expt. 14). Chlorine decomposes sulphuretted hydrogen, combining with the hydrogen therein contained (forming hydrochloric acid gas) and setting free sulphur. If sufficient sulphuretted hydrogen is present in the jar, the action of the chlorine will give rise to the deposition on the sides of a film of solid sulphur, easily visible; whilst by introducing a spoonful of water into the jar and shaking up, a weak solution of hydrochloric acid will be obtained which will redden blue litmus paper (Expt. 57).

The use of "chloride of lime" or bleaching powder as a deodorising agent for evil-smelling drains, &c., is based upon very similar changes thereby produced in the foul gases evolved.

Production of Gases by the Chemical Action of Solids on one another.

As a general rule chemical action does not take place readily between substances unless one or both are rendered fluid, either by applying heat so as to melt the body, or by dissolving in some solvent; but there are some exceptions to this rule.

Expt. 263. To develop a Gas by heating two Solids together.—Take as much *slaked lime* (Expt. 240) as will lie on a shilling and mix with it as much powdered *salammoniac* (chloride of ammonium). Put the mixture into a clean dry test-tube and heat in the Bunsen lamp flame. *Ammonia gas* will be copiously disengaged, as may be proved by smelling cautiously (taking care not to take too vigorous a sniff at the fumes, otherwise they will almost choke you for the moment); or by the turning blue of reddened litmus paper, or other colour change with appropriate test papers as in Expt. 142. All the experiments with ammonia gas above described may be performed with gas prepared thus, instead of by heating strong solution of ammonia as in Expt. 75. Put half an ounce of slaked lime mixed with as much salammoniac in the flask instead of the strong ammoniacal liquor, and apply heat, collecting the gas precisely as before. By leading the gas thus formed into water until the latter has dissolved as much ammonia gas as it can take up, the "liquor ammoniæ" of commerce is produced; for the sake of convenience and cheapness, *ammonium sulphate* is generally used instead of ammonium chloride.

Reduction of Metals from their Compounds.

Expt. 264. To obtain Metallic Lead from Oxide of Lead.—In Expt. 15 we have seen that when molten lead is skimmed bright and then exposed to the air a dull film forms on the surface, and the "dross" thus produced is formed by the combination of the lead with the oxygen of the air, just as magnesium combines with oxygen (Expt. 231), but less energetically. In order to reverse this action and "reduce" metallic lead from oxide of lead, the latter must be heated with some substance that will combine with the oxygen and liberate the lead by an action of single displacement. Such a substance is powdered *charcoal*, if the temperature is high enough, *i.e.*, at a red heat. Mix some *litharge*, or oxide of lead, with a quarter of its weight of powdered charcoal, and press the mixture into a clay crucible; heat this red hot in a kitchen fire, the crucible being well covered; after cooling you will find that more or less of the lead present has melted down to the bottom of the crucible, forming a "button" of metal. The experiment succeeds better if a little oil or tallow is mixed in with the powder.

Expt. 265. To obtain Metallic Bismuth from Oxide of Bismuth and Metallic Tin from Oxide of Tin.—Proceed in exactly the same way as in Expt. 264, using oxide of bismuth or oxide of tin instead of oxide of lead; metallic bismuth or tin respectively will result.

Expt. 266. To obtain Metallic Antimony from Sulphide of Antimony.—Sulphide of antimony consists of antimony combined with sulphur, just as oxide of lead is lead combined with oxygen. Mix powdered sulphide of antimony with half its weight of fine iron filings, and add some *cream of tartar* (bitartrate of potassium) to act as a "flux," *i.e.*, a substance that will promote fusion. Heat the whole red hot in a well covered crucible; the iron will combine with the sulphur of the sulphide of antimony and set free the antimony, so that on pouring out the hot fluid contents of the crucible you will have two kinds of substances, metallic antimony, and the resulting sulphide of iron and flux, &c., which can be easily detached from the antimony when cold.

Expt. 267. To obtain Mercury from Vermilion.—Vermilion consists of *sulphide of mercury, i.e.*, mercury combined with sulphur. When powdered and heated with iron filings the same kind of action takes place as with sulphide of antimony; the iron combines with the sulphur, and the mercury is liberated. As mercury is a volatile metal, this experiment cannot be performed in a crucible; but if carried out in a distilling or subliming vessel where the evolved vapours can be recondensed, globules of quick-silver are easily obtained. Mix powdered vermilion with its own weight of fine iron filings and heat a small quantity of the mixture in a test-tube; a sublimate will form in the upper part of the tube, consisting of fine globules of mercury mixed with a little undecomposed sulphide of mercury, which sublimes unchanged.

Expt. 268. To obtain Metallic Copper from Oxide of Copper by means of Hydrogen.—Inside a piece of hard glass tubing ("combustion tubing"), six or eight inches long and half an inch diameter, place some fragments of oxide of copper, holding the tube horizontally and pushing in the pieces with a pencil or glass rod until they are in the centre of the tube, the ends being clear. Fix in one end of the tube a perforated cork with a bit of quill glass tubing passing through, and by means of a bit of india-rubber tubing connect this with the delivery tube of a hydrogen generator, as represented in fig. 95. Carefully heat the centre of the tube where the copper oxide lies with a Bunsen gas flame or spirit lamp, and when the copper is pretty hot, lead a moderately rapid current of hydrogen through the tube. The hydrogen will act on the oxide of copper, taking away the oxygen therefrom and leaving metallic copper; so that the result of the action will be that water vapour is formed and issues as steam at the far end of the tube, part condensing to liquid water at the cool end, so as to form visible drops; whilst the black fragments of oxide of copper become red, owing to the change in their composition to spongy copper. If the quantity of oxide of copper treated is sufficiently large and the hydrogen supply pretty rapid, the black fragments will become visibly red hot and glow, especially in a darkened room. This shows that the action as a whole is accompanied by the production of heat.

Expt. 269. To reproduce Oxide of Copper from Spongy Copper and vice versâ alternately.—Carry out the last experiment, and when the oxide of copper is almost wholly reduced to spongy metal, detach the hydrogen generator, and substitute for it a tube connected with a gasholder or bladder, &c. (Expt. 177) containing oxygen, so as to send a stream of oxygen gas through the tube instead of hydrogen; the red spongy copper will now combine with the oxygen and reproduce a black mass of oxide of copper in so doing; by alternately coupling on the stream of hydrogen from the hydrogen generator and the current of oxygen from the gasholder, you can bring about at pleasure the two converse actions, reduction of oxide of copper to metal with formation

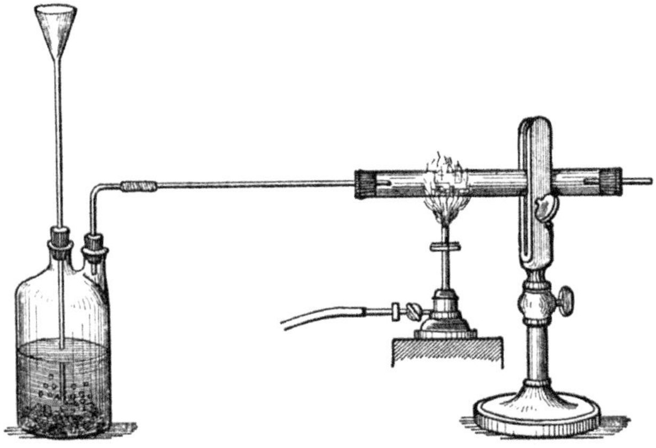

Fig. 95. Oxide of Copper reduced by Hydrogen.

of water, and combination of copper with oxygen to reproduce oxide of copper.

If the copper be fed with oxygen sufficiently rapidly, it will be seen that the spongy metal becomes red hot and glows during oxidation, a considerable degree of heat being obviously produced, although not as much as when magnesium, zinc, or iron unites with oxygen, as in experiments 231, 232, and 233.

In the extraction of many of the metals, &c., employed for various useful purposes in the arts from their natural sources, processes are employed on the large scale, the general principle of which is much the same as that involved in the separation of tin, lead, bismuth, and copper from their respective oxides; and of mercury, antimony, &c., from their sulphides, by the processes illustrated in the above experiments. Thus certain ores of *zinc* are first roasted in order to convert the zinc compounds present into oxide; this oxide is then heated with carbonaceous matter (small coal, coke dust, &c.), when the carbon unites with the oxygen and sets free the zinc; the heat employed is sufficient to volatilise the zinc, so that the metal is obtained by carrying out the process in a large distilling vessel or retort, in the cooler part of which the zinc condenses much as the mercury does in Expt. 267. The metal *sodium* is obtained somewhat similarly from the alkali soda, a still higher temperature being employed to carry on the action and distil the sodium produced. The metals *aluminium* and *magnesium* cannot be easily extracted from their oxides by this kind of treatment; but by heating their chlorides with sodium, a parallel chemical change takes place, the sodium uniting with the chlorine present and setting free metallic aluminium or magnesium.

Various Actions of Double Decomposition.

Expt. 270. To prepare concentrated Nitric Acid.—Grind to powder in a pestle and mortar an ounce of saltpetre (*nitrate of potassium*) and place the powder in a glass stoppered retort. Now pour in carefully through the stopper hole half a wineglassful of oil of vitriol (sulphuric acid), and gently shake the retort so that all the powder may be thoroughly wetted with the liquid, taking care not to splash the mass into the upper portion of the retort. The effect of this is that a chemical action takes place between the sulphuric acid and the saltpetre, resulting in the formation of liquid *nitric acid*, which is volatile and can be slowly distilled by carefully heating the retort, receiving the strong acid which condenses and drops down in a stoppered glass bottle (cork would be quickly corroded by the acid). A nonvolatile substance called *acid sulphate of potassium* is also formed and is left in the retort, from which it can be dissolved out after cooling by pouring in water and allowing to stand.

Expt. 271. To etch Glass.—Glass is a substance capable of resisting the solvent and decomposing action of most kinds of corrosive chemicals, so that they can be retained in glass bottles; but one substance is known which attacks and dissolves glass just as solution of hydrochloric acid would dissolve a zinc vessel (by the evolution of hydrogen and formation of chloride of zinc as in Expt. 10), or as nitric acid would similarly attack a copper vessel (producing copper nitrate and evolving nitric oxide, as in Expt. 116). This substance is termed *hydrofluoric acid*, and is a gas somewhat resembling hydrochloric acid gas, being also excessively soluble in water. To prepare the gas powdered fluor spar (*fluoride of calcium*) is heated with oil of vitriol, when a chemical change occurs somewhat resembling that taking place in the last experiment, hydrofluoric acid and sulphate of calcium being formed by double decomposition. A leaden trough is used, made by bending up the corner of a piece of sheet lead some 6 or 8 inches square, so that the tray produced is about an inch or 1½ inches deep; powdered fluor spar is strewed over the tray to the depth of the thickness of a penny piece, and enough oil of vitriol poured over the powder to wet it thoroughly and make a thin paste. The tray is then placed on an iron tripod and covered over with the plate of glass prepared for etching, and gently warmed with a lamp to bring about the action of the acid more quickly. The glass is prepared by carefully warming it, then dropping on melted beeswax, and causing this to run over the entire surface and allowing it to solidify, so as to form a coating on the glass; when this has set the design to be etched is cut through the wax by means of the point of a needle or penknife or other convenient tool, so as to lay bare the glass where the cuts are made; the design may be drawn on tissue paper with a pencil and then transferred to the wax by pressing on the paper with a uniform gentle pressure and then peeling it off again; if the pencil be of good quality the marks will be more or less completely retained on the surface of the wax. The glass is then set on the leaden trough lid-fashion with the prepared side downwards, so that any vapour evolved in the tray will come in contact with the glass where it is exposed by cutting away the protecting wax; at these points the glass is attacked and corroded, but not at the other portions of its surface covered with wax. When the action has gone on long enough, the glass is removed and warmed and the wax wiped off. Care must be taken not to heat the tray so much as to melt the wax prematurely, and to hide away parts of the etching by the flow of melted wax over the exposed surface of the glass.

Caution.—This experiment should, if possible, be done in a draught place (Expt. 145), or at least in a room well ventilated, and great care must be taken not to breathe in the hydrofluoric acid vapours, as they are even more injurious to the lungs than chlorine or bromine.

Expt. 272. To engrave on Copper or Steel.—Engravings on iron and steel are often executed in a fashion very similar to the above; the metal plate is covered with a coating of "resist" (a substance of resinous and waxy character not affected by the acid subsequently used), and the design traced by cutting away the composition, so as to lay bare the metal; *aqua fortis*, or diluted nitric acid, is then poured on, which "bites" or dissolves the metal where thus exposed; when the etching has gone far enough at any particular spot, further action is stopped there by applying a little "resist" to the lines etched. Finally, the resist composition is cleaned off the plate, and those lines, &c., that require touching up with a sharp tool are then cut further until the plate is in a condition to receive the ink and print off impressions. This process can be easily illustrated by obtaining a polished copper plate (such as those used for printing visiting-cards), covering the polished face with wax, drawing the design, or writing with a sharp needle, so as to cut through the wax and lay bare the copper, and then cautiously pouring on the aqua fortis; but considerable practice is necessary before etched plates fit to print from can be obtained.

Copper rollers with patterns thus engraved on them are largely employed in cotton-dyeing, calico-printing, and similar industries, in order to apply the colouring materials, &c., to the fibre in some parts and not in others (*vide* Expt. 291).

Expt. 273. To produce a Liquid from two Solids.—Rub together in a mortar some *sugar of lead* (acetate of lead) and *Epsom salts* (sulphate of magnesium); the two solids will react on one another and form a wet white mass. The chemical change here occurring is that acetate of lead and sulphate of magnesium form acetate of magnesium and sulphate of lead. The wetness arises from the fact that crystals of Epsom salts, like many other crystallised substances, contain not only true sulphate of magnesium, but also *water* combined therewith, and so to speak rendered solid in the crystal; in consequence such water is termed *water of crystallisation.* As neither the lead sulphate nor the acetate of magnesium combine with the water thus present in the Epsom salts, it is set free, rendering the whole mass wet.

This experiment is only apparently the converse of No. 128, where two solutions mixed together formed a nearly solid mass; there double decomposition also took place, with the result of forming a large quantity of insoluble matter which physically absorbed the water present (as sand would do), but did not combine with it; in the present case a substance insoluble in water (*sulphate of lead*) is also formed by double decomposition, but this is not bulky enough to absorb completely all the watery solution of *acetate of magnesium* also formed, without showing visible signs of the presence of a fluid.

When *plaster of Paris* is mixed with water for taking casts (Expt. 3), the setting or hardening is largely due to the combination of the dry plaster (*sulphate of calcium*) with some of the water to form crystallised *gypsum* (or sulphate of calcium containing water of crystallisation), the crystals of which stick together as a porous mass, which absorbs the rest of the water in its pores, thus becoming apparently *solid* in much the same way as the precipitated carbonate of calcium in Expt. 128. The setting of most kinds of cement and mortar is due to a somewhat analogous change of physical structure of the mass, resulting from some kind of chemical action taking place to begin with.

Expt. 274. To obtain Acids from Sugar.—In Expts. 189 to 193 we have seen that by fermenting sugar *alcohol* is obtainable, and that by the souring or oxidation of alcohol under certain conditions *acetic acid* is formed as vinegar. If sugar be boiled with nitric acid diluted with water for some time a vigorous action is set up; the nitric acid parts with some of its oxygen, and the sugar is partly converted into *carbon dioxide* which escapes, and partly into *oxalic acid*, which can be obtained in the solid state by evaporating the whole to dryness on the steam bath, dissolving the residue in the smallest possible quantity of hot water, and allowing the liquid to crystallise.

Milk contains a peculiar kind of sugar (milk sugar) different from cane and fruit sugars; when milk turns sour an acid termed *lactic acid* is formed, somewhat as acetic acid is from ordinary sugar and fruit sugar.

Expt. 275. To prepare Soap for Household Use.—The different kinds of soap in the market are all essentially substances prepared by the double decomposition of caustic alkalies and fatty or oily matters, resulting in the formation of soap and glycerine. According as the alkali is potash or soda, so, as a rule, is a soft or a hard soap obtained; whilst the nature of the fatty matter also influences the character of the soap obtained. Ordinary household and scouring soaps are generally made from very coarse kinds of fatty matter, such as the grease from slaughtered horses, fatty substances from tannery refuse, the "foots" or thick sediments obtained in refining oils, and such like; toilet soaps are, or at least ought to be, made from superior qualities of fats and oils, subsequently refined, purified, and scented, and finally worked up into tablets by machinery.

Dissolve some solid caustic soda in about four times its weight of water, and make the solution hot; melt down a quantity of kitchen grease or lard mixed with olive oil, taking about six times as much as you have dissolved of caustic soda, or rather more. Pour the strong caustic soda solution in a slow stream into the warm melted fat contained in a convenient basin or pail, stirring vigorously the while; stir for some time till very thoroughly intermixed, and then cover up with a cloth and set by in a warm place for twenty-four hours; at the end of this time, if the operation is successful, you will have a block of moderately solid soap, sufficiently well made to be capable of use for scouring purposes, but probably not of desirable quality for application to the skin; for, when dealing with small quantities, it is almost impossible to make the double decomposition complete, so that the mass ultimately obtained is liable to contain a considerable proportion of unaltered fat and a corresponding amount of unneutralised caustic soda, which excoriates the skin frightfully. Addition of cocoanut oil to the grease treated promotes the soap making action or **saponification**, and furnishes a better product. When working in this way, the glycerine formed together with the true soap during saponification remains mixed up with the soap in the resulting mass; but in manufacturing processes on the large scale, the glycerine is usually separated from the soap, and the latter purified by various devices; in the case of soft soaps (potash soaps), however, the glycerine is not separated.

Expt. 276. Soap made by the Boiling Process.—Most of the soap used for ordinary household purposes is prepared by a somewhat different process, which may be thus imitated on the small scale. Dissolve solid caustic soda in about twenty times its weight of water, and heat the liquor to boiling; then add the fatty matter to be converted into soap (tallow, dripping, olive oil, or a mixture of such substances) to the extent of five or six times the weight of the solid caustic soda used, and keep the whole gently boiling in a capacious saucepan or other vessel for some hours, taking care that the mixture does not froth over; at the end of this time the process of saponification will be nearly complete, when the soap may be made to separate from the watery liquor by throwing in a handful or two of common salt and allowing to stand till cold (compare Expt. 65). A cake of crude soap will then be found to have separated as a soft solid mass at the top, whilst the brine will contain the glycerine dissolved therein; so that the boiling process differs from the one described in the last experiment (called for the sake of contrast the "cold process," being carried out at a temperature below that of boiling water) in that glycerine is separated from the soap in the one case and retained intermixed therewith in the other.

The crude soap thus obtained by the boiling process in practical manufacture on the large scale is separated from the brine before it completely cools and solidifies, and is put through various purifying and finishing operations, after which it is ladled out or pumped into moulds (holding 10 cwt. and upwards), in which it solidifies to large blocks, which are subsequently cut up into bars.

Toilet soaps of superior quality are frequently remelted and further refined and perfumed, again cast into blocks, cut up into small lumps, which are squeezed into the shape of tablets by machinery acting on much the same principle as the presses used for striking coins and medals (Expt. 2); some kinds are ground up in a "mill" to a thick paste, and then formed into tablets by special machinery instead of by remelting and casting, &c.

Expt. 277. To obtain Glycerine from Soap Liquors.—Glycerine is prepared from the brines left in soap boiling by evaporating them down, when salt is crystallised out (used over again for salting out fresh batches of soap), and finally impure glycerine is left, which is refined and purified by various processes before becoming fit for use. Evaporate some of the brine from Expt. 276 to dryness over the steam bath, and then pour some strong alcohol on the residue; this will dissolve the glycerine, but only a small portion of the salt; filter the liquid and evaporate it again until all the alcohol is driven off, taking care that the alcohol vapour does not take fire (Expts. 40, 41). A small quantity of glycerine mixed with a little salt will remain, from which most of the salt can be farther separated by dissolving again in alcohol and evaporating after filtering.

Expt. 278. To make Transparent Soap.—Cut up some lumps of good yellow soap into shavings and expose them to the air in a moderately warm room for some days, so as to get them as dry as possible. Place the dry coarsely powdered shavings in a flask and pour on strong alcohol (methylated spirit will do) to the extent of three or four times the weight of the soap used. Place the flask in a saucepan of water so as to form a waterbath (Expt. 89), and connect a condensing arrangement (fig. 24) with the mouth of the flask, so that when the water-bath is hot the alcohol vapour formed from the boiling spirit in the flask may be condensed and collected for use over again. The soap will gradually dissolve in the spirit, which should be distilled off until but little drips from the end of the condenser; the thick treacly fluid then left in the flask should be poured out into a mould (a small basin will answer) and left to set for 24 hours; on scooping out the mass from the mould in a lump by means of a "spatula" or flexible painter's knife (palette knife), it will be found to be more or less clear, like badly strained jelly; by exposing it to the air for a long time it will gradually dry up and harden, and become somewhat more transparent. If a small quantity of glycerine be added to the alcohol whilst dissolving the soap therein (about 1 part glycerine to 15 or 20 of soap), the resulting product is generally clearer; soap thus clarified, made from a sound quality of stock soap in the first instance, and properly scented, forms the best kind of transparent soap; but the greater part of that extensively advertised and sold is made by substituting sugar for glycerine, a change greatly to the disadvantage of the user, as sugar has a very irritating effect upon some skins.

Another kind of transparent soap of much worse quality is also largely sold, often under the name of "Glycerine Soap," made by the "cold process" (Expt. 275), using castor oil and other oils and fatty matters, with a considerable excess of caustic soda and a liberal addition of sugar, the effect of which is to produce a jelly-like product, about half toffee and half a corrosive soap, which has a very bad effect upon sensitive skins, but is very attractive in appearance, especially when pleasantly scented and prettily tinted by means of a little orange or red dyestuff. *Nimium ne crede colori!*

---

§ 6. Physical Adhesion and Allied Phenomena of Surface Action.

CHAPTER XVI.

Condensation of Gases upon the Surface of Solids.

In the previous pages a number of experiments have been described, illustrating the physical absorption of gases by liquids (no chemical action taking place), forming what are ordinarily termed **solutions**, and somewhat analogous actions taking place between certain kinds of solids and gases (Chapter VII.), often referred to as **occlusion**. In this latter kind of action the gas penetrates into the interior of the solid and becomes pretty uniformly distributed throughout the whole mass; so that if the outside portions were filed off or otherwise detached, and examined separately from the inner part of the solid, no very great difference would be noticeable between the two as regards the quantity of gas taken up by a given weight of solid.

True occlusion of this kind is exhibited by comparatively few solids, and then only in relation to particular gases, at least to an easily measurable extent; but a somewhat analogous, though different kind of action takes place between all solids and all gases to a greater or less degree (under such conditions that chemical action does not modify the effect), the chief difference being that in this more general kind of action *the gas is only attracted to the surface of the solid*, and does not penetrate into the interior except by means of cavities or pores, which are really only prolongations of or convolutions in the outer surface. Thus, if a solid piece of glass or porcelain be heated pretty hot, quickly cooled, and weighed on a sufficiently delicate balance (fig. 49), the moment it has cooled sufficiently to prevent currents of warm air being set up from its surface and so disturbing the action of the weighing instrument, it will be found to weigh perceptibly less than after standing awhile, because during this standing air is attracted to the surface of the object and adheres pertinaciously to it, causing it to increase in weight just as though it had been slightly wetted with water. On heating it again this air will be to a considerable extent driven off, so that, on weighing again after cooling, the weight will be much the same as at first, and perceptibly less than at the second weighing after standing awhile. This condensation of air on the surface of glass and other vessels introduces considerable difficulties in various delicate chemical operations, where substances require to be weighed in containing vessels with a high degree of accuracy; where the surface of the glass is made very large in proportion to its weight (as by drawing out the glass into fine thread or "spun glass" whilst molten by a blowpipe), the bulk of gas thus attracted to the surface is very considerable indeed; and similarly with other solids. The effect of the film of air condensed upon the surface of solids is in some cases to prevent their coming into true contact with liquids in which they are immersed, so that no "wetting" ensues, even though under other circumstances the fluid would wet the solid freely.

Expt. 279. To make Solid Steel float on Water.—A thin bright sewing needle may often be carefully deposited on the surface of a basin of water, and will float thereon without sinking, because a film of air adheres to it, preventing the water from touching it, and buoying it up as a cork would do, but more effectively. If you attentively examine the liquid you will observe that the needle lies in a depression, which is considerably greater in magnitude than the needle itself, the air filling this depression being virtually attached to the needle for the time being, and thus enabling it to displace a much larger column of water than its own bulk, and therefore to float. This action is closely connected with certain phenomena called **capillarity** and **surface tension**, referred to hereafter (Chapter XVII).

In filling barometers the adhesion of air to the glass is a source of error, as this air is apt to creep upwards gradually into the "Torricellian vacuum" at the top of the instrument, and thus spoil the accuracy of its indications; in well-made instruments this air is got rid off by *boiling* the mercury in the tube; the adherent air is partly dislodged by the heat alone, and partly becomes replaced by mercury vapour, and is thus driven out whilst the tube is hot.

In making electrical "vacuum tubes," or sealed glass tubes from which all air or other gas contained is pumped out to the highest possible extent, a similar precaution has to be taken, otherwise the film of air adherent to the glass gradually becomes dislodged, and more or less spoils the vacuum.

Gases thus attracted to the surface of solids bear to gases truly **dissolved** or **occluded** by solids much the same relationship that a film of water adhering to the outside of a gold or silver coin, but not penetrating into its interior, bears to mercury brought in contact with the metal; after a time (Expt. 69) the mercury creeps into the interior of the solid gold or silver; and, if present in sufficient quantity, manifests its presence by rendering the whole brittle; and in the case of gold by visibly whitening its tint, so that the fractured surface is much paler in shade than it would be were no mercury present.

As the adhesion of gases to the surface of solid bodies is only a superficial action, the adherent film not penetrating into the interior (just as a compact solid dipped in water is only wetted externally), it would naturally be expected that if the amount of surface is increased relatively to the weight of the solid by dividing it up into small pieces or powdering it, a given weight of solid would take up more gas the more finely it is divided, just as more water would be taken up by a given weight of fine sand than by the same weight of a compact block of stone. Not only is this the case, but further, the gases thus made to adhere in comparatively large quantity to the surfaces of finely divided solids exhibit the same peculiarity as gases dissolved in or condensed by solids in their interior, *i.e.*, their chemical activity is greatly increased (Expt. 83); and, in consequence, gases thus condensed upon the surface of finely powdered solids, or of solid matters obtained by special devices in highly porous masses like charcoal, often possess the power of setting up chemical changes which the same gases in the ordinary free state (as they would be contained in a jar or bottle) are quite incapable of bringing about. Certain metals, such as platinum, not naturally prone to rusting or "oxidation," can be obtained by particular chemical processes in a fine state of division; and when in this "spongy" condition are well adapted for the illustration of phenomena of this kind. With certain gases, however, a very fine state of division is not requisite, thin wire and foil being capable of attracting the gases sufficiently vigorously to intensify remarkably their chemical activity.

Expt. 280. Absorption of Ammonia Gas by Charcoal.—To do this successfully, a few pounds of quicksilver are requisite, contained in a stout basin; an ordinary mortar of porcelain answers well. Obtain a stout test-tube and a piece of charcoal that will just go inside; heat the charcoal in the flame of a Bunsen lamp or spirit lamp for half a minute, and then plunge it (still hot) under the mercury; the heating expels air from its pores, and the plunging under mercury cools it and prevents its burning away. Fill the test-tube with ammonia gas as in Expt. 75, and then close it with the thumb and bring it to the basin of quicksilver, depressing the thumb and mouth of the tube under the metal so as to prevent air displacing ammonia gas when the thumb is removed. Now insert the piece of charcoal inside the tube, the mouth being still under the quicksilver (fig. 96); support the tube by a clamp or holder and leave it to itself for a few minutes; you will gradually see the quicksilver rise in the tube, the lump of charcoal floating on the top. If the tube was completely filled with ammonia, all air being displaced, the quicksilver will ultimately fill the entire tube; but otherwise a more or less considerable bubble of air will remain unabsorbed.

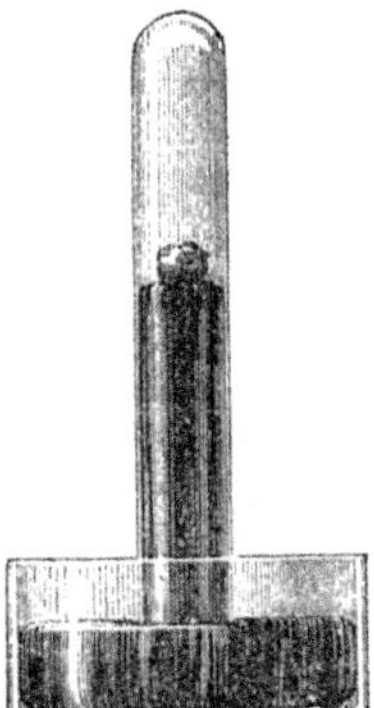

Fig. 96. Ammonia absorbed by Charcoal.

The mercury ascends because the charcoal gradually absorbs the ammonia gas, the pressure of the atmosphere forcing the mercury up as the absorption proceeds, precisely as water was similarly forced up in the analogous experiment (Expt. 76) of dissolving ammonia gas in water. As in that case, too, the experiment shows that some gases, *e.g.*, ammonia, are much more readily absorbed or dissolved (both by porous charcoal and by water) than others, such as air.

Expt. 281. To Cleanse Mercury.—When mercury is used for experiments such as the preceding, it often becomes necessary to clean the surface from dust and dirt. One of the simplest methods of doing this is to make a paper filter in a stout glass funnel (Expt. 56), using a piece of thick paper for the purpose; the dirty mercury is then poured into the paper cone till nearly full, and a fine knitting needle passed downwards through the mercury to the point of the paper cone, so as to make a small perforation; on withdrawing the needle a fine stream of clean mercury passes through the hole and runs down the funnel into a bottle, &c., to receive it; more dirty mercury is poured into the funnel from time to time until all is cleaned, taking care not to allow all to run out before adding more; as otherwise, the mercury running through may carry some dust with it.

Another method is to hold a square piece of chamois leather by the four corners, so as to make a sort of loose open bag or saucer, into which the mercury is put; the corners are brought together and twisted round, so as to squeeze up the mercury like a pudding boiled in a cloth; the mercury is thus forced through the minute crevices of the thin leather, and passes through free from dirt, dropping into a large basin placed underneath to receive it.

Caution.—In all cases where mercury is handled so that it is liable to be spilt and run about, take off all gold and silver ornaments, watch and chain, rings, &c.; otherwise they may be spoilt by the mercury getting in contact with them (*vide*, Expts. 69, 172.)

Expt. 282. Pyrophoric Lead and other Metals.—In Expts. 135, 120, and 199 we have seen that gold, silver, and platinum can be obtained by means of suitable chemical actions in a state of fine division, in which case they constitute *spongy* metals; these metals belong to the "noble" class because they do not readily rust or oxidise in the air* either at the ordinary temperature or on heating.

In this spongy condition they powerfully attract oxygen and other gases to their surfaces, and the gas thus attracted or condensed exhibits a higher degree of chemical activity than before (Expt. 283, *et. seq.*); still the chemical activity is not enhanced so much as to enable the oxygen to combine directly with the metal. It is different when other metals not belonging to the noble class are similarly treated, *i.e.*, obtained in a very fine state of division by chemical means, and then brought in contact with air or pure oxygen; the oxygen attracted by the metal will then often combine directly therewith, evolving heat and light in so doing; so that if the finely divided metal be prepared in a closed vessel in such fashion that no air has access to it, on opening the vessel and allowing the particles of metal to fall out

* Although silver will not combine with oxygen and rust spontaneously, yet it is speedily acted upon and blackened by sulphuretted hydrogen; frequently it will be found whilst experimenting with that gas that silver coins in the pocket will become blackened, whilst the same result is speedily brought about by putting on a shilling a little freshly prepared sulphuretted hydrogen water or a drop of *sulphide of ammonium* (sulphuretted hydrogen dissolved in ammonia water, Expt. 145). On the other hand, although gold is unaffected by either oxygen or sulphuretted hydrogen, it is readily attacked by chlorine (Expt. 229).

into the air, they take fire spontaneously as they fall. Metals in this condition are said to be *pyrophoric*.

Thus, at the ordinary temperature solid lead does not combine quickly with oxygen, especially if perfectly dry; but when obtained as pyrophoric metal, it takes fire instantly on coming in contact with the air. To obtain pyrophoric lead add a solution of *tartaric acid* to one of *sugar of lead* (lead acetate); a heavy precipitate of *lead tartrate* will fall down, which may be washed by decantation (Expt. 120) in a basin, and then dried over a steam bath (Expt. 89). If lumps have formed by clotting together, powder them, and introduce the powder into a long glass tube corked at one end, or (better still) drawn out and closed by melting together the drawn out portions (Expt. 48). Cautiously heat the tube and contents with one or more Bunsen lamps, commencing at the sealed-up end; the tartrate of lead will be decomposed by the heat (somewhat as coal, &c., in Expts. 200, 201, 202), gases being evolved and *metallic lead* in a fine state of division being formed. When gases cease to come off, the open end of the tube must be tightly corked up (or better still "hermetically sealed" by drawing out and fusing together like the other end) and the whole allowed to cool. On opening the tube and shaking out the powder contained, a shower of sparks will be produced, which will be still more brilliant if the particles are allowed to fall into a jar of oxygen.

Many other metals can be obtained in the pyrophoric form by somewhat similar means. In Expt. 268, if very finely divided oxide of copper be used and the hydrogen be made to act on it with as little heating as possible, the particles of copper ultimately obtained are often pyrophoric, especially if shaken out of the tube before they are quite cooled. Very finely powdered oxides of iron and other metals treated in the same way will often similarly furnish pyrophoric iron, &c.

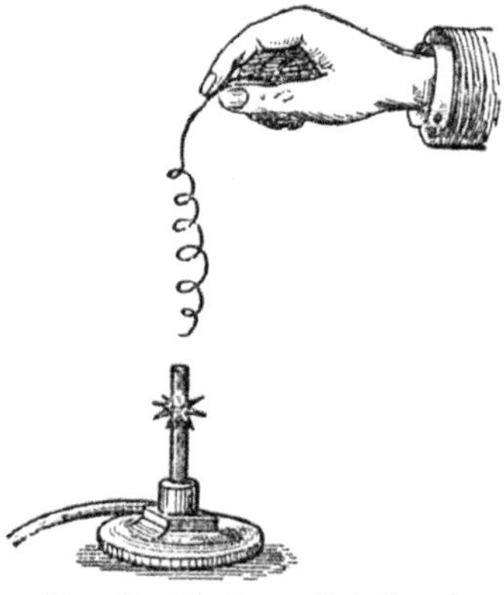

Fig. 97. Platinum Spiral and Bunsen burner.

Expt. 283. To keep Wire Red hot by Means of Coal Gas without Flame.—Obtain a piece of thin platinum wire some 15 or 18 inches long, and coil up round a pencil all but 1 or 2 inches (fig. 97). Light a Bunsen gas burner and hold in the flame the spiral obtained by carefully slipping the pencil out of the coil, using the uncoiled end as a handle. The wire will be heated red hot, and glow brightly. Now turn off the gas so as to extinguish the flame, and immediately turn it on again; a mixture of coal gas and air will pass upwards from the lamp, and coming in contact with the hot platinum wire will be absorbed by the metal, at any rate on its surface. The effect of this is that the coal gas and air can act on each other chemically in the pores of the platinum, just as they do when the lamp burns with a blue flame; and, in consequence, heat is produced which keeps the platinum wire red hot for as long a time as the gas and air come in contact with it. Sometimes so much heat is produced that the wire becomes almost white hot, and lights the burner just as a match would do. By turning off the gas and turning it on again when the wire has so far cooled down that it is not visibly red hot, you can generally succeed in first making the wire glow gently, then more brightly, and finally intensely, so that after a quarter of a minute or so the gas becomes re-lighted. Whether this occurs or not depends upon the exact diameter of the wire, the place where the spiral is held in the issuing gas and air current, whether this current gets blown on one side by the breath or a draught of air, and such like circumstances.

Expt. 284. Perfume Vaporiser.—A kind of perfume vaporiser is sometimes sold acting on the above principles, consisting of a spirit lamp containing scented spirit; a little pin of platinum, shaped something like a clove, is stuck into the wick, the head being made of a sort of miniature cage of fine wire, containing a little plug of "spongy platinum," *i.e.*, platinum prepared by chemical processes in a loose porous condition (Expt. 199). The spirit lamp is lit for a few moments, whereby the pin, and especially its head, becomes heated; the flame is now blown out, when the spongy platinum will continue to glow and diffuse perfume throughout the room. This arises because the spirit passing up the wick becomes evaporated by the heat of the head, a little of the vapour being absorbed by the platinum along with air, giving rise to what is equivalent to burning the vapour without flame, like the gas in the previous experiment, the heat thereby produced keeping the platinum red hot, and consequently maintaining the action as long as any spirit remains in the lamp. To extinguish the vaporiser and stop its action, all that is requisite is to put on the cap of the spirit lamp; this prevents any air having access to the platinum, and consequently stops the action just as an extinguisher puts out an ordinary candle.

Expt. 285. Döbereiner's Lamp.—By using platinum in an

extremely porous condition, and employing hydrogen gas instead of coal gas in Expt. 283, it will happen under favourable circumstances that it is not necessary to heat the platinum before bringing it in contact with a mixture of combustible gas and air in order to make it absorb the two gases and cause them to act on one another evolving heat, and so gradually making the platinum hot enough to light the gas jet; a little instrument was invented by Döbereiner for the purpose of utilising this property as a means of obtaining a light when required, and so superseding matches. This consists of a hydrogen gas-generator, working on much the same principle as those above described, but somewhat different in construction (fig. 98). A cylindrical glass jar, *cc*, is provided with a lid which can be removed if required, but ordinarily fits on pretty closely. In the centre of this lid is a perforation through which passes the narrow end of a funnel-shaped glass tube, *b*, fitted with a valve outside the lid opened by the trigger *e*. When open, hydrogen issues out by a horizontal jet and impinges on some spongy platinum in the little metal box, *f*, thereby causing the platinum to heat, and finally to ignite the hydrogen jet so as to produce a flame.

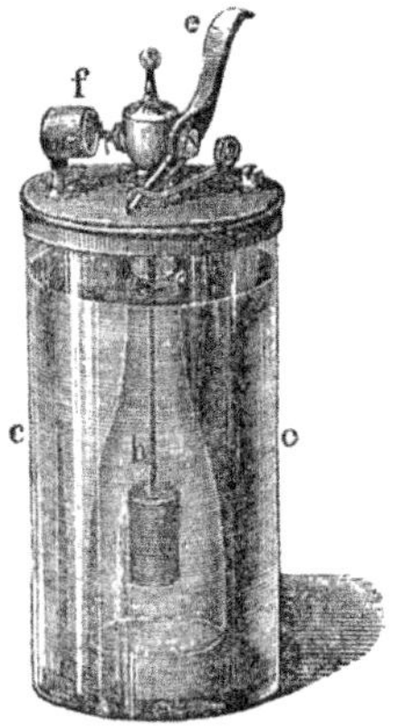

Fig. 98. Döbereiner Lamp.

The hydrogen is generated thus; a lump of zinc is supported by a wire inside the funnel tube, and the jar is filled about half or three quarters full with diluted sulphuric acid by taking off the lid and pouring the liquid in. When the lid is replaced, the zinc dips in the acid so that hydrogen is generated and rises up inside the funnel tube; at first the hydrogen is allowed to escape freely by opening the valve so as to displace all air present; when pure hydrogen issues from the jet, the valve is closed; the gas having no vent accumulates in the upper part of the funnel tube and depresses the level of the acid therein, raising it in the outer cylinder; when the acid is so far depressed that the zinc is uncovered, the action ceases. On opening the valve the pressure of the acid fluid in the outer cylinder forces the hydrogen out at the jet, and the acid rises in the funnel so as to come in contact with the zinc and thus generate more hydrogen. In this way the alternate production of a jet of hydrogen and cessation of production, are effected automatically by opening and closing the valve.

Expt. 286. Use of Charcoal as a Deodoriser and Sanitary Agent.—Although charcoal does not absorb all gases as freely as it does ammonia, it nevertheless does absorb air, and especially the oxygen therein contained, sufficiently readily to enable it to act somewhat after the fashion of the spongy platinum in Expts. 284 and 285; *i.e.*, if also brought into contact with gases capable of being chemically acted upon by oxygen under suitable conditions, it will absorb both and cause them to act upon one another in its pores. In this way deleterious gases from drains and offensive vapours from decaying organic matter that otherwise would escape into the atmosphere and render it injurious to health may be deodorised and rendered innocuous by making them pass through a layer a few inches thick of charcoal broken into small fragments; thus a deodorising charcoal trap to cover sinks and the orifices of drains, &c., is often employed as a remedy for bad smells and injurious emanations; the evil-smelling gases and vapours being absorbed by the charcoal and converted by the oxygen also absorbed into other substances not offensive to the nostril nor injurious to health. The activity of charcoal in this respect is far inferior to that of platinum sponge, so that the action never goes on sufficiently rapidly to cause the charcoal to heat perceptibly.

The carcase of a dead cat or dog may be kept in the open air for weeks in summer without being at all a source of nuisance by thickly covering it with lumps of good charcoal; all offensive products of putrefaction being absorbed by the charcoal and deodorised as rapidly as formed. Coal smoke and soot have the power of exerting a similar action, though to a lesser extent; accordingly, the visible "blacks" in the air of smoky towns in which coal is largely burnt, although most objectionable from many points of view, are nevertheless not wholly without use, as they tend to purify the air by facilitating the natural process of purification by the action of oxygen on deleterious gases and vapours contained in the atmosphere.

Expt. 287. Use of Charcoal and Analogous Substances for Filtration.—Ordinary water (as occurring in rivers, streams, lakes, ponds, &c.) is liable to several sorts of contamination which it is usually desirable to remove as far as possible by various purification processes. One of these applicable to hard spring water containing much bicarbonate of lime consists in adding just so much lime to the water as will convert the soluble bicarbonate of lime into insoluble carbonate (Expt. 156); muddy waters are clarified by filtering them through vast beds of sand or similar material forming the top layer of a "filter bed," the lower layers of which are successively gravel and pebbles; so that as the water percolates through it is finally received in channels and culverts, whence it flows to a suitable reservoir. Most domestic waters, however, contain more or less organic matter not removed by such treatment, but capable to some extent of being removed by filtering again on the small scale through miniature filtering beds made of certain kinds of materials capable of exerting a purifying action by enabling oxygen from the air to act on the organic matter in the pores of the filtering material, as well as effecting mechanical removal of suspended solid particles.

The **chemical purification** thus effected depends on much the same general principles as the use of charcoal for trapping drains and the like; obviously it is essential for the proper action to be carried on that the filtering material should from time to time be exposed to the air, so as to take up oxygen again, to replace that exhausted whilst the purification is going on. This precaution is often neglected, with the result of rendering the filter wholly useless for the purpose for which it is mainly intended, viz., as a means of causing the oxygen to act on the organic matter and destroy it.

Amongst the materials practically used for filtration on the principle of not merely mechanically intercepting solid suspended particles, but of also exerting a chemical purification by oxidation, finely divided charcoal of various kinds and spongy metallic iron are those most largely employed; but fine sand, porous earthenware, and similar materials possess the power of causing a small effect of the kind; in some instances this is partly due to the presence of oxide of iron in the material, this substance acting as a sort of "carrier" of oxygen, *i.e.*, it absorbs oxygen from the air and to some extent parts with it again to the organic matters causing their oxidation. Certain other metallic oxides, such as oxide of manganese, have been used in the same way as ingredients in filtering beds, &c. Fig. 99 represents a simple form of small filter consisting of a block of a kind of porous coke or charcoal cemented into the bottom of a conical jar which stands on the top of a water bottle so that the water must pass through the charcoal before dropping into the bottle.

Fig. 99. Domestic Filter.

A natural purification of impure water as it flows down a river is brought about by the gradual absorption of air by the water, and the action of this dissolved oxygen on the noxious matter; so that some miles down stream from a spot where a river becomes contaminated with sewage or such matters as refuse liquors from tanneries, paper mills, and similar industrial factories, the water is considerably less impure than just below the spot where the contamination occurs. When the water is naturally of a slightly ferruginous character, or where clay, &c., containing iron is disseminated through the water rendering it turbid, this natural purification seems to take place more rapidly, owing to the oxygen carrying action of the ferruginous matter.

CHAPTER XVII.

Mutual Adhesion of Solids and Liquids.

It is an everyday observation that when the hand, stones, and most other solid objects are placed in water, on taking them out again the water **wets** their surface or **adheres** to them; just the same thing occurs with spirits of wine or olive oil; but, on the other hand, if a solid object be greased with a film of oil and then immersed in water, it will not be wetted therewith when taken out again; whilst, conversely, a thoroughly wet finger may be placed in oil without becoming greased in the same way that a dry one would. This arises from the very feeble adhesion between water and oil as compared with the adhesion of either to a solid. On dipping the hand into a basin of clean quicksilver and removing it, no adhesion will be noticed; the same remark applies to a piece of glass if the mercury be pure; but if it contain other metals dissolved therein, it will sometimes adhere to the glass, forming a shining metallic coating. One method of coating the inside of globes and other hollow glass vessels with a reflecting layer of bright metal so as to form mirrors is based upon this property.

Expt. 288. To Silver a Looking Glass.—The ordinary mirrors or looking glasses used for furniture, &c., consist of a flat sheet of glass to which is applied a coating of *tin* and *quicksilver* in such a way that the brilliant surface of the metal is preserved from oxidation or injury by the hinder surface of the glass plate. To silver a large sheet of glass properly requires much practice and dexterity; but a small piece may be coated without much difficulty. The glass must be scrupulously clean and recently polished with a clean leather, &c.; any grease or other matter adhering to the glass will inevitably cause blemishes. Spread out on a flat plain board placed on a table a piece of thick tinfoil about the size of the glass plate, and smooth it with the finger; pour on the centre a few drops of mercury and rub these very gently with a hare's foot, or pad covered with flannel, so as to make the mercury **wet** the tinfoil; gradually add more mercury and extend the rubbing outwards until finally the whole of the tinfoil is coated with quicksilver. Pour on more mercury until the liquid metal forms a coating of the thickness of a shilling. Now gradually slide the glass plate polished side downwards over the tinfoil, so as to push most of the mercury in front of it without allowing any air bubbles to enter between the glass and metal; if this is dexterously done, the under surface of the glass and the mercurialised tinfoil will form a brilliant mirror; but any unsteadiness of the hand in pushing the glass plate will cause particles of scum and dust or air bubbles to get between the glass and metal, causing spots and blemishes. Much of the mercury originally poured on is thus pushed off; more runs off when the board is cautiously lifted on one side so as to tilt the mirror and allow the surplus quicksilver to drain away. After remaining half an hour slightly tilted, the mirror should be placed upright to drain still further; next day the tinfoil will have become entirely penetrated by the mercury that has not drained away, and the tin and mercury together will have set to a solid *amalgam*, which adheres pretty firmly to the glass, but is easily scraped off, being very brittle. The mirror then only requires to be set in a frame with a solid back to avoid the scratching off of the amalgam during use.

Processes by which films of pure *silver* are deposited on glass by chemical action as reflecting agents instead of amalgam have largely come into use of late years, especially for hollow globes and curved surfaces. Expt. 198 illustrates in a rough way how this is effected.

Before these processes were invented such mirror surfaces could only be coated by making a mixture of mercury and other metals nearly fluid when warm, but setting solid or nearly so on cooling, and of such a nature that when shaken about inside a clean glass globe, &c., the mixture would adhere to the glass, wetting it as water would; so that a film would permanently stick to the glass, forming a reflecting coating when the rest of the amalgam was poured out. A mixture of mercury 2 parts, lead 1, tin 1, and bismuth 1 part, answers well for this purpose.

Removal of Substances from Solution by Absorption by means of Solids.—Many solids when brought in contact with solutions gradually remove the dissolved matter from the fluid. This action may be brought about in a variety of ways, the most usual one being where some kind of chemical action is set up with the result of precipitating the dissolved matter in the form of an insoluble compound. The ordinary processes of dyeing are amongst the best examples of this class of action: if textile fabrics (wool, silk, or cotton prepared or "mordanted" in various ways) are placed in a dye vat (*i.e.*, a large tank filled with solution of colouring matter) the fibres of the fabric combine with the colour and remove it from solution, becoming themselves increased in weight and tinted by the operation. But besides this chemical action, some substances, and especially various kinds of charcoal, have the power of withdrawing from solution certain dissolved matters and storing them up in their pores, no chemical action at all taking place, at any rate in the sense in which the term is usually understood.

Thus charcoal, especially "animal charcoal" from bones, &c., when boiled with water containing small quantities of strychnine and similar substances, will absorb them and remove them from solution; on drying the charcoal and boiling it with alcohol the absorbed substances are dissolved out again, and may be obtained from the solution by evaporating off the spirit. This process for separating certain bodies of this kind from beer or such like liquids, is employed in their analytical examination in cases of suspected adulteration, poisoning, &c. Obviously it is closely akin to the process of separation of one substance, dissolved in a solvent, by adding another solvent in which the substance dissolves more freely, the two solvents not being miscible together in all proportions.

Thus in Expt. 62 an aqueous solution of iodine when agitated with chloroform, gives, on standing, two layers of fluid; the upper one water containing but little iodine, the lower one chloroform which has dissolved out almost all the iodine from the water.

Similarly (Expt. 86), when zinc is agitated with a melted mixture of lead and silver, a lighter alloy floats up, consisting of the majority of the zinc with a little lead dissolved therein, and practically all the silver; whilst the heavier alloy consists of the greater part of the lead with a little zinc dissolved therein, but only a very small fraction of the silver originally present. Ether behaves in a similar fashion when agitated with aqueous solutions of certain vegetable products; these substances are removed almost entirely from the water, being dissolved in the ethereal fluid which rises to the top on standing.

The dark colouring matter produced by heating sugar is readily absorbed by charcoal, although pure sugar itself is but little affected;

accordingly the use of animal charcoal for clarifying syrups in the manufacture of refined sugar is an important part of that trade.

Expt. 289. To Decolorise Beer, &c., by Animal Charcoal.—The colour of ordinary beer and porter is mostly due to the presence of substances formed by the action of heat on sugar during the preparation of the malt, &c., just as "Caramel" or burnt sugar is employed by cooks for browning gravies, sauces, and the like. Get a pint of dark coloured beer; put into it an ounce of finely powdered good animal charcoal, and boil the whole for a few minutes in a flask; or in a distilling arrangement (Expt. 38) if it is wanted to save the spirit evaporated during the boiling. Pour the turbid black fluid on to a large paper filter (Expt. 56) supported by a glass funnel, and you will shortly see that the clear fluid that filters through the paper is either colourless, or at least very much paler than the original beer, showing that the charcoal has absorbed most of the colouring matter. By boiling for a longer time, adding distilled water as the fluid evaporates, you can generally succeed in getting a filtrate nearly as white as water from an ordinary beer of only moderately dark tint; with stout or porter the lightening in tint is relatively still more marked, although actually the filtered liquid will usually be somewhat more coloured than that from a lighter ale.

In the refining of sugar the coarse raw brown sugar is dissolved in boiling water and treated with charcoal, whereby the syrup is rendered almost colourless, so that when evaporated down it ultimately forms crystallised sugar in the form of sugar loaves, &c. Precisely the same kind of operation is resorted to in the extraction and purification of many drugs and useful substances extracted from vegetables and other products. Precipitated sulphide of lead (Expt. 158) and many other substances possess the same property to a less extent.

Expt. 290. To dye Wool and Silk various Colours.—Animal fibres, such as woollen cloth and silk, possess the properties of absorbing and combining with various kinds of colouring matters either prepared from natural sources (vegetable like indigo and madder derived from certain plants, or animal like cochineal obtained from a kind of insect), or manufactured artificially by somewhat complex chemical processes from materials generally obtained in the first instance from coal tar, this being a cheap source of the main ingredients. All animal fibres contain *nitrogen* as one of the constituents, a fact easily proved by the test described in Expt. 203; the presence of this nitrogen seems to be connected with the power possessed by such fabrics of absorbing colouring matters and becoming dyed or tinted therewith, inasmuch as vegetable fabrics (such as cotton, linen, hemp, and jute fibres, &c., and goods woven from them) are incapable of so doing (except in the case of a comparatively small number of dyestuffs, mostly artificial), and hence can only be permanently dyed by other processes involving the use of a "mordant" or substance acting as a kind of cement, fixing the colouring matter firmly to the vegetable fibre (Expt. 291).

Get a skein of *white Berlin wool* and wind part of it on a piece of card about 6 or 8 inches long, so as to make a small hank of some 10 or 12 turns; tie the ends firmly with a small piece of the same material to prevent the hank coming unwound, and then immerse it in water tinted with a sufficient quantity of the dyestuff employed; these dyestuffs are now sold in small quantities at a very cheap rate almost everywhere for general household use. Instead of purchasing these artificial colours, *cochineal* solution (such as is often used for tinting jellies and sweetmeats) may be used; or the prepared *indigo* solution described in Expt. 151. Stir the wool about in the dyeing liquid with a glass rod or a stick, for some time; it will gradually take up the colouring matter and become dyed permanently. If too much dye have not been added to the water in proportion to the weight of the wool it will often happen that the wool will take up almost the whole of the colouring matter, leaving the water colourless or nearly so. By using different quantities of dyestuff in different experiments (say 2 drops in one case, 4 in another, 8 in a third, 12 in a fourth, and so on), you will be able to dye different hanks (all of about the same weight) various shades of the same colour, and thus get a gradation of shaded wools varying in tint from a very light shade just perceptible up to the deepest tones possible.

Expt. 291. To dye Vegetable Cloths (Cotton, Linen, &c.).— In order that dyestuffs may adhere firmly to vegetable textile fabrics, so as not to be removed by washing, it is in most cases necessary to prepare the calico, &c., to be dyed by treating it with certain materials termed **mordants**, which have the power of adhering themselves firmly to the vegetable fibre, and also of causing the colouring matter to adhere to, or combine with, themselves; so that mordants act somewhat as a cement, uniting together two things that otherwise would not remain permanently associated together.

The mordants most frequently employed are certain metallic compounds, more especially those containing *iron*, *aluminium*, and *tin*; solutions of certain compounds of these metals are thickened with gum or *dextrin* (converted starch, or *British gum*), and the fluid thus obtained printed on to the goods to be dyed by means of engraved copper rollers (Expt. 272), in such a fashion that the pattern ultimately to appear in colour is first applied to the calico in the form of a nearly colourless fluid. The mordanted goods are then usually put through a process of exposure to air heated by the admission of steam, termed **ageing**, which fixes the metallic compound more firmly on the fibre by virtue of chemical changes brought about by moderate heat and moisture; after which they are passed through the bath of dyestuff dissolved in water, when the colour is absorbed and fixed on the cloth at those spots where mordants have been applied, but not elsewhere; so that, finally, when the cloths are well washed and "cleared," the colouring matter is entirely removed from the body of the fabric, but is retained at the mordanted spots, thus developing the pattern.

One and the same dyestuff can be made to give a variety of colours and shades by suitably varying the nature of the mordanting material, and the quantity of it used relatively to the water, &c., added to dilute it in preparing the fluid printed on to the cloth by the rollers; thus mordanting solutions of *acetate of aluminium* will give various pink and red shades (according to the strength) on calico passed through a dyebath containing the colouring matter of *madder* (either that from the natural root or the artificial *alizarin* now manufactured from coal tar); whilst *acetate of iron* similarly yields shades varying from pale violet and lilac to a very dark purple almost blue black; and mixtures of these compounds together (or with other analogous metallic salts) will give clarets and other shades. By engraving a portion of the total pattern on one roller, another on a second, and so on, up to five rollers; and supplying the first roller with, say, a weak solution of acetate of aluminium, and the second with a stronger one; the third with a weak solution of acetate of iron, and the fourth with a strong one, and the fifth with some appropriate mixture of salts; it becomes possible to produce a pattern with five distinct colours by one immersion only in the same dyebath; the portions of cloth mordanted by the first roller becoming a light red, and those by the second a dark red; those by the third a light violet, and those by the fourth a dark purple; and those mordanted by the fifth roller some other colour. Similarly, patterns in more than five colours can be prepared by suitably multiplying the rollers and varying the mordants.

Pieces of calico thus prepared with a variety of different mordants can be purchased from dealers who supply materials for chemical lecture experiments, as also can alizarin and various other dyestuffs; the process of making and applying the mordant and subsequently dyeing may be roughly imitated on the small scale thus. Dissolve in hot water some alum (or *sulphate of aluminium*), and in a separate vessel some *acetate of lead*, so as to obtain pretty strong solutions when cold; pour the alum solution into the other liquid, whereby a thick white precipitate of *sulphate of lead* will be formed, whilst *acetate of aluminium* remains in solution, these two compounds being formed by double decomposition. The addition of the alum solution should be made a little at a time, the fluid being allowed to stand for the precipitate to subside, so that it can be seen whether the further addition of alum solution produces more precipitate or not; or a sample may be filtered and tested by adding a little alum solution to it. When all the lead is precipitated as sulphate of lead, so that no more is thrown down by adding more alum solution, the whole is to be filtered. Part of the filtered liquid should be diluted with, say, two or three times its bulk of water, and another part with seven or eight times its bulk, and the rest preserved undiluted. Each of the thin liquids is to be thickened with gum arabic or dextrin, so that when a piece of dry calico is written on with the thickened fluid the writing does not run.

By means of a quill-pen, brush, stencil plate, india-rubber stamp, &c., letters or devices are drawn on a piece of calico, using the three liquids according to fancy. The prepared cloth is hung up in a warm room for twenty-four hours, and is then dipped into an evaporating basin containing boiling water in which some alizarin or madder extract has been dissolved, and well stirred about for some minutes; at the end of which time the writing or other devices will be developed, appearing in different shades of red, according to the strength of the mordanting liquor used.

Expt. 292. To discharge Colours.—Some kinds of goods are treated by what is termed the method of "discharge." In this case the fabric is uniformly dyed all over, and, after drying, the pattern to be discharged is printed on (by wooden blocks, &c.), employing solutions of chemicals so chosen as to bleach or otherwise alter the colour. For example, a cambric handkerchief may be dyed red with alumina mordant and alizarin, and a pattern then stamped upon it from an engraved block, &c., using a weak solution of bleaching powder; this will more or less bleach the spots to which the liquid is applied, especially if the handkerchief be "soured" by dipping into water slightly acidulated with a drop or two of sulphuric or hydrochloric acid (Expt. 162); so that the pattern will finally appear in white on a red ground.

In similar fashion calico may be more or less completely dyed black by first boiling it in water that has been previously boiled with chips of *logwood* and strained, and then immersing it in solution of *sulphate of iron*, which develops a purple black colouring matter resembling ink; by repeating the process two or three times, alternately boiling in logwood liquor and steeping in solution of sulphate of iron, the cloth becomes a tolerably good black. Now dry the cloth and print on it from a wooden block any device you please, using as printing fluid a strong solution of *citric* or *oxalic* acid thickened with gum; wherever this is applied the black will be discharged, so that a white pattern on a black ground will result (compare Expts. 139, 140). Silk handkerchiefs are sometimes dyed and then spotted with *nitric acid*, which discharges the dye and forms a yellow colour by acting on the silk.

Expt. 293. To prepare Lakes of various Colours.—The attraction of cotton or other vegetable fibre for metallic compounds is regarded by some rather as a physical phenomenon than as an example of chemical combination, although this view is open to some discussion; but the union of dyestuffs with metallic compounds appears to be wholly of the nature of a chemical change; moreover, the colour of the resulting combination varies with the nature of the metal employed. When suitable metallic compounds are brought into contact with solutions of dyestuffs without the presence of vegetable fibres, they absorb the colouring matters and develop tinted solid matters, which, when collected and dried, can be used as pigments; such compounds are generally called *lakes*.

Dissolve some *alum* in water, and divide the solution into several portions, to each of which add a little solution of dyestuff of various kinds. Now add a little *diluted ammonia* solution to each; this will act upon the alum, precipitating *hydrated alumina* (oxide of aluminium combined with water), so that if no dyestuff be present a white gelatinous precipitate will be thrown down; but if any colouring matter be present it is absorbed by the alumina forming a coloured lake, the shade of which depends on the proportion between the alumina and the quantity of colouring matter taken up, whilst the tint varies with the nature of the dyestuff used. *Crimson Lake*, *Madder Lake*, and a large variety of other lakes made from cochineal, madder, and other colouring matters, natural or artificial, are prepared in large quantities by processes essentially identical with this in principle for the use of artists, house decorators, wall paper manufacturers, and such like users of pigments; by using other metallic compounds in place of or in combination with alum, the tints and shades of the resulting lakes can be modified to almost any required extent.

Expt. 294. To prepare Ink.—Writing fluids of various kinds are in use at the present day, many of which are simply solutions of certain colouring matters thickened with gum, &c., so that the writing will not readily run on ordinary slightly glazed sized writing paper; most of the "aniline inks" and ordinary red ink are of this character. The old fashioned black inks, however, are of a somewhat different nature, being essentially black lakes produced by the combination of iron with certain kinds of colouring matters and allied products, these lakes being in an excessively fine state of division, so as not to settle readily in the bottle, somewhat like Faraday's ruby gold described in Expt. 136. The colouring matter developed in Expt. 292 from logwood and sulphate of iron is one of this class; *gallnuts* boiled with water, and the infusion similarly treated with *sulphate of iron* and thickened with *gum* or *dextrin*, forms the basis of one of the best kinds of ordinary writing ink. Iron inks of this class often write pale, but darken greatly after a while; this is because the iron compound present absorbs oxygen from the air, and develops a much darker tint in so doing, the action being somewhat analogous to that occurring when white indigo solution is exposed to the air, whereby oxygen is absorbed and insoluble blue indigo precipitated in the pores of the cloth (Expt. 151); the ink, as at first used, is chiefly a mixture of various substances all truly in solution; but under the influence of the oxidising action of the air an insoluble lake is formed and precipitated in the body of the paper, so that water will not wash out the tinting substance produced.

A good black ink may be thus prepared. Finely powder some gallnuts and boil the powder with water for three hours, using about 7 parts by weight of water to 1 of galls (or about 3 ounces of galls per pint of water), and adding more water from time to time to supply that lost by evaporation; strain the decoction tolerably clear, and then add to it a solution of sulphate of iron dissolved in three or four times its weight of hot water, and another of gum of the same strength; 5 parts of sulphate of iron and 5 of gum being used for every 12 of galls; so that in all the proportions will be about—

Gallnuts, ...	12	parts by weight.
Sulphate of iron, ...	5	,,
Gum, ...	5	,,
Total water, ...	120	,,

Copying ink chiefly differs from ordinary writing ink in containing *sugar* or *glycerine* intermixed, which prevents the ink from drying in so completely, and enables some of it to be transferred more readily to damp tissue paper by pressure in a copying press.

Pencils are sometimes used, consisting of some soluble colouring matter in the solid state mixed with gum and other materials, so that when the writing is moistened a solution of the colouring matter contained in the pencil marks is formed, which tints the paper as a soluble ink would do; or by pressing damp tissue paper on the writing a copy is obtained, the colouring matter dissolving in the water moistening the tissue paper.

In Expt. 140 directions are given for removing from paper, &c., ink stains produced by inks of the gallnut and iron class.

Expt. 295. Capillary Attraction and Repulsion.—The difference between the relative amounts of adhesion of various liquids to given solids, and the cohesion of the liquids themselves, gives rise to the difference observed in the behaviour of various liquids when one and the same solid is dipped into them and removed; thus a glass rod dipped into *water* or *alcohol* comes

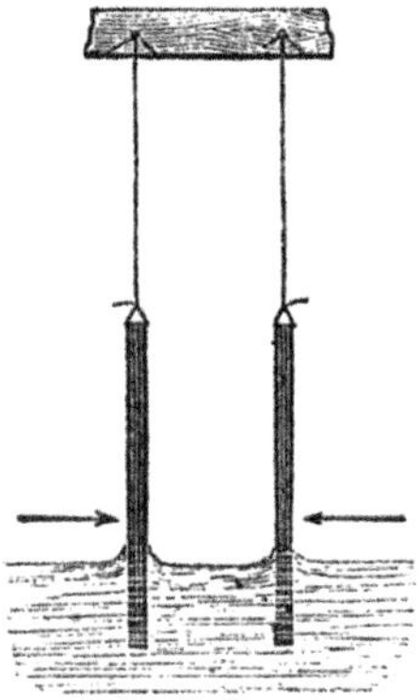

Fig. 100. Rods of Glass dipped in Water.

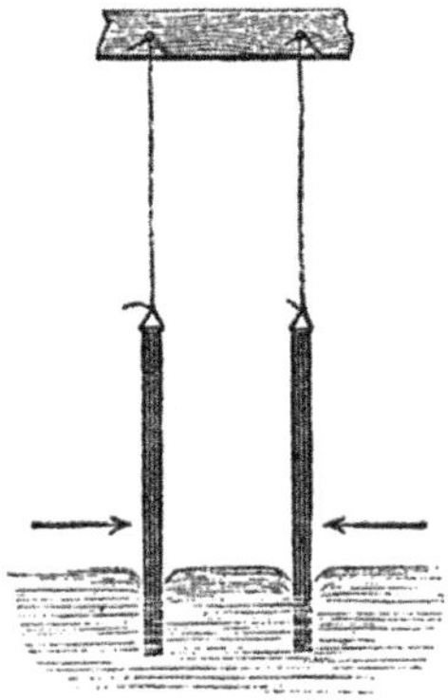

Fig. 101. Rods of Glass dipped in Mercury.

out **wetted** with the fluid; *i.e.*, a portion of the fluid adheres to the glass so strongly that the adhesion overcomes the cohesion of this part of the liquid for the rest, so that when the rod is lifted out from the liquid this portion is removed too. If, however, a similar rod be dipped into clean pure *mercury*, no metal whatever will stick to it on lifting it out, the adhesion of the mercury to the glass being less than its cohesion towards the portions of mercury adjoining. As a consequence of the nature of the surface actions producing this difference, the surface of a basin of water into which a freely suspended rod of glass (or any other body wetted by the water) is dipped, will be found to be elevated in a curve towards the surface of the glass, as in fig. 100; whilst the opposite action (fig. 101) will be manifest when mercury is used instead of water, or if the glass be well greased, so as not to be wetted by the water. For the same reason water contained in a narrow vessel of about $\frac{1}{3}$ to $\frac{1}{4}$ inch internal diameter will have a **concave** surface; whilst mercury in a similar vessel will have a **convex** surface. A rise of the liquid inside a narrow tube will similarly be observed on dipping it into water, the amount of ascent being the greater the narrower the tube (fig. 102); whilst if dipped into mercury, the quicksilver will stand at a lower level inside the tube than outside, the difference of level, as before, being greater the narrower the tube. Obtain two flat plates of glass and tie them side by side together, with a thin strip of cardboard or glass between them at one side, so that the space enclosed between them is wedge-shaped. Place the plates in a vessel of water, and a rise upwards of the fluid will be observed between the plates, greatest at the narrow end of the wedge, and least at the wider end, so as to trace out a curve. If placed in mercury an analogous sunken curve, due to depression,

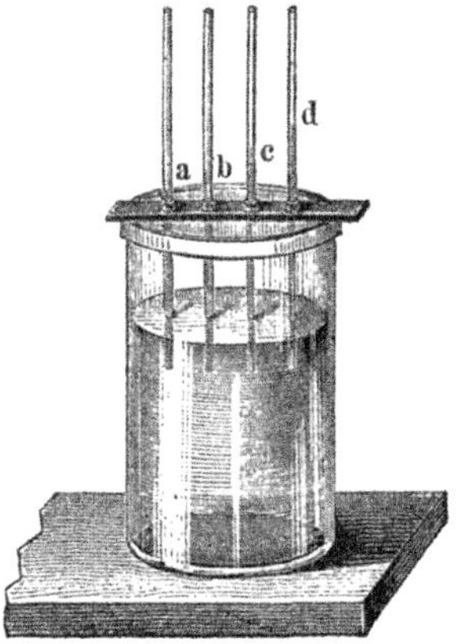

Fig. 102. Capillary Tubes and Water: *a*, *b*, *c*, *d*, height of Water.
Water.

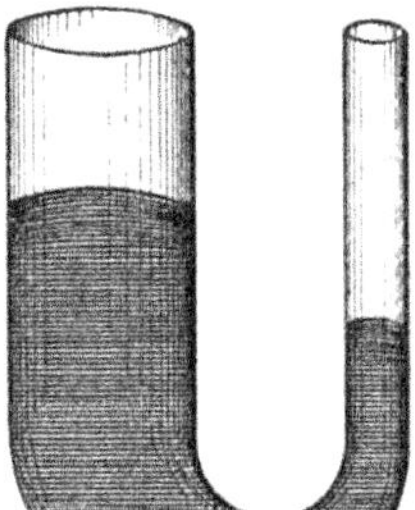
Fig. 103. U Tube containing Mercury.

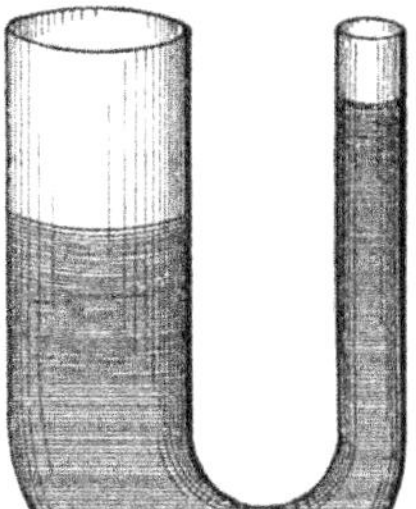
Fig. 104. U Tube containing Water.

will result, less easy to see on account of the opacity of the mercury.

Expt. 296. Capillary Action in Narrow Tubes.—Get a piece of glass tubing some 6 inches long and $\frac{1}{4}$ to $\frac{1}{3}$ inch bore; draw it out in a lamp flame or gas jet (Expt. 48), and whilst hot bend it up into a U shape (fig. 103), so that one limb of the U is much narrower than the other. If quicksilver be poured in it will be found that the mercury will stand highest in the wider limb; whilst the opposite will be noticed if alcohol or water be used (fig. 104). In the first case the surface of the quicksilver will be convex upwards in both limbs; whilst in the second both water surfaces will be concave.

Expt. 297. Apparent Attraction and Repulsion between Floating Bodies.—A curious result follows from the nature of the curved surface of the water or other fluid in which a solid body floats; if the fluid wets the body, as with a piece of cork floating on water or a glass rod suspended therein, the curved surface is concave; and, in consequence, two such floating or freely suspended objects placed close together will tend to approach one another and will apparently attract one another (fig. 100). The same thing will be observed if the two floating objects are *not* wetted by the water, *e.g.*, two pellets of wax or two suspended well-greased glass rods, in which case the curved surfaces of the water are convex (fig. 101). But if one object be wetted by the water and the other not, as when a piece of cork and a lump of wax are used together, or a clean glass rod and a greased one, the curvatures of the water surfaces surrounding the two are not alike, one being concave and the other convex; in this case the two objects will not approach one another, but if floated on the water close together will separate, apparently repelling each other (fig. 105).

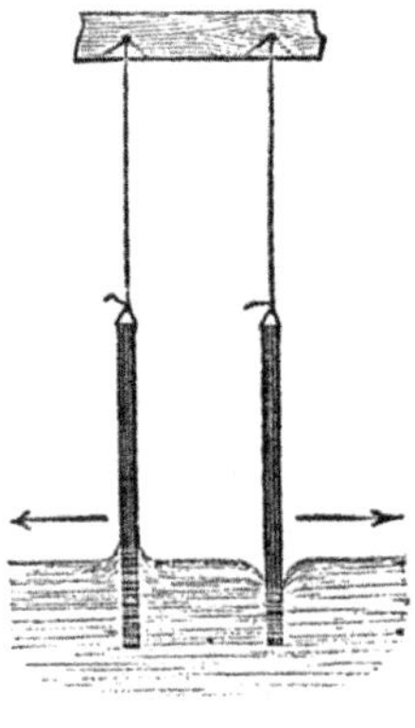

Fig. 105. Greased and clean glass Rods in Water.

The floating of a comparatively heavy small object, such as a sewing needle, on water (Expt. 279) is partly due to non-wetting of the polished steel by the water; a well-oiled small fragment of stone, or other greasy heavy substance not too large, will often behave in the same way. But if the needle be washed with alcohol, or the greasy stone with solution of caustic potash, and placed on the surface of a basin of water whilst still wet, it will immediately sink, because the fluid is now enabled to wet its surface, and consequently no air cushion, so to speak, is formed between it and the water buoying it up. Water beetles and many other insects skim over the surface of a pond without sinking into it, because their feet are not wetted by the water, and, consequently, depressions are produced which buoy them up like the needle.

Expt. 298. To Empty a Basin of Water by means of a Hank of Cotton.—If a porous substance, such as a piece of calico, be hung so that one end dips in water, the water will gradually travel upwards and wet the whole of the piece; the ascent of the oil in the wick of a lamp or of the melted tallow or wax in the wick of a candle is due to this action, which is substantially the same as that of the ascent of a fluid in a fine capillary tube (Expt. 296).

If a piece of lampwick, a twisted towel, a hank of cotton, or similar material be hung over the side of a basin of water so that one end is in the water and the other hangs over outside, water will be found to drip down from the outside end, being gradually raised out of the basin by this wick-like action. In this way, a wineglass full of water may be gradually emptied without touching it by placing a prawn or shrimp on the rim so that the tail end is inside the glass and the head outside. If the water in a basin be turbid from the presence of a little fine clay or other suspended matter, it may be filtered perfectly clear whilst passing through the "Capillary Siphon" formed in this way by a hank of cotton, &c.; but if it be coloured by some soluble coloured substance this will usually not be abstracted from the water by the passage through the siphon.

If, however, the capillary siphon be made of white wool (previously dipped in water and wrung out), and certain kinds of dyestuffs be dissolved in the water, the wool will in many cases remove these from the water whilst passing through, so that the water dripping down at the far end will be colourless or nearly so, whilst the wool itself becomes more or less tinted or dyed. This kind of action is sometimes exerted to a small extent upon dissolved substances by paper. If a drop of a solution of a coloured salt (such as *sulphate of copper*, or *perchloride of iron*) be carefully placed on a piece of white blotting-paper, it will "run" and produce a circular wet spot. If another drop be placed in the centre of this the circle will extend; and by repeating the process a wet circle of a couple of inches or more in diameter may be produced. In many cases it will be found that the outermost rim of this wet circle is colourless, consisting of paper wetted with little but plain water, the dissolved solid having been mostly removed from the solution by passing along the porous paper fibres from the centre outwards; so that whilst the central part of the wet spot is highly coloured, the outer edges are colourless, or at any rate much less strongly tinted. If the outermost edge be marked off with a pencil and the paper dried in a horizontal position (over a hot plate, &c.) it will often be visible that the coloured spot left when dry does not extend quite up to the pencil mark; and by dipping the paper in some chemical fluid that will react with the substance absorbed by the paper (*e.g.*, *ferrocyanide of potassium* with perchloride of iron or sulphate of copper, *sulphuretted hydrogen solution* with the latter, and so on), it can be rendered still more obvious that the salt dissolved in the fluid originally dropped on to the paper has been more or less removed from solution and retained in the central parts, whilst a weaker solution, and finally little but water has passed on to the extreme edge of the circle.

CHAPTER XVIII.

SOAP BUBBLES AND FILMS.

One of the most familiar observations is that whereas ordinary pure spring or distilled water will not develop a permanent froth, or permit of bubbles being blown therewith, water in which a little soap is dissolved possesses the property of adhering together in such fashion that it can be obtained in the form of thin flat films (as when a ring of wire is dipped in soap water and taken out again), or as nearly spherical shells with very thin walls, or "bubbles," as when a tobacco pipe is dipped in soap water and the flat film formed at the opening of the bowl enlarged by carefully blowing into the stem and finally dislodging the bubble by a dexterous jerk. Films and bubbles can also be obtained with many other kinds of liquids besides soap water, more especially with "collodion" solutions (guncotton dissolved in ether) such as are employed to form a coating on glass plates for photographic purposes.

Expt. 299. Analogy between a Soap Bubble and an India-rubber Balloon.—Blow a good-sized bubble by means of a tobacco pipe, but do not dislodge it by shaking from the pipe; remove the stem of the pipe from the mouth and hold it near a lighted candle; you will now see that the candle flame is blown on one side; a current of air issues from the pipe stem, and at the same time if you watch the bubble you will see that it gets smaller or "collapses:" in short, the film of soap water behaves exactly as if it were a thin india-rubber ball, capable of expansion when blown into, but collapsing again when the air is allowed to escape.

This property of the bubble is due to the different attractive action of the particles of the liquid forming the outsides of the walls of the bubble on one another as compared with their action on the adjacent air, thus leading to a kind of strain in the soap liquid termed **surface tension**, the effect of which in causing the film to collapse is similar to the tension or strain in the elastic rubber ball when expanded. Like the rubber, if the soap bubble be expanded to too large dimensions by continually enlarging it by blowing, the strain on its walls becomes greater than its cohesion can support, and the ball bursts.

Expt. 300. Analogy between a flat Soap Film and a stretched Membrane, such as a Drum head.—Prepare a flat ring with a handle by bending a piece of copper wire, as indicated in fig. 106, and dip it into the soap water; a thin sheet of soap water like a fine skin or membrane will then adhere to the ring, and can be watched for a long time if the ring be held steadily by sticking the handle end into a cork fixed in the mouth of a bottle, or in a candlestick, so as to form a firm support, or by holding it by a clamp such as that shown in fig. 12. Such a ring, moistened with soap water, may also be used to catch and support a bubble blown by the pipe, and so prevent it from being carried away by currents of air; the bubble is blown over the ring, holding the pipe bowl downwards, until the bubble touches the ring; by carefully jerking the pipe the bubble can be removed from the bowl without separating from the ring, and is thus left sticking to the ring and supported by it. Or the pipe may be held bowl upwards and the ring held above the bubble so as to catch it at the top.

Fig. 106. Wire Ring held by corked Bottle, and supporting a Film of Soap.

Just as a soap bubble is comparable with a hollow elastic ball, swelling when blown into and contracting and expelling air again when allowed to take its own course, so is a flat soap film analogous to a stretched drum head or *tambourine*. If you press upon the centre of a tambourine, or on the tightly stretched paper cover of an unopened jam pot, the membrane will yield more or less to the pressure and cease to be quite flat, becoming somewhat hollowed or *concave*; or if you pass a knotted string upwards through a minute hole in the centre of the tambourine so that the knot catches the membrane, and then pull the string, the membrane will again yield to the pulling force, and become *convex*, or projecting in the centre. In exactly the same way a flat circular soap film can be bent as it were downwards if a bit of paper moistened with soap water be carefully dropped on its centre, and a gentle pressure be then exerted on the paper by touching it with a pencil or thick knitting needle; or it may be pulled outwards if the paper have a knotted string attached to it to serve as a sort of handle (fig. 107).

Expt. 301. Solutions suitable for Soap Films.—When the soap water serving to form a bubble or film is of proper strength the permanence of the film is very considerable, and its tenacity or power of cohesion very remarkable. The strongest films of this kind are made from a clarified soap solution containing glycerine, such as that obtained in the following way. A piece of good soap (oil soap, castile soap, or even ordinary yellow soap, is usually preferable to transparent kinds for this purpose) is shaved with a knife, and the shavings well stirred up with about 30 or 40 times their weight of warm water till all is dissolved or nearly so; half an ounce of soap to a pint of water is about the proportion, distilled water being preferable; or failing this, the softest rain water you can obtain. The solution is then filtered through blotting paper, and some glycerine added, about half its bulk being usually sufficient.

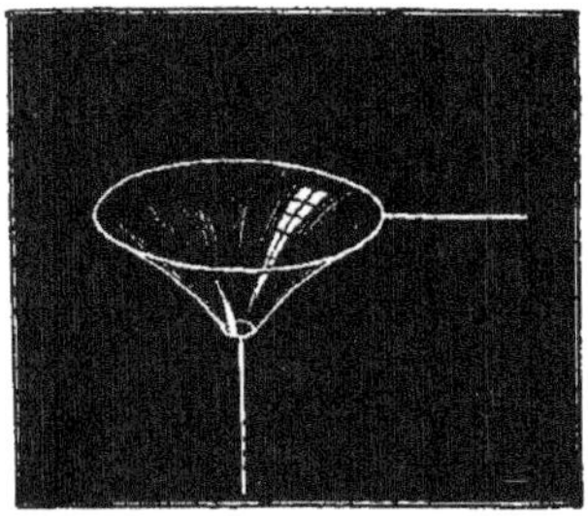

Fig. 107. Flat Soap Film supported by Wire Ring, and pulled downwards, forming funnel-shaped Film.

The following receipt for making exceedingly permanent soap films is due to M. Plateau, a celebrated French physicist. *Marseilles soap* (or better, *pure oleate of soda*) is dissolved in about 40 parts of warm distilled water and the solution filtered and mixed with $\frac{3}{4}$ of its bulk of pure *glycerine* and allowed to stand for a week; the vessel containing the fluid is then immersed in fragments of ice for six hours and again filtered to remove the deposited matter, care being taken to prevent the liquid in the filter from becoming warmed again (which would redissolve the deposit) by placing inside the filter a little corked bottle filled with fragments of ice. Films and bubbles made with this solution will often last 12 hours and more if kept under a glass shade in slightly moist air. Instead of a tobacco pipe a piece of glass tube about $\frac{1}{4}$ inch bore may be conveniently used as blowing tube.

Collodion films may be prepared by means of the following

liquid. Dissolve 1 part of photographic guncotton in a mixture of 1 part absolute alcohol and 16 of ether, and allow the liquid to subside if not quite clear; decant off the clear solution and add to it about three-fourths of its volume of pure castor oil; extremely tenacious films can be produced from this fluid.

Expt. 302. Tenacity of thin Films.—The great strength and tenacity of soap films made from a proper solution may be illustrated by the following experiments. Blow a good-sized bubble and make it adhere to a ring of wire so as to rest on the ring (Expt. 300); carefully bring to the top of the bubble a circular bit of paper, through the centre of which a knotted thread or thin string has been passed (fig. 108), the surface of the paper being wetted with soap liquor; the paper will then adhere to the bubble; by pulling the thread the bubble can be elongated and pulled upwards to a considerable extent without breaking; conversely, if the bubble be made to hang down from the ring, and the paper disc and thread be attached below, and a little paper cage or tray be tied to the thread, a number of small shot can be put in the cage and their weight suspended by reason of the tenacity of the film of soap liquor due to its "surface tension"; a light wire ring with a cage suspended therefrom by three threads (fig. 109), may be used instead of the paper disc and knotted thread. Exact numerical experiments made with soap films show that the amount of weight which is required to increase the *area* of the film to a given extent is directly proportionate to the increase in area thus brought about. A soap film may thus be used like the *spring balance* used for weighing letters, and similar objects, where the amount of extension or compression of the spring used is proportionate to the weight acting; and where, consequently, the amount of weight can be read off on a scale by means of an "index" or pointer attached to the spring.

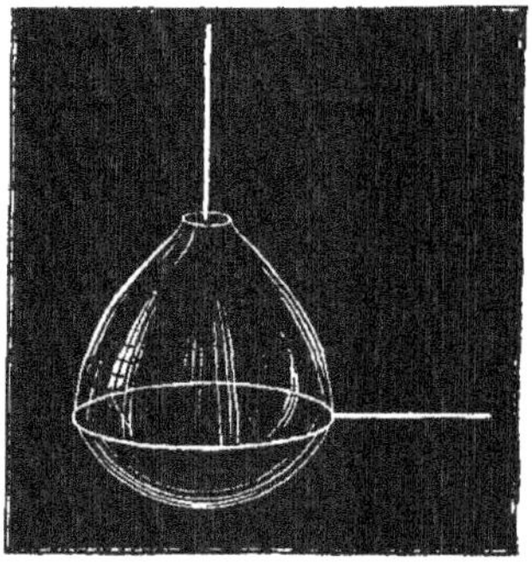

Fig. 108. Soap Bubble supported on Wire Ring, and pulled upwards into a conical shape by means of a Thread.

Fig. 109. Soap Bubble supported by Wire Ring and pulled downwards to a conical shape by attached Car and Weights.

Expt. 303. Soap Film Spring Balance.—Procure a small ordinary drawing-slate and break away the slate and one end of the wooden frame so as to leave the other end and the two sides grooved where the slate fitted in (*a a a*, fig. 110); then get a straight knitting needle (*b*), and with the aid of a file cut it to such a length that it can slide up and down in the groove parallel to the end of the frame without sticking or passing out sideways. Attach a loop of thread (*c c*), to the wire supporting a small scale pan (*d*), or bucket; and similarly, fix a loop of string (*e*) to the wooden frame, so that it can be hung up and supported by the loop from a wire projecting out of the cork of a bottle filled with water to make it heavy and steady, or from a stand (such as fig. 12). The frame and wire being dipped into the soap liquor, by gently pulling the loop attached to the wire a rectangular film is formed; by carefully putting weights (sand, shot, &c.), into the scale pan, or removing them, it can be shown that the wire descends or ascends so as to increase or diminish the area of the film, according as the weight is increased or diminished, precisely as the index of a spring balance moves under similar conditions. In short, the surface tension of the soap film causes it to behave *precisely as though it were an elastic membrane* like a sheet of thin rubber.

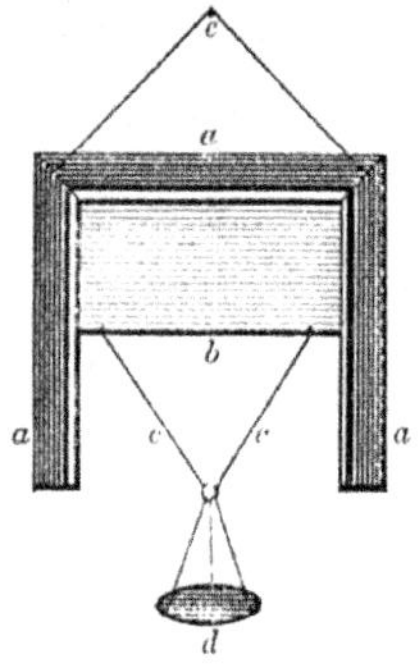

Fig. 110. Flat Soap Film supported by Frame of Wood and Cross Wire with attached Pan and Weights.

Expt. 304. Ring-shaped Soap Film.—Make a flat circular soap film, as in Expt. 300; carefully place on it a bit of thread, the ends of which are tied together, and which is moistened with soap liquor. Now touch the film *inside* the thread loop with a hot wire or needle; this will break the film inside the loop, but will not affect that part of it which lies outside. The effect of the pulling force brought into play by the tension of the film will be to pull the thread loop into a perfectly circular form lying concentrically inside the circular wire supporting the film, as in fig. 111.

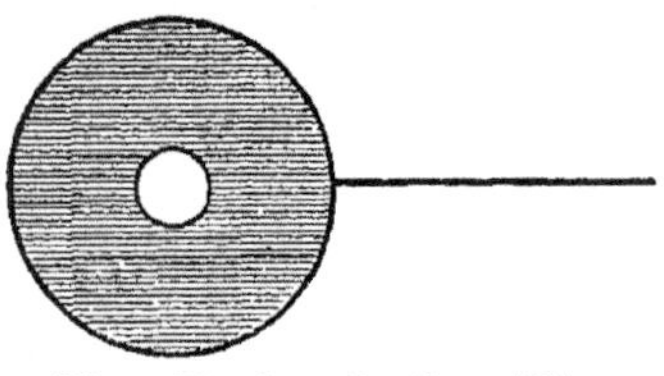
Fig. 111. Annular Soap Film.

Expt. 305. Two Soap Bubbles will not touch each other readily.—Blow a soap bubble, but before shaking it free from the pipe or glass tube, &c., used for blowing, bring it in contact with a wire ring previously moistened with soap solution and mounted in a corked bottle, so that the ring is vertical. A dexterous shake will now free the bubble from the blowing tube, and will leave it sticking to the ring. Now blow another bubble and bring it near to the side of the first one; you will find it extremely difficult, if not impossible, to make the two unite in one large bubble; pressing the two together sideways will flatten them out somewhat, but they will not run together or coalesce into one. If, however, a stick of warm dry sealing wax be briskly rubbed with warm flannel, so as to "electrify" it, and be then brought near, the two bubbles will often join into one, which sticks to both of the supports. The reason for this is that the "surface tension" prevents the two bubbles from actually coming into contact under ordinary circumstances, the surface of the one exerting a sort of repellant action on that of the other, and preventing their coalescence; somewhat as two drops of water on a dusty plate or leaf will not run together, contact being prevented by the film of dust sticking to each drop. The presence of an electrified body, however, modifies the surface tension in such a way as to render coalescence easy.

Expt. 306. One Bubble inside Another.—Blow a bubble and mount it on a horizontal wire ring as in Expt. 300. Now dip the blowing tube again in the soap liquor, and by a quick movement pass it through the top of the bubble so that the end of the tube is an inch or more inside; you will now be able to blow a second bubble *inside the first one* (fig. 112.) Whilst you are so doing you will notice that the outer bubble increases in size, the fact being that the bulk of the air contained in the space between the two bubbles remains much the same, so that as the inner bubble grows the volume of the outer one must simultaneously increase to about the same extent. By a careful shake you can dislodge the inner bubble from the blowing tube, which can then be removed, leaving the two bubbles one inside the other. As in Expt. 305 the two bubbles cannot be made to touch one other absolutely, and consequently will not coalesce; but in this case an electrified body will not cause them to unite in one when brought near. The

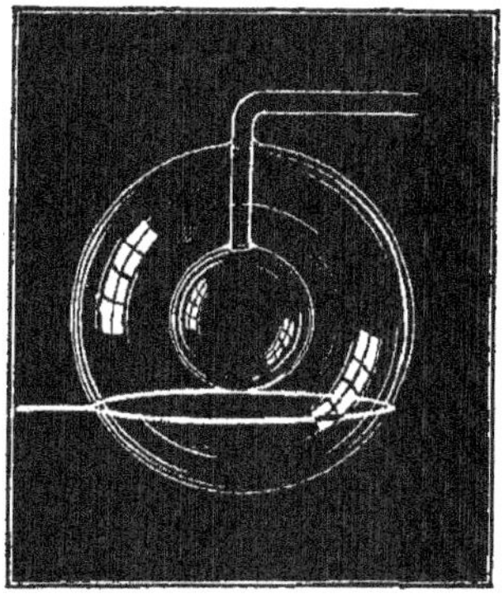

Fig. 112. One Bubble blown inside another.

reason for this is that the kind of electrification here in question *only affects the outer surfaces;* and the inner bubble is consequently entirely unaffected, as to all intents and purposes it is an internal object protected against electrical influences by the outer coating of the larger bubble. The most delicate electrical indicators exhibit no sign of electrification when they are similarly situated, *i.e.*, when they are completely enclosed in some conducting outer coating.

Expt. 307. Soap Bubble Balloons.—Provide a bladder, into the neck of which a piece of ordinary ¼ inch brass gas piping is fixed air tight, a tap being attached to the piping (Expt. 177). By means of a piece of india-rubber tubing the bladder may be connected with the burner of an ordinary gas jet, all air having been first squeezed out of the bladder; on opening the taps the bladder will become slowly inflated with coal gas. If the bladder thus filled be held under the arm and gently squeezed, the coal gas will be forced out of it when the tap is opened; by means of the india-rubber tubing the blowing tube can thus be supplied with coal gas instead of air from the lungs. If you then blow a bubble filled with coal gas and detach it by a shake from the blowing tube, you will see that it ascends up to the ceiling more quickly than a bubble simply filled with the breath; this is because, whilst slightly warm air is lighter than cooler air, bulk for bulk, coal gas is lighter still, and is therefore more buoyant when used to fill a bubble or balloon.

Provide a light wire ring with three threads attached to it at equal intervals (fig. 113); moisten the ring with soap solution, blow a coal gas bubble and make it adhere to the ring, and detach the blowing tube by a shake; if the ring is not too heavy, the bubble will be light enough to lift its weight and a small paper disc or car attached to the threads, thus forming an actual miniature balloon. A light thread attached to the car then converts the whole into a "captive balloon," which can be allowed to ascend or pulled down again by the thread at will.

Fig. 113. Soap Bubble Balloon with attached Car.

After some time the gas escapes through the wall of the soap bubble by osmosis, and air more or less completely takes its place (*vide* Expt. 79); so that the ascending power becomes considerably weakened or even entirely lost.

Caution.—In carrying out this experiment care must be taken not to allow the bubble filled with coal gas to come in contact with a chandelier or sunburner, &c., otherwise a slight explosion may be caused by the firing of the gas, and glass shades may be thus broken; the bubbles at the ceiling may be fired without any danger by means of a long taper tied to a stick, provided they are nowhere near to curtains, hangings, or other inflammable objects.

Fig. 114. Balloon with one Bubble blown inside another.

Expt. 308. Buoyant Bubble inside a Heavier one.—Blow a bubble with air and fix it on the ring as in the last experiment. Now dip the blowing tube again in the soap solution, and by a quick motion pass the blowing tube through the top of the first bubble, as in Expt. 306, and connect the blowing tube with the bladder of coal gas. You will now be able to blow one bubble within another, the innermost containing coal gas and the outermost air. The ascensional power of the innermost one will be sufficient (if the ring is not too heavy) to lift up both, the outer one with the car attached being lifted up by the lighter internal bubble (fig. 114). If the car be held by the thread and the outer bubble broken by touching with a dry stick, &c., the ring and car will drop on account of the buoyancy being no longer exerted upon it, whilst the inside bubble will escape and rapidly rise to the ceiling. The outer bubble in this case is comparable with the netting thrown over the covering of the ordinary balloon, from which netting the car is actually suspended.

Fig. 115. Soap Bubble floating in Tub half-filled with a heavy Gas.

Expt. 309. Illustration of Density of Carbon Dioxide.—Get a large basin or small tub and strew powdered chalk or whiting over the bottom inside; now pour on a tumblerful of strong vinegar, or some diluted hydrochloric acid;

carbon dioxide gas will escape vigorously (Expt. 99) and gradually fill the basin. Blow a soap bubble with cold air from a bladder, or render one blown by the breath slightly heavier than air by attaching to it a light car, as in Expt. 307, so that it will sink when detached from the blowing pipe; let it fall on to the centre of the basin of carbon dioxide, when it will remain suspended, floating on the sea of heavy gas in the basin, just as a hollow india-rubber ball would on the surface of a tub of water (fig. 115).

---

§7. Effects of Heat upon Bodies other than Change of State and Production of Chemical Action.

CHAPTER XIX.

Expansion and Contraction.

When a body becomes warmer it generally becomes larger, although there are some exceptions to this rule; the ascent of the quicksilver in a thermometric tube on heating it has already been explained (Expt. 28) to be due to the fact that whilst the glass case becomes larger or **expands** by heat, the quicksilver expands more rapidly, and therefore occupies more space.

Expt. 310. To show that Metals Expand on Heating.—A simple arrangement for showing this consists of a brass or copper ball (fig. 116) suspended by a wire and turned to such a size that when cold it will just drop through a horizontal ring fixed so as to stand on three legs. Heat the ball for a short time in the flame of a Bunsen lamp or spirit lamp; it will then be found too large to pass through the ring. If left a while standing on the ring it will by and by drop down through it, partly because as it cools it becomes smaller, and partly because the ring becomes heated by touching it, and consequently expands and becomes wider. Fig. 117 represents another arrangement for illustrating the same thing, consisting of a bar of metal with a gauge of such size that the bar can just fit in when cold, but is too large to do so when hot.

Fig. 118 illustrates an arrangement for making a determination of the amount of increase in the length of a bar when it is heated; a bar of metal (t) is held in a flame in such a way that one end is firmly clamped by a screw (v) whilst the other presses against a lever or index (h) near to its centre, this lever in turn

Fig. 116. Ring and Ball (expansion).

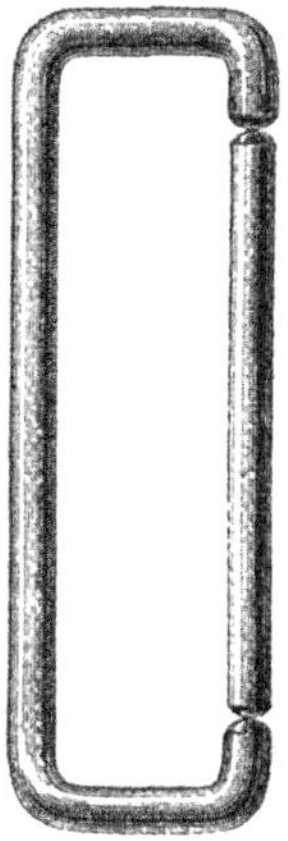

Fig. 117. Bar and Gauge (expansion).

pressing upon a second (p). When the bar is heated by a large

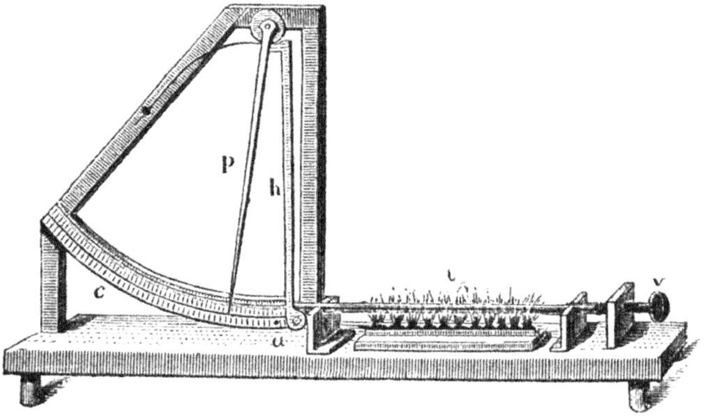

Fig. 118. Measurement of Expansion (bar).

flame of spirit underneath it grows longer, and consequently presses upon the lever and makes it move so as to register on the scale (*a*, *c*) the increase in length.

Expt. 311. Apparent Contraction on Warming.—Fit into the mouth of a 6 or 8 ounce flask or bottle of thin glass a perforated india-rubber cork, through which a piece of quill glass tubing (*a*, *b*, fig. 119) passes air-tight. Fix a paper scale to the tube, take out the cork and tube, fill the flask quite full with water tinted with indigo, black or red ink, or other convenient coloured liquid, and insert the cork again so that the liquid is pressed up in the tube to some convenient point on the scale, *a*. Provide a saucepan full of hot water and dip the flask into it, holding it by the glass tube; the first effect of the heat will be to make the level of the liquid in the stem or tube drop a quarter of an inch or more, making it appear as though the heat produced contraction. What really takes place, however, is that the heat first causes the glass to expand, so that the flask becomes a little bigger and holds more fluid; after a few seconds the heat penetrates through the glass into the water and causes it to expand, when the liquid begins to rise in the stem towards *b*. On taking the flask out of the hot water the opposite effects are produced; for a few seconds the water *rises* in the stem after withdrawal, because the chilling action of the air first affects the glass of the flask, causing it to contract and squeeze the liquid upwards; but as soon as the water inside begins to cool, the liquid begins to sink again in the stem.

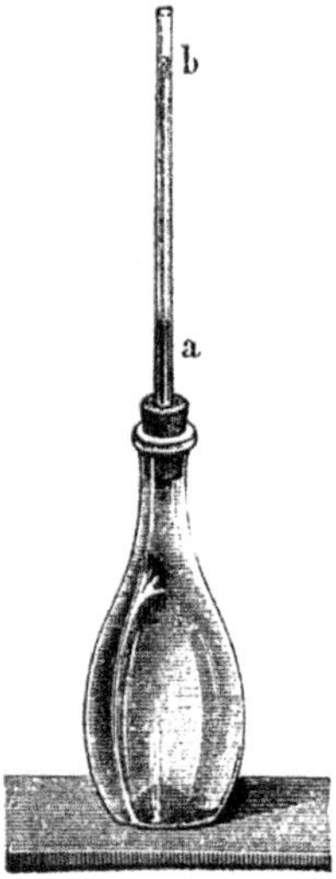

Fig. 119. Measurement of Expansion (liquid).

Expt. 312. To show that Solids do not expand all alike.—Just as quicksilver and glass do not expand equally, so is there observed a greater or less difference in the rate of expansion of all solids. Thus bars of iron, copper, glass, and other substances, when heated in the same way in the arrangement described in Expt. 310 (fig. 118), will cause the index to move over very different amounts of space; the increase in length which a copper rod will undergo will be found to be about half as much again as that indicated with an iron rod, which again will be greater than that shown by a glass rod; on the other hand, a rod of lead or zinc (not heated so highly as to melt) will expand nearly twice as much as one of copper heated to the same extent.

Expt. 313. Curving of Compound Metallic Bars on Heating.—

Obtain two strips of sheet copper and iron respectively, some two inches wide and two feet long, of pretty stout material; have them firmly riveted together so as to form a flat double bar; on heating this with a lamp the bar will curve, because, as the copper expands more than the iron, it must either break the rivets or else take the *outside* of a curve so as to occupy a greater length than the less expanded iron on the inside of the curve. On cooling down the curved compound bar will gradually become flat again; if chilled in a freezing mixture the bar will again curve, but now with the copper *inside* the curve, since the copper contracts the most. In laying railroads, building iron joists into walls, and such like operations, the differences in length caused by the variations of temperature between day and night, summer and winter, &c., have to be carefully allowed for, and space left to permit of expansion; were this not done the rails would become curved, the walls thrust out of the perpendicular, and similar serious derangements brought about. A long iron girder bridge will often become measurably longer if a gleam of sunshine fall upon it on a cloudy day, or shorter if a cool wind suddenly spring up and blow upon it; with some structures the amount of growth and decrease can be readily watched at the ends by the eye under favourable circumstances.

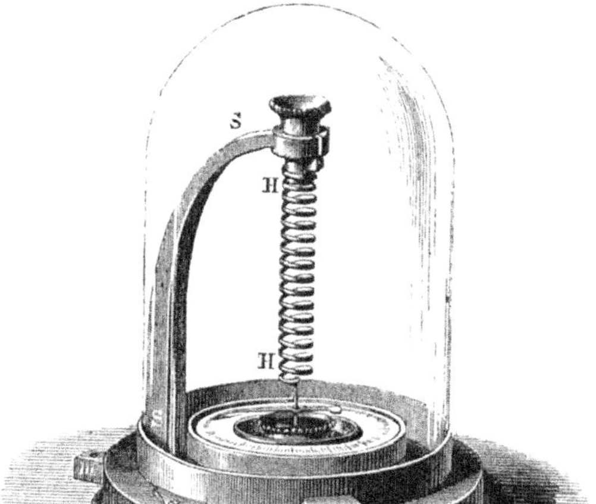

Fig. 120. Breguet's Thermometer.

Breguet's metallic thermometer (fig. 120) is an instrument depending on the different rates of expansion of different metals. A long compound spiral, HH, is formed of two or more metals with the most expansible one inside; usually a triple strip is employed with silver inmost, gold next, and platinum outside. As the temperature rises, the silver expands most and the platinum least, so that the spiral uncoils; conversely, on cooling the spiral twists up closer. The upper end of the spiral is fixed to a firm support, S; to the lower end a long index is attached, so that the temperature is read off by the position of the index on a circular scale, somewhat like the hand of a watch on the dial.

Expt. 314. To show that Liquids do not all Expand alike.—A series of bulbs of equal size blown on tubes of the same bore are arranged side by side (fig. 121) with the bulbs in a trough, so that they can all be heated simultaneously to the same extent by pouring warm water into the trough. The bulbs and tubes are first filled to convenient heights with different fluids (alcohol, water, benzene, oil of turpentine, ether, &c.), the levels being read off on scales fixed for the purpose. On heating the bulbs in the trough, the different liquids will be found to ascend in the tubes to very different extents. Thus, water expands much less than alcohol for a given rise of temperature, whilst benzene expands more than alcohol, and ether more still. Instead of purchasing bulbs and tubes, the apparatus may be readily constructed at home by selecting a number of small flasks all of as nearly as possible the same size (best ascertained by measuring the quantity of water that each will hold), and fitting them with perforated corks and pieces of quill tubing of the same bore for each.

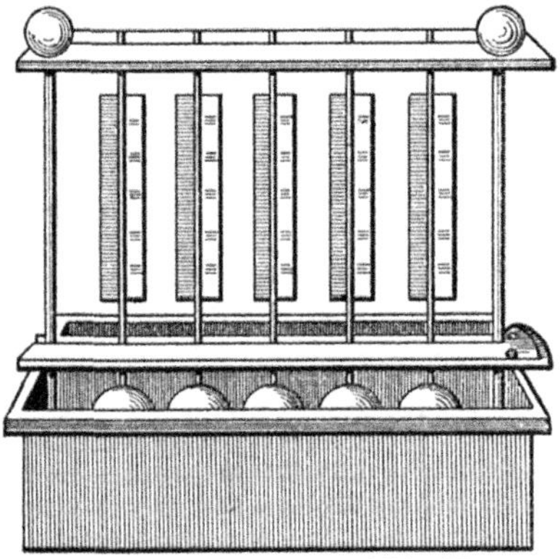

Fig. 121. Expansion of Liquids.

Expt. 315. Irregular Expansion of Water.—A remarkable peculiarity in the expansion by heat is exhibited by some few substances, and more especially by water and bismuth. If a block of ice be melted so as to form water at 0°, and this water be allowed to rise in temperature, it will be found *to shrink* in volume until the temperature reaches about 4° C., after which it begins to expand as the temperature rises. In consequence, a given quantity of water occupies the least possible space at 4°; or otherwise, a given space filled with water will hold a larger weight of water at 4° than at any other temperature. Accordingly 4° is spoken of as the *temperature of maximum density of water*.

A bucket of water in a cold room during frosty weather furnished with two thermometers, one indicating the temperature of the water at the bottom of the bucket and the other that at the top, will register different temperatures at top and bottom as the water cools. Suppose the water is at 10° to start with; as the cooling goes on the top thermometer will at first indicate a higher temperature than the bottom one, because the cooler water is heavier and tends to sink to the bottom; by and by the two thermometers will indicate the same temperature, 4°; after which the top thermometer will register a lower temperature than the other, because now the cooler water is higher and rises; finally, ice will form on the top. Thus, if readings of the two thermometers be taken every 5 minutes, numbers will be obtained something like the following:—

Time in Minutes.	Top Thermometer.	Bottom Thermometer.	Difference.
5 minutes.	9°·5	8°·5	+1°·0
10 ,,	9°·0	7°·0	+2°·0
15 ,,	8°·0	5°·5	+2°·5
20 ,,	7°·0	4°·5	+1°·5
25 ,,	5°·5	4°·25	+1°·25
30 ,,	4°·0	4°·0	0°
35 ,,	3°·0	3°·5	−0°·5
40 ,,	1°·5	3°·0	−1°·5
45 ,,	0°	2°·0	−2°·0
50 ,,	Ice at surface.	1°·0	−1°·0

Fig. 122 represents a form of arrangement for trying this experiment in an ordinary room; a large cylinder containing water has a sort of gallery arranged half way up, filled with fragments of ice, so as gradually to chill the water, when numbers will be obtained on taking readings of the two thermometers somewhat as above. It is probable that all substances that expand in solidifying, like water, exhibit the phenomenon of "maximum density;" but in most cases it is very difficult to prove that this is so, as the point of maximum density lies but little above the melting point.

Expt. 316. Expansion of Gases.—Fill a bladder nearly, but not quite, full of cold air with a pair of bellows, and then keep it in a warm place, *e.g.*, in front of the fire, but not so near as to burn and shrivel the membrane; the bladder will soon become plump and fully distended, and if not very strong may even burst from the increased pressure, due to the tendency of the air inside to expand by heat.

Fig. 123 represents an arrangement illustrating in another way the development of pressure by this action. A glass bulb with a long cylindrical stem (a piece of wide tubing attached to a flask will answer) is provided with a tightly fitting piston like a syringe; on heating the bulb the air expands and presses on the piston, forcing it outwards; on cooling the bulb again the air contracts, and the pressure is relieved, so that the atmospheric pressure outside forces the piston down again.

Motive power is obtained by means of "gas engines" in this way. A chamber containing a mixture of coal gas (or other combustible gas) and air is fired, so that much heat is developed by the chemical action taking place; this heat largely expands the products of combustion, producing pressure upon a piston working in a cylinder, and so causing it to move; by means of rods, cranks,

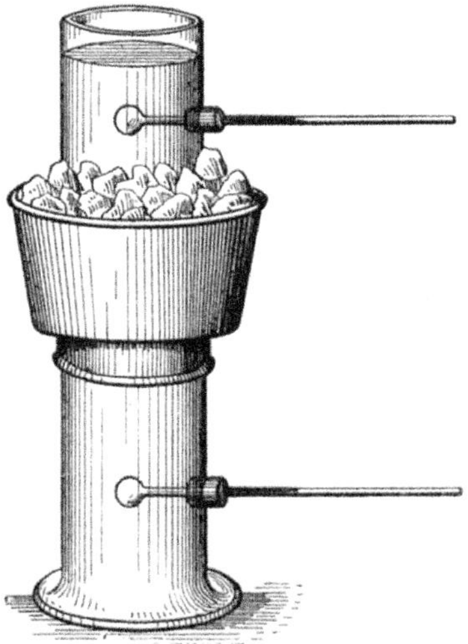

Fig. 122. Maximum Density of Water.

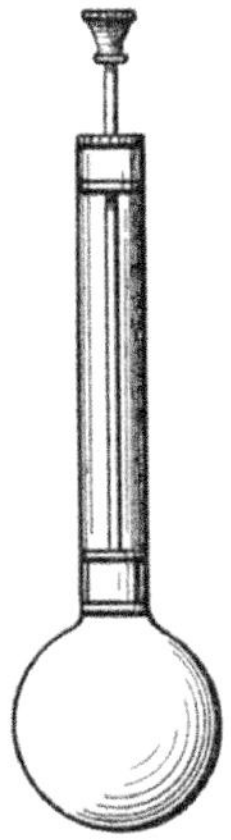

Fig. 123. Expansion of Air moves Piston.

shafts, &c., this motion is communicated to the machinery to be driven by the engine. A steam engine works after somewhat the same fashion, only here the power is produced by the expansion of steam produced in a pressure boiler.

Expt. 317. Air Thermometer.—Into the mouth of a 6-ounce flask fix a cork air-tight, with a piece of quill tubing passing through the cork, like the arrangement used in Expt. 311. Hold the tube and flask with the latter upwards, and place the lower end of the tube in a bottle half full of water. On warming the flask the air inside will become expanded and will partly bubble out through the water in the bottle (like the air from the diffusion apparatus,

Expt. 109, when increased in volume by the passage inwards of hydrogen). On allowing the flask to cool again, the atmospheric pressure will force the water up the tube, where it will stand at a height above that in the bottle, depending on the difference between the temperature to which the flask was heated and that to which it has subsequently cooled; as the temperature of the room alters, so will the level of the water rise or fall, somewhat like the quicksilver in an ordinary mercurial thermometer. Fig. 124 represents a more neatly made form of air thermometer, where the flask and quill tube are replaced by a large bulb blown on a piece of stout glass tubing, the whole being mounted in a flask containing some coloured liquid by passing the tube through a hole in a cork, taking care that this does *not* fit air tight into the flask. A scale for reading off variations in the level of the liquid in the stem is also provided.

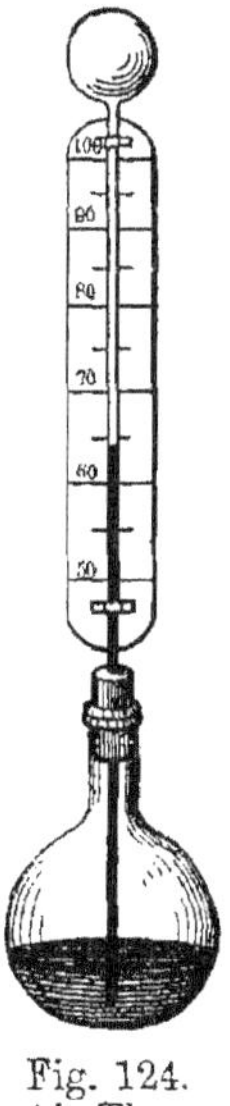

Fig. 124. Air Thermometer.

Expt. 318. Ascending Currents of Hot Air.—An important result follows from the fact that gases expand largely in bulk when heated; viz., that if they are not confined in an inexpansible envelope (such as a bladder when fully distended) they become lighter, bulk for bulk, than previously. A soap bubble blown with warm air from the lungs consequently ascends, notwithstanding that the soap water itself is heavier than air, and that the expired breath is also heavier than air would be at the same temperature, on account of its containing more carbon dioxide from the chemical changes going on in the lungs (Expt. 153): this extra weight is more than compensated for by the increased bulk, and consequently superior lightness, communicated to the breath in virtue of its warmth as compared with the surrounding air; consequently the bubble ascends, just as one filled with hydrogen or coal gas would do (Expt. 307), or as a cork held at the bottom of a tumbler of water would rise to the surface if let go. Hence, over any heated object an ascending current of warm air is necessarily set up, because the air in contact with the hot body becomes warmed and expands, and consequently rises, cooler air flowing in to supply its place.

Expt. 150 illustrates this effect; here the source of heat is the chemical action taking place between a lighted candle and the air in which it burns; ventilating appliances, where foul air is removed and fresh supplied, mostly work on this principle, the hot vitiated air being removed in virtue of its tendency to ascend and so set up a current of air. The ascent of smoke in a chimney is due to the same cause.

When a current of air strikes upon a surface inclined to its direction in a certain way, pressure is produced on the surface, tending to make it move; the turning round of the sails of a windmill, and of several children's toys which revolve when a gentle wind blows upon them, is thus brought about. In the same way, if the top of a chimney be fitted with a revolving head where the issuing smoke and hot air pass out through orifices radiating out from the centre and bounded by flat surfaces fixed obliquely to the direction in which the current issues, a continuous rotation of the head is set up; by connecting the revolving head with a spit in front of the fire an automatic turning round of the joint to be roasted is brought about.

Trace a spiral line on a piece of card or thin sheet metal and then cut through the card along the line with a sharp knife, &c.; pull the centre of the spiral outwards so that the whole presents a conical appearance, the surface of the card winding round and round screw-wise, but each successive turn being larger in diameter. Balance the central point of the "flat spiral" thus formed on a needle stuck point upwards in a cork in the neck of a wine bottle, or held by any convenient form of stand; on bringing a candle underneath the spiral, the ascending current of hot air will impinge on the surface of the spiral, producing the same effect as wind on the vanes of a windmill, setting the spiral in rotation. Toys based on this principle are often met with, continuous revolution being set up by fixing the toy near a fire over the edge of a mantelpiece, with a candle underneath, or in some other situation where an ascending current of warm air can play upon the vanes.

Expt. 319. Differential Air Thermometer.—Provide two small flasks of equal size and two tightly-fitting india-rubber perforated corks, with a piece of quill tubing about 2 feet long fitting air-tight into the corks when fixed in the mouths of the flasks. Bend the glass tube to a double right angle so as to form a large U, so that each limb may be about 8 inches long and the two limbs about 8 inches apart; when cool pour into the tube enough water (tinted with indigo or ink, &c.) to fill the limbs about 4 inches high with fluid. Now fix on the flasks at the two ends of the tube, so as to form a double-bulbed air thermometer. If one bulb be warmed by the hand or a flame, &c., the air therein will expand, and will press the column of coloured fluid in the U tube down on the warmed side and up on the other; so that any difference of temperature between the bulbs is indicated by a variation in the relative levels of the surface of the fluid in the two limbs. Fig.

125 represents a more neatly finished form of differential air thermometer where the two bulbs and connecting stem are blown in one piece.

A better instrument results if the corks are doubly perforated and the second perforations supplied with short bits of glass tubing about an inch long, the two being connected by a piece of thin india-rubber tubing; there is thus a direct communication between the two bulbs through the india-rubber tube, so that any inequality of level that might otherwise accidentally exist in the liquid in the U tube is obviated by the equalisation of pressure thus brought about. When the thermometer is to be used, the india-rubber tube is closed by means of a spring pinch cock or screw clamp (fig. 126), so that the direct communication between the bulbs is cut off. A further improvement is to use a longer glass tube and bend over the ends again at right angles, so that the flask bulbs may depend downwards, and so be capable of being placed in different vessels of water, &c. When the bulbs and tube are blown in one piece the connecting rubber tube and clamp are replaced by a cross piece of glass tubing provided with a tap, so that when the tap is open the two pendent bulbs are in direct communication, and when closed the thermometer is ready for use; this form is known as "Matthiessen's differential air-thermometer," the one with upright bulbs being "Leslie's" form of the instrument.

Fig. 125. Leslie's Differential Air Thermometer.

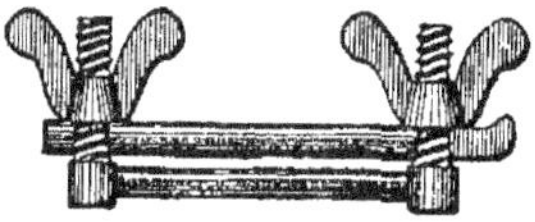

Fig. 126. Screw Pinch Cock.

CHAPTER XX.

Convection and Conduction of Heat.

Expt. 318 shows that when a gas is heated under such circumstances that it can expand and move about freely, it becomes lighter and ascends bodily, thus setting up currents. This process of **conveying** heat from one spot to another by causing a heated mobile fluid to move as a whole is termed **convection**, and is applicable to liquids as well as to solids. On the other hand, if a poker is put in the fire, by and bye its other end becomes more or less warm. The process by which the heat travels along the length of the poker from fire to handle end is termed **conduction**. Yet another way in which heat is propagated from spot to spot is by the process termed **radiation**, best exemplified by the heat of the sun, or by the effect produced by holding a red hot poker a foot or so from the face and *level with it*, so that no currents of hot air can reach the face by convection; the sensation of warmth on the skin and the effect produced on the eye by the emission of light are usually spoken of as manifestations of **radiant energy**. Some sources of radiant energy are also capable of setting up chemical action; as, for example, the effect of light in causing the combination of chlorine and hydrogen gases (Expt. 215), or in decomposing certain silver compounds (Expt. 196).

Expt. 320. Convection Currents in Water and Gases.—Obtain a large glass shade (or a flask or bolt-head) nearly full of water, and set it over a large gas burner or spirit lamp, so as to heat it up gradually; throw into the flask some bran or finely chopped up wooden shavings, when the formation of currents of heated water rising from the bottom to the top and of descending currents of cooler water will become visible by the motion of the particles of bran. Another way of rendering the currents visible is to throw into the flasks some lumps of solid litmus, fragments of the "lead" from an aniline dye pencil, or other material that will tint the water without dissolving too quickly; the flow of differently tinted streams will become easily manifest.

The ordinary processes in use for heating houses, churches, &c., by hot water pipes is based on this property of convection in liquids. A boiler or other heating arrangement with a fire attached is fixed in the vault or basement, and connected with the supply pipe for hot water at the top of the boiler; the return pipe, bringing back the cooler water after it has done its work and warmed up the air in contact with the coils of piping arranged for the purpose, joins the boiler at the bottom; this construction evidently necessitates that, as the water in the boiler becomes heated, it will rise upwards and flow through the supply pipe, its place being taken by the cooler fluid entering by the return pipe.

Instead of plain water, solutions of various saline matters are sometimes employed, these being capable of being more highly heated than water only.

When it is required to *cool* the air of a building, an analogous arrangement is adopted; a cooling engine is provided (usually involving the rapid evaporation of highly volatile liquids, such as liquefied ammonia or sulphur dioxide gas) in place of a boiler; the coils of pipe intended to cool the air of the apartment are filled with brine, or glycerine, or some other fluid that does not easily freeze, and a continuous current of cold fluid is thus kept circulating, flowing outwards and downwards from the base of the cooling vessel (which should now be placed at the highest elevation), and returning at a somewhat higher temperature to the top of the cooling vessel. When artificial ice is to be made, water is gradually frozen into blocks by cooling the vessels containing it below zero by means of pipes containing highly chilled brine. By placing a horizontal network of pipes on the floor of a room and sprinkling them with water, a crust of ice can be formed on the pipes; which, by cautiously repeating the addition of water, can be finally converted into a solid flooring of smooth ice, thus producing an artificial ice skating rink, such as is termed in various parts of the country a *glaciarium*. In the manufacture of paraffin wax from petroleum and shale distillates, &c., and in many analogous industries, chilling appliances of this description are largely employed to cause the solidification of the more valuable portions, which are then removed by pressing out the still fluid part by means of powerful hydraulic presses (Expts. 26, 331).

Anything that mechanically hinders the setting up of convection currents in the air surrounding a given body will prevent its rapidly altering in temperature if materially hotter or colder than the air. The naked skin is soon chilled down, even in a room with door and windows closed, so that there are no draughts or horizontal currents of air; partly because heat is lost by radiation, but principally because convection currents are set up all over the warm body, which continually bring fresh cool air in contact with the skin, and thus rapidly abstract heat; but a loosely fitting garment greatly retards this motion of the warmed air; whilst a series of garments one above the other, especially if wadded with wool or quilted with eider down, &c., or lined with fur, effectually prevent the formation of convection currents and consequent chilling action. Such clothing literally "keeps one warm" by preventing escape of heat; but it in no way produces heat, so that the phrase "warm clothing" is not exactly correct. For exactly the same reason that materials of this description will prevent a hot body from cooling quickly, they will equally prevent a cold body from becoming warmed by the air; thus to preserve a block of ice from melting rapidly in a hot room, or in summer, it suffices to envelop it in several folds of flannel or loosely wrapped newspaper, which prevents the access of warm air in this case, just as it would keep the human body from chilling in a cold atmosphere.

"Norwegian stoves," for keeping food hot a long time, are simply boxes lined with a thick coating of felt or similar porous material, which will thus prevent chilling. Ice chests are similarly coated outside with a double jacket, either containing air only, or better still, a layer of felt, dry sawdust, or similar material. Similarly, fireproof safes are protected by a corresponding jacket filled with sand, asbestos, mineral wool, or analogous material, capable of preventing the easy passage inwards of heat, whilst not being itself susceptible of charring.

Expt. 321. Conduction of Heat.—The property possessed by an iron poker thrust into the fire of becoming more or less heated at the handle end (even though this be screened from the radiant heat of the fire) by the conduction of heat along its length, is shared to a greater or less degree by all solids, whilst liquids and gases possess it only to an extremely minute extent. If a copper rod and another one of iron, both of the same dimensions (say 2 feet long), be placed with one end of each in the fire, it will be found that the far end of the copper rod will soon become too hot to grasp comfortably with the hand, whilst the iron rod will be much less affected, showing that the copper conducts heat better than iron. Fig. 127 represents another mode of illustrating this. Two similar bars of different metals (such as copper and iron) are fastened together, end to end, and a number of light wooden balls cemented by means of melted beeswax to the under portion of the compound bar at equal intervals (say 1 inch apart). The central part where the two ends meet is then heated by means of a lamp; heat is conducted along each metal, causing the wax to melt and the attached balls to drop off; the better conducting bar allows the heat to pass more rapidly and to a greater distance than the other; so that in the case of copper and iron the balls attached to the copper rod will drop sooner, and more of them will be ultimately detached than will be the case with the iron rod.

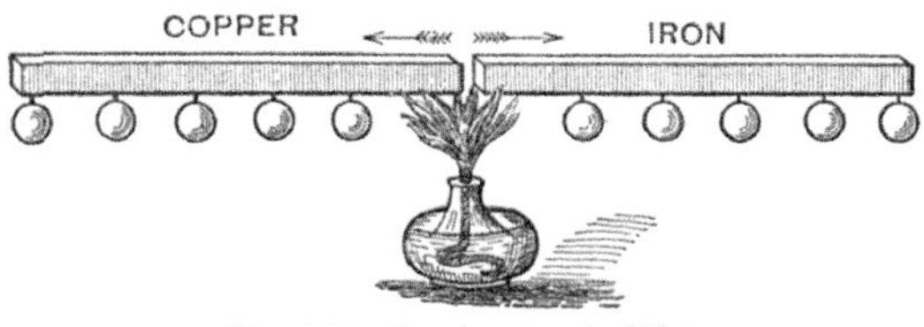

Fig. 127. Conduction (solids).

Expt. 322. Another Illustration.—Fig. 128 illustrates another way of showing the same thing. A box is provided (a cigar box made water tight with putty will do), in the side of which are bored a number of holes in the same horizontal line. In each hole is fixed a cork, through which passes a rod of metal, &c., a few inches in length. One rod may be of copper, another of brass, another of iron, another of glass, and so on, all being as nearly as possible of the same dimensions; each rod is so arranged that equal lengths always project outwards from the box, whilst an inch or more is inside it. The external portions of the rods are coated over with melted beeswax by a brush, and small balls of wood cemented on to the far ends by the same means; after cooling and standing to harden the wax, boiling water is poured into the box so as to cover the ends of the rods inside; this heats the rods and the heat is conducted along them outside, melting the wax, and finally causing the ball to drop off at the far end, if the rod is of sufficiently good conducting material. It will then be seen that copper conducts much better than iron, iron better than zinc or tin, and glass or slate very little; so that with these substances the heat travels far enough to melt only a very small quantity of wax.

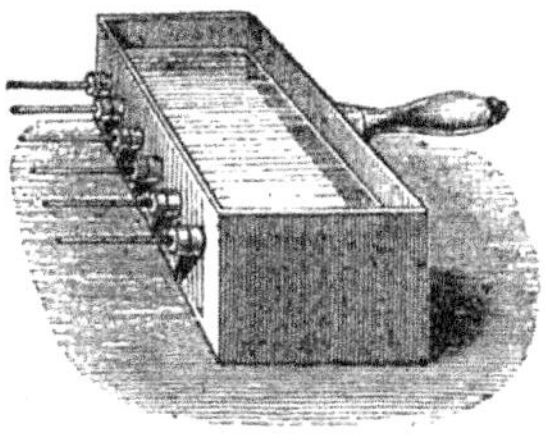

Fig. 128. Conduction of Heat (solids).

Owing to the low conducting power of glass, slate, &c., these substances may be rendered extremely hot by means of a flame at one spot, whilst only a very short distance away the material remains cool enough to hold in the hand without inconvenience. Thus, by the aid of a blow pipe a candle flame may be directed on the end of a piece of glass rod held horizontally so as to heat it red hot and even to melt it; whilst less than an inch away from the softened glass the rod may be readily held in the fingers.

On the other hand, iron, and especially copper and brass, conduct heat so readily that tools made of these metals require to be fitted with wooden handles if the tools are intended for use in a heated condition, so as to prevent the hand being burnt; *e.g.*, the soldering tool referred to in Expt. 47, or an ordinary pair of ladies' hair curling tongs. For a similar reason the flat irons used by laundresses for smoothing starched collars, &c., and kettles of boiling water, are grasped by means of a folded piece of cloth or similar "holder" to prevent the heat passing to the hand so rapidly as to scorch it; an extremely hot plate may be handled safely at table if the hand be protected by a glove or by catching hold of the plate with a napkin; whereas, if directly grasped, pain would be produced by the heat. The handles of metal teapots and coffeepots are frequently made separate from the body of the vessel, and united thereto with rings of ivory, or ebony, &c., between the ends of the handle and its attachment to the pot, so as to diminish the heating of the handle by the conduction from the hot pot, the wood or ivory being comparatively an extremely bad conductor.

It is somewhat remarkable that those bodies which conduct heat best (silver, copper, &c.,) are also those which conduct electricity the best, and *vice versâ*.

Expt. 323. Different conducting power of Wood and Metal.— The difference in conductivity of heat subsisting between wood and metal, especially copper, may be readily illustrated in a simple way. Provide a cylindrical rod of wood some inches long and $\frac{3}{4}$ to 1 inch in diameter (part of a broomstick will do). Slightly cut away the outer part of one half, and bend a piece of stout sheet copper over the remaining wood at the cut end, so as to cover the wood with copper, keeping the outside diameter about the same as at first (fig. 129). Strain a piece of white writing paper over the prepared rod, holding it so that one fold of paper covers both the wooden end and the other metal-coated half, and then place the rod and paper in the flame of a spirit lamp or Bunsen gas lamp, so that the flame may impinge on the paper at about the centre. In a few seconds that part of the paper in contact with the wood will be charred and set on fire, whilst the other part in contact with the metal will remain unaffected by the heat; the reason being that the metal conducts heat away rapidly, and consequently cools the paper and prevents its becoming heated to the charring point for some time, whilst the wood is incapable of thus protecting the paper, on account of its bad conducting power.

Fig. 129. Conduction Wood and Metal compound Bar.

Expt. 324. Effect of Wire Gauze in stopping the passage of Flame.—The conducting power of metals in cooling down combustible substances, and so extinguishing flame by preventing the temperature continuing high enough to propagate the chemical action of burning, is utilised in the construction of various forms of *miners' safety lamps* for the purpose of diminishing the liability to dangerous explosions of fire damp (Expt. 208) in collieries and underground workings. If a mixture of oxygen and hydrogen be made in a bladder, in the mouth of which is fixed a piece of $\frac{1}{4}$ inch metal gas pipe some few inches long, on applying a light to the end of the pipe the issuing explosive mixture will be fired, and the flame will pass through the pipe into the bladder, causing its contents to explode; just as a train of gunpowder would allow the flame of burning powder to pass along it from one end to the other, or as a piece of burning touch paper (Expt. 258) will allow the smouldering combustion to go on, gradually but regularly, till the whole is consumed.

If, however, the piece of touch paper be placed between two thick flat copper plates with a portion projecting, on lighting the part sticking out it will burn as usual; but the combustion will not penetrate far between the copper plates, the conducting power of these being great enough to cool down the burning paper so as to extinguish it.

In much the same way, by employing a long fine metal tube, instead of a comparatively short wide piece, in the mouth of a bladder of mixed oxygen and hydrogen, it is possible to prevent the flame passing back through the tube and firing the contents of the bladder; the sides of the tube act here as the copper plates with the touch paper, conducting heat away so rapidly that the flame cannot pass along, the chemical action of burning being stopped by the cooling. What is termed a *safety jet* based on this principle is sometimes used to produce an oxyhydrogen flame (Expt. 211), the two gases being mixed and made to pass through a tube filled full of pieces of wire, packed longitudinally in the tube as close as possible; the interstices between the wires allow the gases to pass to the jet at the end, where they are burnt together, but act as the long narrow tube with the bladder, preventing the flame from passing back by the cooling action exerted in virtue of conduction.

When the tubes are fine enough, a very short length suffices to enable this cooling action to be sufficiently rapid to prevent an ordinary candle or oil lamp flame from passing; so that a piece of fine wire gauze may be pressed down on the flame almost to the base without allowing the inflammable gases and vapours that pass through to take fire; or, conversely, if a gas jet be turned on, and a piece of wire gauze held over it, the gas that passes through the gauze may be lit (fig. 130), but the flame will not pass downwards through the wire meshes. This property of wire gauze is utilised in the construction of various forms of gas burner, where air and gas are mixed and burnt together so as to produce a blue

Fig. 130. Wire Gauze and Flame.

nonluminous flame, like that of a properly arranged Bunsen burner.

Expt. 325. The safety lamp.—Fig. 131 represents a form of safety lamp, based on much the same principle, consisting of a small oil lamp surmounted by a wire gauze cage. If the lamp be lighted and the cage fixed on, the lamp may be held in a mixture of coal gas and air (such as that issuing from a large unlighted Bunsen burner) without firing the mixture, although a blue flame is formed inside the cage if sufficient inflammable gas passes through the meshes to burn *inside.* In actual use underground, the presence of a dangerous amount of firedamp in the atmosphere is indicated by a peculiar "cap" or fiery appearance inside the lamp, due to this action; in such a case the lamp ought to be extinguished, as, if kept alight, the gauze is very apt to be heated red hot, when it might permit the flame to pass through, on account of the cooling action being no longer sufficiently exerted.

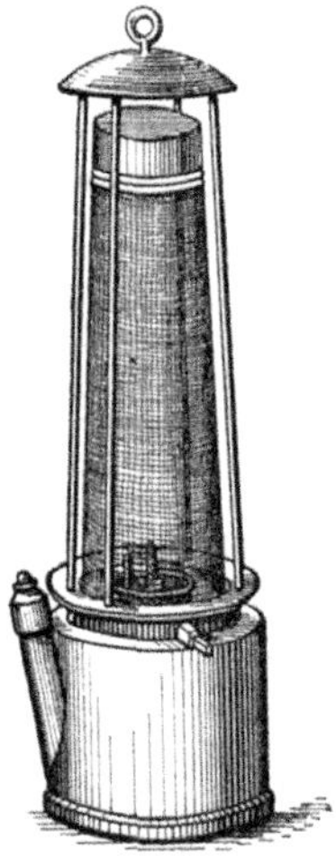

Fig. 131. Safety Lamp.

Any sudden shock or impulse given to the air (as by firing a blast, a sudden fall of rock, &c.) is liable to drive the flame of even the best safety lamps through the gauze for an instant, the cooling action being in such a case not sufficiently rapid to chill the flame down below the **point of ignition**, or temperature sufficient to cause the chemical action of burning to commence. Should this occur, of course any explosive atmosphere outside the lamp would be forthwith fired; hence, although safety lamps have proved a great boon to colliers by diminishing the risk of explosions, they are by no means perfect safeguards, especially when used carelessly and incautiously.

Expt. 326. Illustration of bad conducting Power of Water.—Water possesses only an extremely low power of conducting heat, although from its fluidity and mobility convection takes place very readily, not only with pure water, but also with all aqueous fluids and solutions, unless rendered extremely viscid (like treacle), by the presence of very large quantities of dissolved matter, Tie a piece of lead or brass to a small lump of ice, or put both into a small bag of cambric or netting, and place them in a large test-tube filled with ice cold water, the piece of metal being sufficiently heavy to cause both to sink to the bottom. Now heat the upper part of the test-tube by means of a lamp, holding the tube in a sloping position (fig. 132), and taking care that the flame of the lamp does *not* heat the bottom part; the water in the upper part of the tube will soon begin to boil, so that *ice cold water and solid ice are contained at the bottom of the tube, whilst the water at the top is boiling vigorously;* the heat of the boiling water is so slowly conducted downwards that the lump of ice will remain for some time unmelted. If, however, the bottom part of the test-tube be warmed instead of the top, the ice will rapidly melt, and will all disappear long before the water in the upper part of the tube becomes heated to any considerable extent.

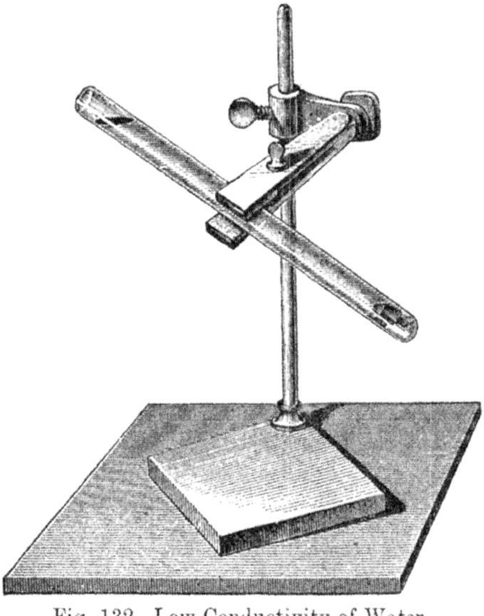

Fig. 132. Low Conductivity of Water.

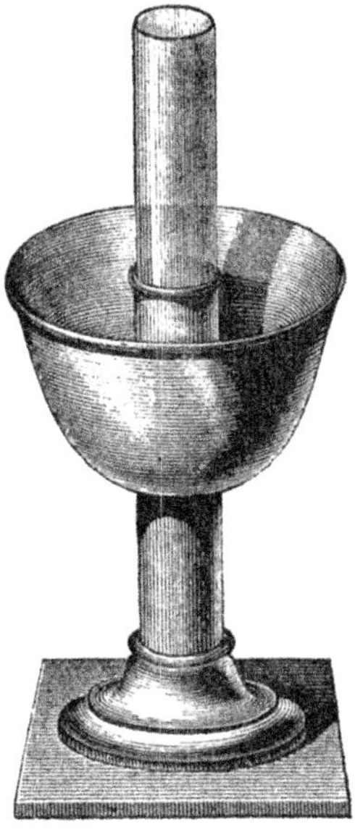

Fig. 133. Low Conductivity of Water.

Instead of applying a flame to the outside of the tube, a better plan is to place in the upper part of the test-tube a thin spiral wire of platinum, through which an electrical current can be passed so as to heat the wire considerably; in this way the upper portion of the water can be made to boil without the risk of heating the lower part of the tube by currents of warm air, &c.; so that if the bottom part be wrapped in dry flannel or cotton wool, &c., to keep off the heating action of the air of the apartment, a much longer time will elapse before the ice becomes melted. Fig. 133

illustrates another way of heating the top part of a vertical tube containing water by means of an outer cup, which is filled with hot oil.

Expt. 327. Another Illustration.—Provide a small differential air thermometer (Expt. 319), and weight it with lead so that it will sink in water; place it inside a large bell jar turned upside down and filled with water, the jar being so large that when the thermometer is turned on its side, with one bulb near the bottom of the jar and the other higher up, the upper bulb is still some inches under water. Preferably the differential thermometer may be constructed with a U shaped glass connecting tube, the legs of which are of unequal length. Now place a large red hot poker in the upper part of the jar so as to heat up the top layer of water. It will be seen, on repeating the application of the hot poker several times, that the upper layer of water will ultimately become very hot, far too hot to bear the hand in; but the heat will travel downwards by conduction so slowly that very little difference of temperature between the two bulbs will be evident, the index column of fluid moving but little, if at all. Obviously, if the heat travelled readily, the upper bulb would become much more rapidly warmed than the lower one, and a considerable difference of temperature would be indicated.

It is somewhat difficult to carry out experiments with gases in reference to their power of conducting heat, on account of their mobility and the ease with which convection currents (Expts. 318, 320) are set up in them. When these are prevented by employing loose highly porous substances (such as eider down, feathers, loosely piled cotton wool, &c.) the conducting power of the entire mass is found to be extremely low; so that, as the passage of what little heat does pass in such cases is partly due to the solid matter present, and partly to the not entire absence of convection currents, it is obvious that the conductivity of gases by themselves is very small.

Unless a current of air (wind, &c.) be set up, the abstraction of heat from a well-clothed person by even very cold air is but slow, so that intensely cold frosts and Arctic climates can be borne without any particular inconvenience when the air is still; but a much more moderate degree of cold is far less bearable if the air is in motion. At extremely low temperatures contact of the bare skin with substances of high conductivity, such as metals, produces local frostbite almost instantly, and corresponding injury to the animal tissues thus affected; although mere contact with the air or other nonconducting substances at the same temperature produces no such result (*vide* Expt. 328).

Certain remarkable results follow from the difference in power possessed by different substances as regards the passage of heat along them by conduction, one of which is that when heat passes through, or is evolved at or near the junction of the surfaces of two substances of different conducting power, a portion of the heat becomes transformed into electricity, whence the term *thermoelectricity* applied to this kind of action.

CHAPTER XXI.

Measurement of Heat—Specific and Latent Heat.

In the complete measurement of most natural kinds of manifestations made known to us by the senses, there are two distinct qualities, both of which are involved in the phenomenon examined, and both of which require to be measured before a complete knowledge of the matter is gained; thus, in the case of light and bodies rendered visible thereby, two different impressions are made on the eye, one of which is termed **colour**, and the other **brightness**, **brilliancy**, or **illumination**; thus a candle and a powerful oil lamp give out light of much the same colour in each case, but of very different degrees of brilliancy; whilst by holding various kinds of glass between the eye and the candle, &c., a long series of different colours may be obtained, not differing greatly in their relative brightness. In the same way, in reference to audible sounds and more especially musical notes, there are the two corresponding qualities, viz., the **Pitch** of the note (whether high or low, treble or bass) and the **Loudness** of the sound.* Similarly, in the consideration of a projectile force, such as a stone thrown by the hand, an arrow propelled by a bow, &c., there are the two separate ideas involved, viz., the **direction** in which the body is made to move, and the **rate** at which it moves, or the **velocity** imparted to it. In the case of the power or energy possessed by a moving body of water, such as a hydraulic engine supplied by a stream of water in a pipe, there are again two measurements requisite, viz., the degree of **pressure** exerted by the water, dependent on the height of the cistern or reservoir out of which

* In the case of musical notes, a third quality also comes into play, called **timbre** or *quality of sound*: the same note played on a violin, cornet, or clarionet, &c., has a different effect on the ear in each case, although the pitch may be the same throughout and the loudness not widely different.

it flows, and the actual **current** or quantity of water passing through the pipe in a given time. In like manner, the phenomena of heat, when considered to the fullest extent, involve an analogous pair of notions, viz., the degree of manifestation of heat termed **temperature**, to some extent correlative with the pressure of the water current; and the **quantity of heat** present, corresponding with the quantity of water flowing. Similarly, in relation to electric current energy, two analogous ideas are requisite, viz., that of **potential** or **tension**, and that of **quantity of electricity**.

Variations of temperature are indicated by instruments termed *thermoscopes*, and are usually measured by means of *thermometers* (such as the mercurial thermometer, Expt. 28), dependent for their indications on the alteration produced in the bulk of a given substance or object by its expansion and contraction as the temperature varies, the starting point being determined by some well marked natural change, always produced at one and the same temperature, such as the melting of ice or boiling of water under definite conditions. The **quantity of heat**, however, cannot be thus measured; for it is found that very different results are obtained with different substances under the same conditions; and therefore some one substance must be selected as a standard. Thus, suppose two precisely similar glass flasks to be heated in exactly the same way over a lamp or other source of heat; in one water is put, so that the effect of the passage of the heat into the water by conduction through the walls of the flask is to heat up the water, *i.e.*, to cause its temperature to rise. Suppose some other fluid, such as mercury, to be placed in the other flask and this to be heated in the same way; obviously the rate at which heat passes through the walls of the two flasks will be the same; but, nothwithstanding, it will be found that the mercury heats far more rapidly than the water if equal weights of the two be used; whilst if the quantities of water and mercury are so adjusted that they both rise in temperature at the same rate (say 5° per minute, or so) it will be found that the mercury weighs about 30 times as much as the water. This difference is expressed by saying that mercury has a much smaller **capacity for heat** than water; so that a given quantity of heat will produce some 30 times the effect in heating up mercury that it will when made to heat up an equal weight of water.

Expt. 328. To show that the Human Body will not always distinguish Temperature accurately.—At first sight it would seem that nothing is more simple than to distinguish heat from cold by the sensation produced on the body; but in point of fact such is not the case. In the depth of winter in the Arctic regions very cold metallic substances produce the same effect on the skin as hot ones; *i.e.*, they produce pain and a "burn" or blister, on account of the destructive action of unduly low or high temperatures on the living tissues*; but even without going to such extremes as these, it may be easily shown that the human body will have a sensation of warmth or coldness communicated to it by one and the same object according to the condition in which it has been placed previously. Arrange three basins on a table side by side; in the left hand one place water with some lumps of ice, so as to chill it as much as possible; in the right hand one put water as hot as the hand will bear without being scalded; in the middle one put a mixture of equal bulks of the hot and cold water, so that the temperature here may be intermediate. Now stand with the left hand in the left hand basin, and the right hand in the right hand basin; and after a minute or two, when each hand respectively has become somewhat accustomed to the water in which it is immersed, lift them out of the basins and place them both in the middle basin; the lukewarm water therein will feel quite cool to the right hand that has previously been in much hotter water; but to the left hand it will seem distinctly warm by contrast with the ice cold water in which it was previously immersed. With three buckets of water the feet and legs may be similarly used.

Thermometers and such like non-vitalised instruments, however, show no such irregularity; a thermometer placed in the hot water will sink on transference to the middle basin, whilst one previously immersed in the ice water will rise on being placed therein; but each of the two will register the same temperature when they have stood a while in the lukewarm water, which will not seem warm to the one and cold to the other as it does to the human nerves of sensation.

The following Table indicates some of the most remarkable centigrade temperatures connected with various phenomena, ranging from the lowest to the highest:—

Absolute zero (theoretical), ...	− 273° C.
Greatest recorded natural cold of Arctic regions, .	− 50°
Freezing point of mercury, ...	− 40°
Fahrenheit's zero (temperature of snow and salt), .	− 17°·8
Melting point of ice, ...	0°
Point of maximum density of water (Expt. 315), .	+ 4°
Blood heat, ...	36° to 37°
Boiling point of water at normal pressure, . .	100°
Do. mercury, ...	350°
Do. sulphur, ...	420°

* "The parching wind burns frore (*i.e.*, freezingly), and cold performs the effect of fire."—Milton's *Paradise Lost.*

Red heat just visible in diffused daylight,	500° to 550°
Full red heat (cherry red),	700° to 800°
Melting point of silver,	near 1000°
Intensely white heat,	1300° to 1500°
Melting point of platinum,	near 2000°

Expt. 329. To show that Bodies differ in their Capacities for Heat.—Into three similar basins or dishes put exactly equal quantities of water measured in a graduated vessel; say half a pint = 10 ounces in each; if the water have been taken out of the same jug, obviously each one of the three quantities will have the same temperature to begin with, which equality of temperature may be readily proved by means of a thermometer.

Provide a lump of *lead* weighing say 4 ounces, and tie a string to it, so that it can be lowered into a saucepan full of boiling water; when it has got hot (after some three or four minutes), lift it out by the string and quickly plunge it into one of the basins of water; stir up this water for two or three minutes, and take its temperature by means of the thermometer; of course the water will have become a little hotter through plunging the hot lead into it; if the original temperature were, say 15° C., the temperature after the hot lead was plunged in will rise to something like 16°, or a little higher. Now repeat the experiment, using instead of the lead a lump of *tin* of exactly the same weight as the lead, and plunging the hot tin into the second basin; the thermometer in this case will rise considerably more, say from 15° to about 17°, showing that the tin at the temperature of boiling water parts with more heat in cooling down than does the lead; whence evidently more heat would be required to heat up the tin than the lead, *i.e.*, the tin has a greater capacity for heat than the lead—roughly speaking, about twice as much—the rise in temperature of the water in the basin being nearly twice as great with the tin as with the lead.

Into the third basin plunge similarly a lump of *iron* of the same weight as the lead or tin previously used, and similarly heated in boiling water; the rise in temperature of the water due to the heat parted with by the hot metal in cooling will now be greater still, amounting to about 4°; showing that the capacity for heat of iron is nearly twice as great as that of tin, or four times as great as that of lead.

If a piece of *aluminium* of the same weight as before were used, the rise in temperature of the water would be again nearly doubled, since the capacity for heat of aluminium is nearly double that of iron, quadruple that of tin, and about 8 times that of lead; whilst, if instead of a piece of hot metal an equal weight of *boiling water* were employed, the rise in temperature would be still greater, the capacity for heat of water being greater than that of almost all other known substances.

In measuring quantities of heat, what is taken as the "unit of measurement" (as a foot, yard, or mile is taken as unit for measures of length; a pint, gallon, or cubic yard, &c., for measures of bulk; a second, hour, or year, &c., for measures of time; and so on) is *the quantity of heat which suffices to raise the temperature of a unit of weight of water* (*e.g.*, 1 lb. of water) *through the range of* 1°.* Hence the capacity for heat of a given substance as compared with that of water may be represented as being the increase in temperature that would be produced in a given weight of water by the quantity of heat that would raise the temperature of the same weight of substance 1°. This value is spoken of as the **specific heat of the substance.** For instance, the quantity of heat that would raise the temperature of a given weight of copper or zinc by exactly 1° C., would only raise the temperature of the same weight of water about one tenth of a degree; so that the capacity for heat of copper or zinc is only about one tenth of the capacity for heat of water; and the **specific heats** of these metals are consequently expressed by decimal fractions close to 0·1.

Expt. 330. Latent Heat of Fluidity.—Weigh up a pound of crushed ice and put it in a basin, and into another similar basin put a pound ($\frac{4}{5}$ of a pint) of water chilled down to 0° C. by means of pieces of ice, but not containing any pieces of actual ice floating about in it. Into each basin pour one pound of hot water, measured or weighed out in a flask, and heated over a lamp until it is just at the temperature 80° C. Mix up the water thoroughly by stirring in each case, and then take the temperature with a thermometer; in the first case it will be found that the addition of the hot water will suffice to melt all the ice after sufficient stirring; but the temperature of the resulting 2 lbs. of water (one added as such, the other produced by the melting of the ice) will be very little above 0°; whilst in the second case the temperature of the resulting 2 lbs. of water will be about 40° C. Taking a pound as the unit of weight, 1 lb. of water at 80° will contain 80 units of heat more than the same water at 0°; none of this is lost by simply mixing hot and cold water together, since 2 lbs. of water at 40° will still contain $2 \times 40 = 80$ units of heat more than the same water at 0°; but when ice is melted by hot water, the

* Strictly speaking, the amount of heat requisite to raise the temperature of a given weight of water from 0° to 1° is not quite the same as that requisite to raise the temperature from 10° to 11° or from 50° to 51°, and so on; but the difference is very slight.

heat disappears entirely; the 80 units of heat contained in the original hot water are said in this case to become **latent**, whilst 1 lb. of ice is melted; or otherwise, the quantity of heat rendered latent during the fusion of ice (**latent heat of fusion**) is said to be 80 units.

The reason why snow and salt and other mixtures of saline matters with ice or snow (Expts. 21, *et seq.*) become chilled, is simply that by virtue of their chemical or physical action on one another the ice becomes melted and dissolves the solid, forming an aqueous solution; heat is necessarily rendered latent during the fusion of the ice, and consequently the temperature of the liquid falls greatly.

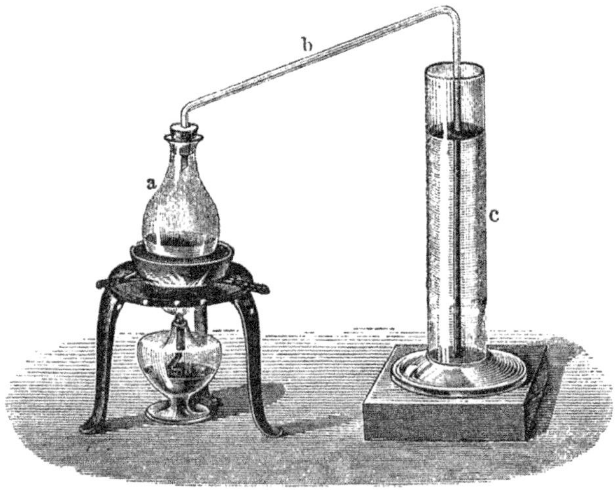

Fig. 134. Latent Heat of Steam.

Expt. 331. Latent Heat of Vaporisation.—Into the mouth of a flask (*a*) fix airtight a perforated cork with a piece of quill glass tubing (*b*) bent as shown in fig. 134, so that if water be boiled in the flask the issuing steam can be blown into a cylinder (*c*) containing water, whilst any water mechanically spirted upwards will run back into the flask by the sloping tube (*b*). Measure into the cylinder any convenient quantity (say $\frac{1}{2}$ pint = 10 ounces) of water chilled down to 0° by pieces of ice, but not containing any solid lumps of ice. Now blow a current of steam into the ice cold water by boiling water vigorously in the flask, and keep the ice water constantly stirred with a thermometer; you will observe that the temperature continually rises until the thermometer marks 100°, after which the steam simply bubbles through the hot water as a current of air would do, without becoming condensed as it did at first. When this occurs, remove the cylinder and weigh or measure the quantity of hot water now contained in it (more conveniently, cork it up and wait till cool enough to measure without danger of being scalded). It will be found that the original 10 ounces of water has become increased to something like $11\frac{6}{7}$ ounces, the extra $1\frac{6}{7}$ ounces having been produced by the condensation of the steam. It thus results that whilst $1\frac{6}{7}$ ounces of steam have become condensed to water at 100°, a large quantity of heat previously **latent** in the steam has become **sensible**, producing the effect of raising the temperature of 10 ounces of water from 0° to 100°; *i.e.*, taking now 1 ounce as unit of weight, $10 \times 100 = 1000$ units of heat have been rendered sensible, or have passed out of the latent state during the conversion of $1\frac{6}{7}$ ounces of steam at 100° to water at 100°; so that during the condensation of 1 ounce of steam, $\frac{1000}{1\frac{6}{7}}$ = nearly 540 units of heat have become sensible.

Hence *the latent heat of steam is said to be* 540 *units;* that is, in order to convert water at 100° into steam at 100°, 540 units of heat must become latent. In precisely the same way heat is rendered latent during the conversion of all liquids into vapours or gases by volatilisation; the more quickly this operation is performed the greater will be the fall in temperature of an evaporating liquid, owing to the conversion of sensible into latent heat, just as with a freezing mixture. Thus in Expt. 25 the rapid evaporation of ether produces a sufficient degree of cold to freeze water and other substances; by the use of more volatile substances still (such as ammonia or sulphur dioxide gases reduced to the liquid state by cold and pressure) an extremely quick evaporation can be produced, more especially when the gas or vapour formed by the evaporating liquid is rapidly removed by a powerful pump; the modern forms of powerful freezing machines (Expt. 320) are produced by employing these or other analogous agents, waste being prevented by recondensing the evolved vapour by compression and cooling in another part of the apparatus, and using the recovered liquid over and over again.

§ 8. Radiant Action: Visible Light.

CHAPTER XXII.

EMISSION OF VISIBLE LIGHT—ILLUMINATION OF OBJECTS NOT SELF-LUMINOUS—COLOUR AND ABSORPTION—REFLECTION AND REFRACTION AT PLANE SURFACES.

The various sources of visible light known to us may conveniently be divided into two chief classes, viz., those of origin **outside the world**, and those of **terrestrial origin**. The first class includes the **heavenly bodies**, such as the sun, moon, planets, fixed stars, comets, and meteors, &c., the study of which in the main belongs to that branch of science known as **astronomy** rather than to general elementary physics, excepting in so far as the light from these sources obeys the same general laws as the artificial and natural lights producible on the earth. Of late years, by means of an instrument termed the *spectroscope*, much valuable information has been gained respecting the various heavenly bodies, especially when photography (Chapter XXV.) is applied to the indications of the spectroscope, so as to obtain permanent records, instead of having to trust to the comparatively fleeting results of eye observations; by the examination of the light reaching us from the sun, for example, and comparing it with artificial lights produced in various ways by employing chemical or electrical action with different kinds of elementary substances (*e.g.*, when elements such as metals are burnt in contact with oxygen, or when electric sparks are made to pass between pieces of metal) a variety of results are deducible, leading amongst other things to the conclusion that many forms of matter known to us here on earth as elements are also present in the sun; analogous remarks also apply to various of the fixed stars.

The terrestrial sources of light are again divisible into two kinds, viz., those **spontaneously produced in nature**, and those **developed by artificial means.** To the former class belong the light emitted in volcanic regions by "burning mountains" and natural highly heated substances, this source of light being simply due to the existence of sources or stores of heat sufficient to cause the heated bodies to become luminous, just as when a poker made "red hot" in a fire becomes visible in a dark room by virtue of the light emitted whilst hot. Electricity developed in nature also produces visible light under suitable conditions, *e.g.*, the **lightning flash** and the **aurora borealis**; whilst chemical action also occurs naturally under such conditions as to develop visible light; in some instances, in consequence of great heat being produced (*e.g.*, where natural petroleum or other combustible matter is fired); in others without the production of any notable amount of heat, *e.g.*, in the case of the glow-worm and "phosphorescent" animal and vegetable decaying matter, such as fish, certain kinds of wood, marshy emanations (Will of the wisp), &c., &c.

Of the artificial sources of light the two chief ones employed are, firstly, those where light is produced *by means of electricity*, itself generated either *by chemical action* (directly, as in the voltaic battery; indirectly, as when fuel is burnt and motive power then developed by means of a gas engine, steam engine, or other analogous appliance), or *by utilising natural sources of mechanical power*, such as the flow of water from a higher to a lower level, the tides, the pressure of the wind, &c.; the mechanical power being transformed into electricity by one or other of a variety of forms of a machine now generally known as a "dynamo." Secondly, those produced by the agency of *heat developed by chemical action*, ordinary candle flames, oil and gas lamps, furnaces and fires, the lime light (Expt. 211), and fireworks, &c., being all cases of this kind. Sparks produced by **percussion** (such as the **flint and steel mill**, where a feeble light is developed by the continual striking together of pieces of steel and fragments of flint, &c.), and **phosphorescent action** produced by friction, &c. (as when quartz pebbles are rubbed together or loaf sugar broken up in the dark, or on heating fragments of the mineral termed *fluor spar* on account of its use as a flux in metallurgical operations—*vide* also Expt. 271), are also processes by which visible light is developed without the occurrence of chemical changes. Certain bodies possess the property of absorbing and storing up light, so to speak, so that when exposed for a short time to a tolerably bright light, and then removed to the dark, they glow and become luminous, emitting again the light absorbed during the previous exposure. Many articles are now sold prepared with a "luminous paint" of this kind, consisting of a solid phosphorescent body (usually a preparation containing *sulphide of calcium*) ground to powder and mixed with oil, so that the articles will glow at night in virtue of having been exposed to light during the day.

Certain kinds of radiant energy which are capable of exerting energetic chemical action (Chapter XXV.) are not perceptible to the ordinary human eye; but various substances have the power, when exposed to this kind of influence, of absorbing (so to speak) the invisible light, and emitting it again in a form perceptible by the eye; so that a solution of *sulphate of quinine*, glass containing the somewhat rare metal *uranium*, a decoction of *horsechestnut bark*, and many other substances, will apparently become luminous and emit light when placed in the track of a beam of invisible light of this kind. Here an alteration obviously takes place in the quality of the light, that emitted being not identical in quality with that absorbed; to this property the term **fluorescence** is applied, it having been first observed with certain kinds of fluor spar.

Whatever may be the source of light, whether terrestrial or extra-terrestrial, whether natural or artificial, certain points are common to all lights, whilst certain differences exist in particular cases. The ordinary eye distinguishes two qualities in reference to light, viz., **brilliancy** or **illuminating power** and **colour** (*vide* Chapter XXI). The light of *the sun*, especially in cloudless tropical climates, the intensely bright electric *arc light*, and in a somewhat lesser degree the light emitted by *intensely heated quicklime* (lime light, Expt. 211) and other substances, and by *burning phosphorus* (Expt. 234), &c., represent sources of light of the highest degree of brilliancy. When none of these or other less brilliant sources of light are active, ordinary objects are invisible to the eye; but when illuminated in presence of a source of more or less bright white light, such objects become visible, and in many cases show the phenomena of difference in colour; thus the grass appears *green*, the poppy flower *red*, the sunflower *yellow*, the violet a tint approaching to *blue*, and so on. Objects thus become visible because they act on the light illuminating them somewhat after the fashion of sulphate of quinine, &c., upon invisible light, in so far as they *absorb the light falling upon them and re-emit it*, but not altogether in the same way; as a rule, they do *not* emit light of a different quality from that falling upon them, the differences in colour, &c., resulting from the fact that ordinary bright white light is, so to speak, a conglomeration of all kinds of colours, of which a given object illuminated therewith *can re-emit one or more kinds more easily than the others.*

Thus the grass appears *green* in sunlight, because sunlight contains rays of all kinds of colours, which when mixed together produce upon the eye that effect called "whiteness" (Expt. 350); whilst of these different kinds of rays those producing the effect termed "greenness" are re-emitted by the grass much more freely than the others. The poppy appears *red* for the analogous reason that this flower has the power of re-emitting the rays producing the sensation termed "redness" more freely than the others; and similarly in all other cases. If the grass be illuminated by light containing no green rays, it will not appear green any longer, but dingy grey or even darker; similarly, if the poppy be illuminated by light containing no red rays, it will no longer appear red; and so on in other cases. The reason for the peculiar appearances seen in Expt. 245, where the light used is that of spirits of wine tinted by a salt of sodium (such as common salt, or chloride of sodium), is that the light employed is purely *yellow*, so that objects which in white light would appear of other colours cannot show these colours when exposed to the sodium spirit flame, because rays of those colours are not contained in the light for them to absorb and re-emit.

The light emitted from the surface of a given object, not self-luminous but illuminated from other sources, is spoken of as **scattered light**, being propagated equally in all directions, so that wherever a beholder may be situated this scattered light will equally meet his eye, so as to enable him to perceive the object in question; if the object possesses the power of scattering all kinds of coloured rays to an equal extent, it will appear *white* when illuminated by white light; or, if illuminated by any particular coloured light, it will appear of that same colour; the larger the proportion of the light originally falling upon or **incident** on the body from the source of illumination is thus scattered the brighter will the object appear, the various effects of light and shade (apart from colour) in producing whiteness, greyness, and blackness of different objects simply depending upon the relative powers of thus scattering incident light possessed by the objects respectively. Thus, a sheet of white note paper, a grey felt hat, and a black cloth coat may be all three equally illuminated, but they scatter the incident light to very different extents; the paper scatters largely, the felt hat but little, and the cloth coat practically not at all.

Unless light meet with substances exhibiting differences in their chemical and physical nature or conditions whilst passing onwards, no scattering takes place. A chamber filled with air or any other gas or vapour absolutely free from all suspended solid particles, such as motes, or liquid vesicles (minute fog or water spray, &c.), appears perfectly dark when beams of the strongest light are passed through it; if, however, liquid or solid particles be present, these more or less absorb and scatter the light, so that the track of the beam of light is more or less illuminated and rendered visible by this action, which thus serves as a physical test of the purity of a given atmosphere as regards particles of suspended matter, such as ordinary dust and germs of plant life, disease germs, and similar organised matters.

Besides the light thus scattered, in virtue of which objects generally become visible in daylight or when artificially illuminated, certain kinds of objects affect incident light in other ways; thus a flat polished silver surface **reflects** the light, the light thus reflected being different from scattered light in that it is *propagated only in one direction*, dependent solely on a certain geometrical consideration embodied in the **law of reflection** (Expt. 332). A pane of windowglass, on the other hand, not only reflects some of the light incident upon its surface, in addition to scattering a part thereof, but also allows a considerable portion to pass inwards into the glass from the outside exposed to the illuminating influence, of which portion a considerable fraction emerges again at the other side of the pane; so that part of the original light is scattered, part reflected, and part *absorbed in the glass*, whilst the remainder *passes through*.

Bodies which thus allow light to pass through are said to be **transparent**. If the different coloured rays of incident white light are all equally absorbed in passing through, the transmitted light is also white, but of diminished brilliancy in proportion to the amount of the original light lost by scattering, reflection, and absorption conjointly. If, on the other hand, certain rays are absorbed more readily than others, the light that passes through is defective in these rays as compared with white light, and consequently produces a sensation of colour, the character of which depends upon the nature of those rays that pass through most readily. Thus, a piece of glass stained dark yellow (such as is used for photographic purposes) absorbs all rays producing blue and violet coloured effects (and also, and more particularly the chemical or actinic invisible rays—Expt. 349), so that the light that does pass through is deprived of these rays, and produces on the eye much the same effect as the light reflected or scattered from a yellow object. On the other hand, glass tinted deep blue violet with cobalt, stops yellow and green rays readily, but allows blue and violet rays, also the actinic ones, to pass pretty readily. Alum, on the other hand, allows light of all colours to pass through equally readily, and hence does not appear of any particular colour; but it exerts a most powerful absorbent action on in visible rays of low refrangibility, or heat rays (Expt. 377).

When light passes into a transparent body, or **optical medium**, from another medium outside, it generally suffers an alteration in the direction in which it is propagated, somewhat after the fashion of reflected light, but in accordance with a different law, known as the **law of refraction** (Expt. 333), or alteration of direction in passing into one transparent medium from another. The law of reflection does not involve the nature of the reflecting body, but only the relation between the position in space of its surface and the direction of the incident light; the law of refraction, on the other hand, involves the nature of both media, the direction taken by the refracted ray being influenced by each of them, and being dependent on certain differences existing between the power possessed by each of transmitting light, as well as on the relation between the surface of the body on which the light is incident and the direction in which the light falls.

Expt. 332. To Illustrate the Law of Reflection.—Place a flat reflecting surface (such as an ordinary lookingglass, or preferably a polished plate of white metal, such as silver, or a vessel of clean mercury (Expt. 281) upon a table, and arrange a candle (fig. 135) in such a way that the **reflected image** of the flame can be seen in the bright mirror; it will be found that, no matter how the candle (A) may be placed in reference to the mirror (B) in order to see the reflection, the eye must be placed at some point (E), which is determined by the following geometrical considerations.

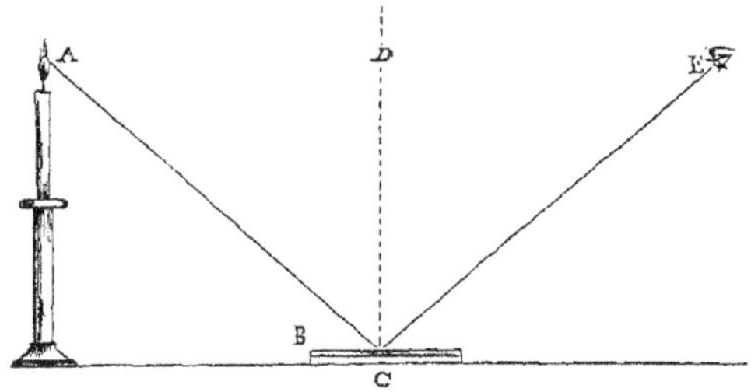

Fig. 135. Reflection of Light.

1. The line AC (or *incident ray*), the line CE (*reflected ray* meeting the eye at E), and the line CD, drawn perpendicular to the surface of the mirror (termed the *normal at the point of incidence* are all in one plane.

2. The incident ray and the reflected ray lie on opposite sides of the normal at the point of incidence, and are so situated that each ray makes an equal angle with the normal. Or, in other words, the angle ACD is equal to the angle ECD, which is expressed by saying that *the angle of incidence is equal to the angle of reflection.*

This is equally true whatever may be the nature of the reflecting surface. A flat plate of glass will reflect the candle exactly in the same way as a plate of polished silver or a basin of mercury at rest, so far as direction is concerned; but in each case the brilliancy of the reflected light is different, *i.e.*, a different proportion of the incident light is reflected. With glass, when the light falls nearly perpendicularly, so that the angle of incidence is acute, very little is reflected, the majority being transmitted; but if the light fall almost glancingly along the surface, so that the angle of incidence is not far from a right angle, very little is transmitted and almost the whole reflected. The power of glass to reflect as well as to transmit light is utilised in the optical illusion commonly known as "Pepper's Ghost" (Expt. 340), and in Amici's Camera Lucida (Expt. 338).

Expt. 333. To Illustrate the Law of Refraction.—Place a coin, such as a penny or shilling, at the bottom of a shallow cup or ordinary basin, and then move away until the coin just disappears from view, being hidden by the rim of the cup or basin. Now let the vessel be filled with water by some one else, yourself remaining perfectly stationary; as soon as the surface of the water comes to rest, you will see that the coin is distinctly visible, so that the presence of the water in the cup enables you, as it were, *to see round a corner.* Fig. 136 represents the track of two of the

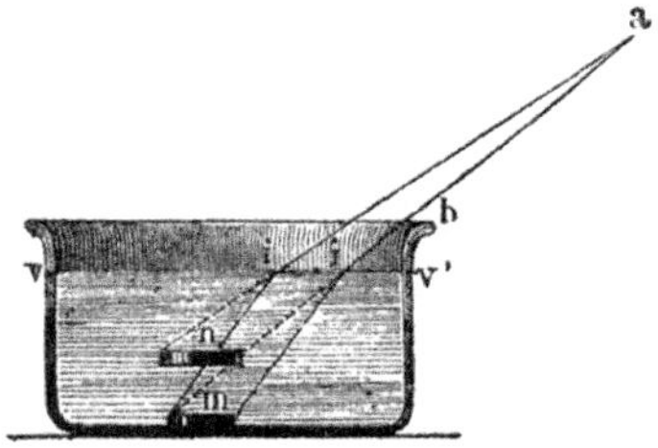

Fig. 136. Refraction of Light.

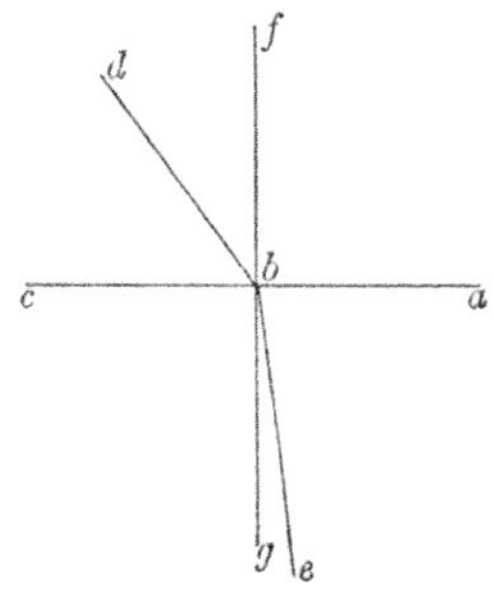

Fig. 137. Law of Refraction.

rays of light scattered from the surface of the coin which meet the eye and enable the coin to be seen; whilst traversing the water they pass in straight lines, *mi*; after passing out of the water at *ii*, they again travel in straight lines, *ia*, but the direction of these lines is not the same as before; at *ii*, where the rays pass from one medium (water) into another (air), they suffer "refraction" or deviation from the original paths, so that the directions before and after refraction are inclined to one another at an angle, instead of being one and the same line. Accordingly the coin is seen as though it were situated at *n*. The angle thus made depends on the nature of the media, always obeying a particular geometrical law, which may be thus described. Let the line *abc*, fig. 137, represent the common surface where the two media join, and let the light from a candle or other source of light, *d*, strike this surface at the point *b*, so as to pass after refraction along the line *be*; then the relative positions of the source of light *d*, the refracted ray *be*, and the **point of incidence**, *b*, are determined by the following considerations:—

1. As in the case of reflection, the incident ray, *db*, the refracted ray, *be*, and the normal at the point of incidence, *fbg*, lie in the same plane, the two former being respectively situated on opposite sides of the last.

2. The incident ray and the refracted ray are so situated with reference to the normal, that the sine * of the angle of incidence, *dbf*, bears to that of the angle of refraction, *ebg*, a constant ratio, no matter whether the angle of incidence be acute or obtuse.

The ratio of the sine of the angle of incidence to that of the angle of refraction is called the **refractive index** relatively to the two media; thus, when light passes from air into water the ratio is nearly $\frac{4}{3} = 1{\cdot}33$; when from air into glass, about $\frac{3}{2} = 1{\cdot}5$. Conversely, when light passes in the opposite direction, the values are the reciprocals of these, viz., from water into air about $\frac{3}{4} = {\cdot}75$ from glass into air about $\frac{2}{3} = {\cdot}67$. In general, when light passes from a rarer into a denser medium the refractive index is greater than unity, so that the refracted ray is bent **towards the normal**; whilst, if the light pass from a denser into a rarer medium, the refractive index is less than 1, and consequently the ray is bent **away from the normal.**

One result of this alteration of direction of light, on passing from one medium to another, is that any object under water viewed from

* By the *sine* of an angle is meant a trigonometrical value, depending only on the magnitude of the angle, and thus determined:—Let the angle be enclosed by the lines, AB, CB, fig. 138. From some point, D, on one of these lines let fall a perpendicular, DE, on the other line; then the ratio of the length, DE, to the length, DB, is called the sine of the angle, DBE. Calling this angle α, the relationship is usually written thus:—

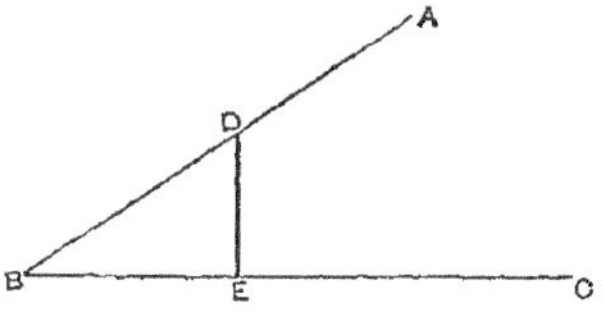

Fig. 138. Sine of an Angle.

$$\text{Sin } \alpha = \frac{DE}{BD}$$

In similar fashion, the ratio of the length BE to the length BD is called the *cosine* of the angle DBE; written thus:—

$$\text{Cos. } \alpha = \frac{BE}{BD}$$

the outside (above the water) *appears raised in position*: *e.g.*, the experiment with a coin in a cup of water just described. On looking somewhat obliquely into a pool of clean water, the stones, &c., at the bottom will be similarly affected, so that the water appears considerably less deep than it really is; a matter that should be borne in mind by persons who cannot swim before plunging into water, as it will generally happen that the water is deep enough to drown, if, when looked at from the bank, it appears to be only breast deep or shoulder deep when no allowance is made for refraction.

Expt. 334. Another Illustration.—A long straight stick or pole, thrust obliquely into clear water and viewed from above, will appear to be bent at the surface of the water to an angle, on account of the apparent raising of the submerged part. When, by reason of refraction, reflection, or otherwise, an object is apparently seen to be in a position different from its true one, in virtue of the rays that actually meet the eye passing thereto in directions different from those in which they were actually emitted, what is seen is termed **an image.*** Thus the appearance of the coin in the cup of water (fig. 136) is due to the **formation of an image** of the coin at **n**; similarly the apparently raised bottom of a pond viewed from above, and the seemingly bent stick partly immersed in water, are due to the formation of images of the stones, &c., and the stick.

Expt. 335. To illustrate the Absorption of Light.—Whenever light is reflected at the surface of, or passes through, a medium, more or less is always absorbed during reflection or passage; those media which absorb all kinds of light so greatly that practically none at all passes through are said to be **opaque.** Unless the thickness of the layer of such a medium traversed by the light is extremely small, so little passes through as to be invisible to the eye; thus a plate of gold or silver is absolutely opaque, even when no thicker than ordinary writing paper; but if beaten out extremely thin into gold leaf or silver leaf, it becomes **translucent,** allowing some light to pass. In the case of gold the light which passes is of a greenish tint, showing that the absorption of rays other than those producing the sensation of greenness is greatest. In the same kind of way, a *coloured transparent medium* (such as clear claret or port wine, aqueous sulphate of copper or indigo

* Certain kinds of lenses and curved mirrors (Expts. 363 and 366) are capable of forming images that can be received on a screen like the glass plate of a camera, or the strained sheet used for a magic lantern; these are spoken of as *real* images; images that cannot be so received on a screen (like the reflection in a plane mirror) are called *virtual* images.

solution, a ruby, emerald, or topaz, or a piece of tinted glass, &c.) is one which absorbs certain kinds of light more readily than others, but does not absorb any one kind to any great extent as compared with merely translucent or opaque substances.

If white light be made to pass through a sufficiently thick layer of a given coloured medium, it is possible to absorb practically all rays except those of some particular colour, depending on the nature of the medium; if the light thus treated be now passed through another differently coloured medium, it will often happen that the second medium rapidly absorbs those rays which were able to pass through the first one, so that but little of any kind, or even none at all, passes through the two media jointly. Thus, it is possible to select two pieces of coloured glass (say a bright emerald green and a pure red) which when put together *will almost entirely stop all light*, so that nothing can be distinctly seen through the two together, although most objects would be pretty readily visible through either separately. If certain shades of purple and deep orange be used together, the light passing through the two jointly will be *red;* the purple glass allows red and blue with intermediate shades to pass, absorbing chiefly yellow and green; the orange absorbs green and blue rays, but transmits mostly red and yellow ones, so that the two together filter out, as it were, all colours excepting red.

Certain media have the peculiar property of stopping all rays visible to the eye, but allowing other kinds of radiant energy to pass freely, or *vice versâ.* Thus, a strong solution of *iodine in bisulphide of carbon* has such a powerful absorbing action on all kinds of visible light that if a stratum of sufficient thickness be used nothing whatever perceptible to the eye passes through, but radiant heat (Chapter XXIV.) passes freely; on the other hand, yellow glass allows much visible light to pass through, but stops all actinic rays (Chapter XXV.), whence its use in photography. A plate of solid *alum* cut from a large crystal, or a trough of alum solution, allows all visible rays to pass through equally freely, so that white light emerges; but such a plate absorbs radiant heat very rapidly. Various kinds of ordinary white glass act in much the same way. On the other hand, a plate of clear *rock salt* allows radiant heat to pass just as readily as visible light, being equally transparent to all kinds of rays.

Expt. 336. Absorption during Reflection.—A beam of light impinging on a flat polished plate of silver is mostly reflected (Expt. 332), but little being scattered; whilst, if a piece of black cloth or a well-smoked fragment of glass or metal be substituted for the polished metal plate, scarcely any light at all is either reflected or scattered. In the latter case, practically all the incident light is absorbed by the cloth or lampblack without being re-emitted in the form of visible rays; even with the most highly polished reflecting surfaces some amount of absorption always takes place, the brilliancy of an object illuminated by reflected light being less than it would be if directly illuminated from the same source at such a distance as would cause the rays of light falling on the object to traverse equal amounts of space in each case. Silver and white metals generally absorb much the same amounts of all kinds of coloured rays, so that the reflected light is substantially white; though, even with such metals as polished aluminium, silver, tin, and zinc, slight but perceptible differences in whiteness are observable, aluminium and zinc giving a bluer shade, and tin a yellower one, than silver. Polished copper absorbs red and orange rays less readily than green and blue ones; so that when white light falls upon copper the reflected rays contain a larger proportion of red and orange than white light, and consequently appear coloured; and similarly with gold and other coloured metals. By causing the rays to be reflected several times in succession from two or more polished surfaces, the absorption of certain coloured rays is intensified and a much deeper colour is produced. Thus, a gold cup looked into obliquely appears of a much deeper orange red at the bottom than at the rim, on account of the greater absorption, owing to repeated reflection taking place with rays from the lower part. A polished sheet of copper rolled up into a short wide tube, and looked into obliquely, exhibits an analogous deepening in tint for the same reason.

Expt. 337. Total Internal Reflection.—A somewhat remarkable result is brought about when an incident ray is inclined at a particular angle to a refracting surface, when the ray tends to pass from a denser to a rarer medium. As shown in Expt. 333 the refracted ray will be bent away from the normal in a direction given by the law of sines. Now, suppose that the index of refraction relatively to the passage from the denser to the rarer medium is represented by the letter n, obviously $n<1$; if the angle of incidence be such that its sine is $\frac{1}{n}$, then the sine of the angle of refraction will be $n \times \frac{1}{n} = 1$; that is, the angle of refraction is a right angle, so that the emergent ray passes into the rarer medium at right angles to the normal at the point of incidence, that is, glances along the common surface of the two media. But now, suppose that the incident ray falls upon the surface at such an angle that the sine of the angle of incidence is *numerically greater* *than* $\frac{1}{n}$; it would result from the law of sines that if the ray could be refracted outwards under these circumstances, it would traverse a path such that the sine of the angle of refraction would be greater than $\frac{1}{n} \times n$, that is greater than 1. Since this is simply impossible from the definition of a sine (Expt. 333, footnote) it results that the ray *cannot pass out of the denser medium.* What happens is that it is entirely reflected (in accordance with the law of reflection) from the surface back again into the denser medium; whence the term **total internal reflection**, no absorption of light taking place with this kind of reflection which is not the case with an ordinary mirror, as above shown (Expt. 332).

The angle where $\frac{1}{n}$ represents the greatest possible value of the sine of the angle of incidence compatible with refraction outwards is termed the **critical angle**, and has a value of about 42° from glass to air, and 48½° from water to air. When swimming on one's back under water, with the eyes open, certain portions of the bottom of the bath and other submerged objects may be seen apparently suspended in the air, being *reflected from the upper surface of the water* in contact with the air to the eye of the observer below the surface; these objects are so situated that the angle between a vertical line (normal to the horizontal surface of the water) and the line drawn from any given submerged object to this normal at the point where it cuts the surface exceeds 48½°. light passing along this line cannot be refracted out of the water upwards, and is consequently reflected downwards again, thus meeting the eye under water, and giving the appearance of being derived from an object situated *above* the surface of the water.

The greater the value of the refractive index, the smaller is the critical angle; consequently highly refractive bodies possess to a great extent the power of reflecting light by internal reflection, and thus appearing extremely brilliant. This is especially the case with the *diamond*, which owes most of its lustre to its high refractive power; by cutting the surface into "facets," or flat planes slightly inclined to one another, the tendency to internal reflection of light from one or other of the facets is considerably increased, with corresponding increase in the brilliancy of an ornament, such as a necklace consisting of a number of stones so cut.

Fig. 139 represents a simple illustration of total internal reflection from the surface of water. A glass beaker, *b* (or, better still, a rectangular glass trough), is filled with water and placed on a convenient table or stand, with a candle or other source of light in front of it and a mirror, *a*, lying flat on the table, below the level of the surface of the water. A beam of light reflected from this mirror will pass along the line indicated; it will be refracted on entering the water at the first surface, totally reflected at the upper surface; and again refracted on passing out into the air at the other side.

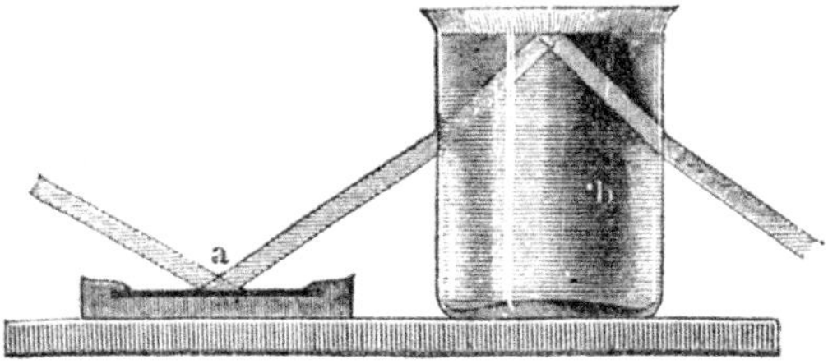

Fig. 139. Total Internal Reflection.

Fig. 140 represents an analogous case; the glass vessel is filled half full of water, and benzene poured on the top; these two liquids will not mix, whilst the refractive index of benzene is much greater than that of water; accordingly, a ray of light, striking the side of the beaker slantingly downwards, will be refracted, and then totally reflected from the common surface of the water and benzene, and finally refracted outwards again as indicated.

Expt. 338. The Camera Lucida.—The *camera lucida* is a small instrument intended for making drawings of various objects, so arranged that the eye sees simultaneously the paper and pencil point used for the drawing and the reflected image of the object, formed by one or more internal reflections from a suitably shaped mass of glass. Fig. 141 represents *Wollaston's camera lucida*,

Fig. 140. Total Internal Reflection.

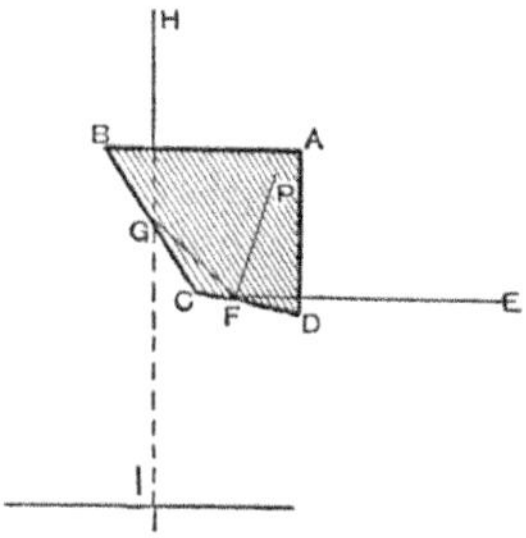

Fig. 141. Wollaston's Camera Lucida.

where the glass "prism" is so shaped that the angle A is a right angle, and the angle C is 135°, so that both the angles B and D are

67½°. A ray of light proceeding from any given object along the line EFG is totally reflected at F; because the angle of incidence, EFP, is 67½°, and therefore greater than the critical angle from glass to air (Expt. 337). For similar reasons the ray is again reflected internally at G, so that it emerges and meets the eye at H, which therefore sees the image formed at I on the paper where the drawing is to be made. A lens is usually arranged at H so as to enable the eye to distinguish clearly at once both the image and the pencil point. Fig. 142 represents *Amici's camera lucida*, consisting of a right-angled glass prism, ABC, and a plate of glass, DE, so arranged that the line AB is perpendicular to DE. The incident ray proceeding from an object at F enters the prism at G, is refracted to H, where it is internally reflected, emerges by refraction along IK, and at K is reflected from the glass plate to the eye at L; so that the image is seen at M, whilst the pencil is also perceived *through* the glass plate DE.

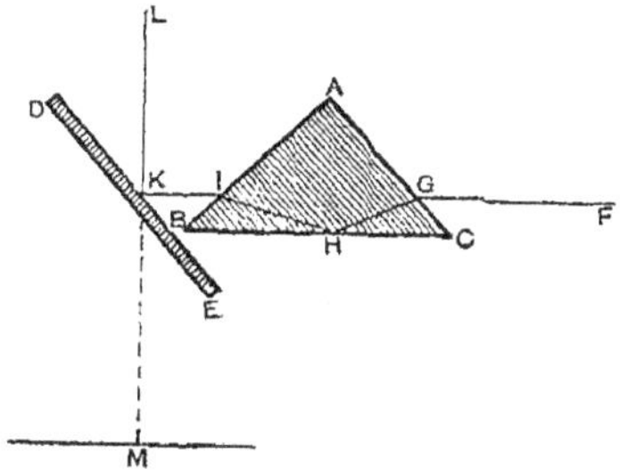

Fig. 142. Amici's Camera Lucida.

Expt. 339. The Mirage.—The *mirage* is a peculiar natural optical illusion noticed most frequently in hot sandy plains; the ground in the distance becomes invisible and in its place is seen, apparently, a lake of water; any objects standing up in the distance are usually seen inverted, as though reflected in the lake. Fig. 143 represents the course pursued by a ray of light from such

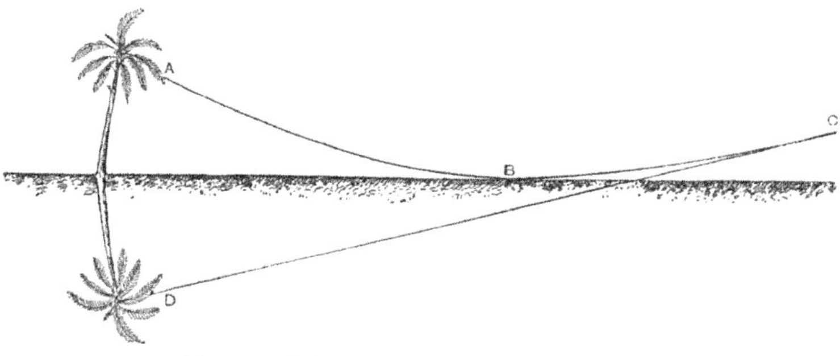

Fig. 143. Palm tree seen inverted by Mirage.

an object (a palm tree); the air in contact with the heated ground and the successive layers above it form a succession of media of gradually less and less refractive power the nearer the ground; hence the ray, ABC, becomes continually refracted to a direction more nearly horizontal, until at a point, B, it becomes totally internally reflected (Expt. 337); it then passes upwards, suffering successive refractions in the inverse order as it gradually traverses more and more refractive layers; and finally, it meets the eye of the observer at C, producing the same effect as though it had been emitted from the point D. An inverted image is thus seen, just as though reflected from water, the appearance of water itself being caused by the similar reflection of the sky.

It is difficult to reproduce this effect on a small scale, but it may sometimes be seen when a large plate of hot metal is allowed to cool, *e.g.*, when an armour plate for a warship is being rolled in a factory for the preparation of such articles, or when a blast furnace for smelting iron is "tapped," so that the molten metal runs out into a series of moulds in a large bed of sand, and there solidifies to "pigs" of cast iron; at a certain stage of the cooling of the whole "pig bed," the phenomenon of mirage is sometimes clearly discernible.

The contrary kind of refraction and reflection is sometimes observable at sea, inverted images of ships being formed in the air *above* the real object. In the case of the mirage of the desert, the lowest layers of air are the hottest, and therefore the least refractive, convection (Chapter XX.) not taking place sufficiently rapidly to interfere with this general disposition; hence the inversion takes place as represented in fig. 143; but in the sea mirage (fig. 144), the lower layers of air are the coolest, and, consequently, the bending of the rays takes place in the opposite direction, an ascending ray, ABC, being gradually bent towards the horizontal direction, then reflected at B, and passing downwards again with a gradually increasing slope, so that an observer at C, sees the image as though it were situated at D. Sometimes the coast line, or a vessel, &c., below the actual horizon becomes visible in an inverted position above the sea level owing to this action.

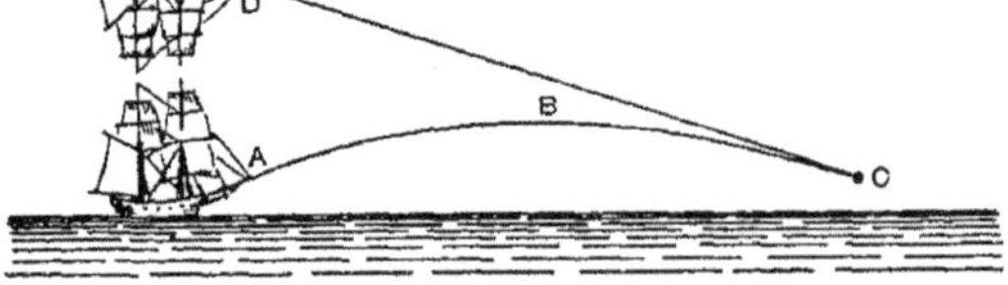

Fig. 144. Ship seen inverted by mirage.

In making astronomical observations, as also in the trigonometrical surveying of countries and similar operations, the effects of refraction, owing to the passage of light through layers of air of different densities, have to be allowed for.

Thus, when the sun apparently sinks below the horizon (say at a seaside spot with westerly sea view), the disappearance is delayed to a notable extent by refraction; after the sun has sunk to such an extent that no rays passing in a perfectly straight line could reach the observer's eye, the disc will still be visible above the horizon, the rays traversing a curved path, somewhat like that in the case of sea mirage (fig. 144), whilst passing through the lighter exterior layer of the atmosphere towards the denser layers at the sea level, the action here being similar in general effect to that illustrated in Expt. 333 with the coin and cup of water, whereby the observer is enabled to see round the corner, as it were, in virtue of refraction.

Expt. 340. Pepper's Ghost.—This beautiful illusion substantially depends on the same principle as that involved in *Amici's camera lucida* (Expt. 338), viz., that a plate of glass can act simultaneously as reflector and as a transparent medium, through which objects can be rendered visible by suitably illuminating them. As worked in public exhibitions a large sheet of glass is arranged at the front part of the stage, sloping forwards at an angle of about 45° towards the spectators. In front of this the stage is cut away (fig. 145), so as to form a kind of well, which is invisible to the audience; this well is lined with black cloth and furnished with powerful illuminating lamps, which can be turned up and down as required; similarly the scenes and actors on the back part

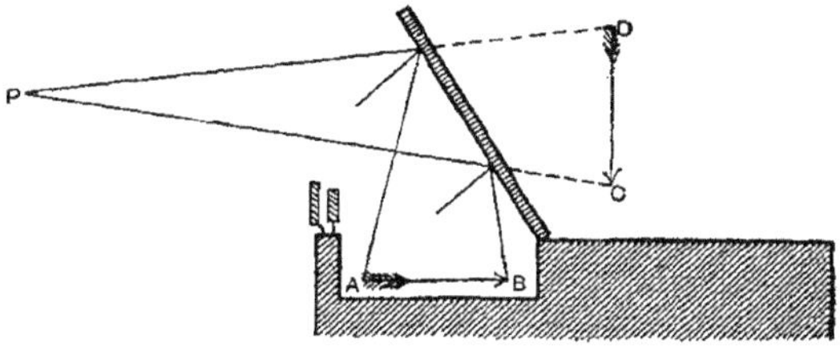

Fig. 145. Pepper's Ghost.

of the stage can be brightly illuminated or thrown practically into darkness by working another set of lights. When the lamps lighting the well are turned down, nothing in the well is visible to the audience, either by direct vision or by reflection; if the other lights are more or less turned up, the actors and scenes behind the enclosed glass plate are seen *through* the glass, precisely as in the case of an ordinary theatre, saving that with this latter there is no intermediate glass plate. When, on the other hand, the ordinary stage lights are turned down and those illuminating the well are turned up, the scenes behind the glass become invisible for want of light, whilst any object or actor, AB, placed horizontally in the well is seen by a spectator in front at P by reflection from the front surface of the inclined glass plate, apparently standing on the stage at CD.

As in all cases of reflection, the right and left hand become *inverted; i.e.*, if the real actor extend his *right* arm, the virtual image seen appears to move his *left* arm, and so on; so that to make the illusion complete, the actor whose image forms the "ghost" must be careful to do all actions with his left arm that would ordinarily be done with the right, *e.g.*, drawing a sword, fencing, &c. If the actor lie with his feet nearest the glass and his head away therefrom, the image will be erect as usual; but if he lie in the reverse direction, the ghost will be inverted and will apparently stand on its head. If in the former case the feet be at a suitable distance from the glass, the image will appear level with the stage flooring, but if further away from the glass the ghost will be proportionately elevated in mid air; so that by drawing the actor away from the glass (by allowing him to rest on a board, &c., covered with black cloth and furnished with castors), the image will apparently rise in the air.

Ordinarily, the lights are so arranged that both the back of the stage and the well are illuminated together, so that the spectators see the real actors *through* the glass and the spectral illusion by reflection from it simultaneously. By turning down suddenly the lights illuminating the well the spectre disappears, whilst the real actors remain visible.

A model "Pepper's Ghost" can be easily made at very little cost by fixing a large inclined pane of glass (such as an ordinary large window pane) in front of a miniature stage with the ordinary pasteboard characters, &c., and placing under the sloping glass a deep tray, painted dull black or lined with black cloth. One or more powerful dark lanterns must be provided, throwing a beam of light into the tray when the dark slide of the lantern is opened, but not otherwise visible to the spectators. The stage itself must be illuminated with candles, &c. (all the rest of the room being darkened) to such an extent that the "characters" *on* the stage can be distinctly seen, whilst at the same time the reflection of another character placed flat on the tray and illuminated by the lanterns can also be distinctly seen. With a little patience and ingenuity a very effective ghost illusion in miniature can be thus arranged. Scenes from "Faust," where Mephistopheles suddenly appears and disappears like a phantom, and such like incidents, can be thus rendered very prettily, care being taken to hide all the accessories from the spectators by means of curtains or screens so arranged that only the stage itself with the pane of glass in front is in view. The spectators must be placed as much as possible *in front* of the show, and not too much at the side; otherwise the reflection may not be properly visible to those at the extreme sides.

Expt. 341. The Bodyless Lady, or Talking Head.—Another popular illusion worked by reflection is one which is capable of being shown in many different ways, but generally as a head and bust suspended in mid air without any attached body, or apparently decapitated and lying in a dish, &c.; the head obviously belongs to a live person, as the eyes can move and the lips and mouth can talk, but no other part of the body is to be seen.

One way of showing this illusion is to fit up the stage (preferably in the form of a square cabinet, in the centre of which the "talking head" is to be seen) with a carpet of such a pattern as to present just the same appearance, whether viewed one way or the reverse, whether right side up or down, whether from left to right or *vice versâ*, some kind of tesselated device being usually chosen. The walls of the apartment are papered with some analogous kind of paper, the object being *to prevent the spectator from distinguishing whether he is looking at the real carpet or wall, or is only viewing a reflection thereof.* Two glass mirrors are then arranged vertically on the floor, one edge of one being in the right hand back corner of the square room, and one edge of the other being in the left hand back corner, the other edges meeting towards the centre of the room, so that the vertical surfaces of the mirrors are exactly at right angles to one another. Fig. 146 represents a plan of the room, AB and BC being the lines on which the mirrors stand. The mirrors are of such a height that the person whose head is to be seen can comfortably sit, stand, or recline (according to the position of the head required) *behind* the angle formed by the two mirrors, the head appearing above their upper edges, but the rest of the body being hidden. A wreath of flowers or some other kind of elegant border or setting is provided, so arranged as to form a sort of top to the angle of the mirrors. If the shape of the cabinet is properly proportioned to the pattern of the carpeting and wall papers, and these are properly applied and illuminated by hidden lights, an observer situated in front of the mirrors and with his eyes about level with the "phantom" head, will see the head, &c., above the top of the mirrors, just as if nothing intervened between his eyes and the rest of the body;

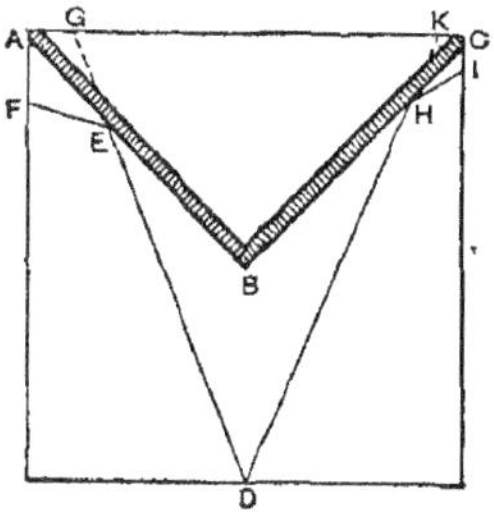

Fig. 146. Bodyless Lady.

but all below the neck, or bosom, or dish (according to the nature of the illusion exhibited) is hidden behind the mirrors; and if these reflect the pattern of the carpet and papering accurately in position, the observer apparently sees the carpet and paper, &c. *underneath the head*, the reflected images formed in the mirrors of the *front* corners and junction of carpet and paper, &c., being coincident in position with what would be the back corners and back wall, &c., of the room, were these not hidden by the mirrors. Thus an observer at D sees the rays emanating from F and I as though these points were situated *behind* the mirrors at G and K respectively, owing to the reflections at E and H.

As with Pepper's Ghost, this illusion can be readily arranged at

Fig. 147. Bodyless Lady.

home on a small scale (or even on a larger life sized one) with a little care and patience in getting the mirrors and patterned papers, &c., exactly into position. For a miniature representation, the flat plates of silvered glass contained in small framed mirrors or looking-glasses a few inches square will do, but proportionately larger and more expensive mirrors are requisite for life size arrangements.

Another way of working the illusion is as follows:—

A square mirror of suitable size has a portion cut away from its centre large enough to enable the head and neck of the real individual seen to pass through. The plate is then arranged in the cabinet, sloping up backwards at an angle of 45° (AB, fig. 147). The head is then passed up through the cavity, and a wreath, &c., arranged as in the previous experiment. Obviously, nothing but the head of the performer (or whatever portion projects upwards through the cavity) can be seen by a spectator at P in front; whilst, if the wall papering of the sides of the cabinet *above* the level of the edges of the mirror and of the *roof* of the same are properly arranged, a reflection of them will be seen in the inclined mirror, giving the impression that the flooring and walls are visible *below* the level of the head, and hence making it appear that the head is suspended in mid air without any body being attached. Thus the observer at P sees at E and F respectively the reflections of the points C and H, formed by the rays CD and HG being reflected at D and G.

Expt. 342. Wizard's Mirrors.—If two flat mirrors be arranged, suitably situated with reference to one another, a ray of light, emanating from some bright object in front of one may be reflected therefrom to the surface of the second, and thence reflected a second time to an observer's eye. Thus, if two mirrors, AB and BC, be arranged at right angles, as indicated by fig. 148, an observer at P will see the reflection of the object at EF as though it were situated at GH; whilst the actual object is hidden from him by the screen, K.

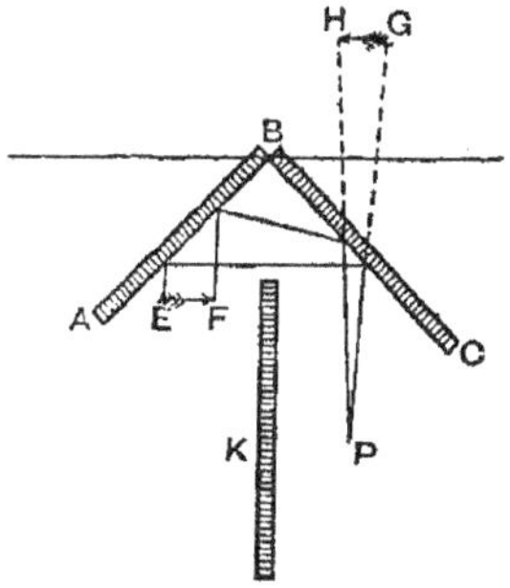

Fig. 148. Magic Mirrors.

A highly effective drawingroom illusion may be arranged after this fashion by means of two large looking glasses; preferably the one looked into should be supported on its side so as to swing on a vertical axis. By the aid of a screen and properly arranged lights,

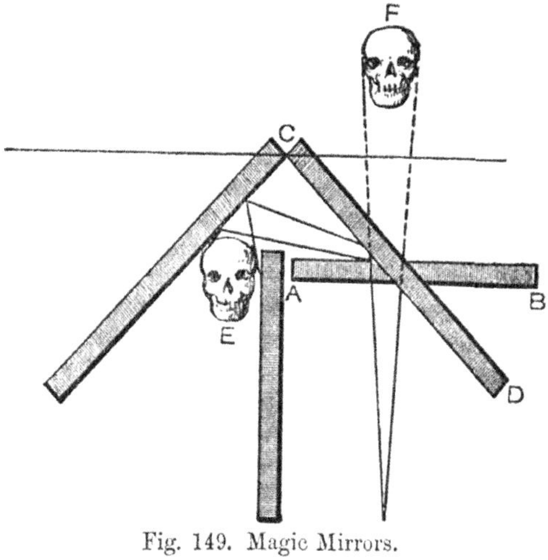

Fig. 149. Magic Mirrors.

the effect may be produced (fig. 149) that an observer, looking into the mirror, sees his own reflection at first, but suddenly this changes to a grinning skull or something equally horrifying. This is effected by suddenly changing the position of the mirror (by swinging it on its axis) from that directly fronting the observer, AB, to the inclined position marked by the dotted line CD, when the reflection, F, of the skull hidden behind the screen at E is seen instead of the observer's own face. The skull should be supported on a table, &c., draped with black cloth, so that nothing else can be seen, and should be well illuminated, the lights being so placed as not to be directly visible to the observer.

A somewhat similar illusion was formerly largely used in their calling by so-called necromancers and magicians, consisting of a flat mirror swinging on a vertical axis like a door, (*e.g.*, like the mirror CD, fig. 149), so as to give the reflection of the person looking into it, when placed in one position, and that of something else, otherwise hidden from view, when placed in another position. When the wizard was consulted and made use of the magic mirror to "reveal the future," the person to be impressed was first placed in front of the mirror in the ordinary position, and consequently saw his or her own reflection. His attention being momentarily called off by some appropriate "business" (as an actor would say) or by-play, such as extinguishing the lights in the room, the mirror was noiselessly placed in the second position, and the light illuminating the picture supposed to be supernaturally revealed gradually turned up; accordingly, the expectant spectator saw gradually developing in the mirror a vision representing the answer to his questions. By and by, when the full effect on his imagination was produced, the first mirror was restored to its ordinary position fronting him and the lights relit, when everything in the apartment would resume its usual appearance. A magic lantern was often employed to produce the vision, the picture being at first produced out of focus and "misty," but subsequently becoming clear and definite as the lenses were focussed, and finally fading away again and disappearing, the light being cut off by means of a dark slide.

Expt. 343. Formation of two Images by ordinary Looking Glasses.—A "looking glass" or flat mirror consists of a plate of glass, behind which is affixed a layer of reflecting substance, such as a film of metallic silver deposited by chemical processes (Expt. 198) or a sheet of tinfoil and mercury amalgam (Expt. 288); consequently there are *two* reflecting surfaces present in such a mirror, whereas there is only one in the case of a polished plate of metal or *speculum*.

In the ordinary way of using a looking glass the incident rays are not far from perpendicular upon the surface, so that the angle of incidence is very small; consequently the two reflections are nearly superposed; whilst, moreover, the one from the metal coating at the back is far brighter than the one from the front surface of the glass plate, so that only one image is noticed.

But it is otherwise when an object is looked at obliquely in such a mirror, more especially if the plate of silvered glass is tolerably thick. Thus, fig. 150 represents a candle flame seen obliquely in such a mirror; one ray, AB, is reflected from the outer surface of the glass at B and meets the eye at C, producing the same effect as though the candle had been situated at G; in other words, forming a **virtual image** at G. Another ray strikes the glass at D and is refracted along the line DE; at E it is reflected from the common surface of the metal film and the back portion of the glass, so as to traverse the line EF; at F it emerges again into the air and is refracted along FC, meeting the eye at C, and producing the same effect as though it had originated from the point H, *i.e.*, forming **a second virtual image** at H. Hence the eye perceives *two* reflections of the candle; the more obliquely the rays fall (*i.e.*, the greater the angle of incidence), the greater is the amount of light passing along the track ABC relatively to the amount passing along the other track, ADEFC. With a small angle of incidence the reflected image, G, is so faint compared with the other one, H, that it is practically imperceptible; but as the angle of incidence is increased by suitably moving the candle flame, and the position of the eye correspondingly varied so as to see the reflections, it will be noticed that the image G grows brighter and brighter relatively to H which gets fainter; by and by with considerably oblique rays G becomes the brighter of the two, and ultimately H becomes so faint as to disappear entirely (*vide* Expt. 345).

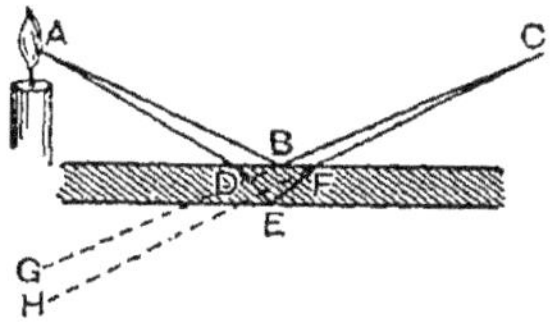

Fig. 150. Double Image of Candle Flame.

Expt. 344. The Kaleidoscope.—In Expt. 341, the effect of two flat vertical mirrors inclined to one another at right angles was obviously to reflect the walls of the cabinet in such a way as to produce virtual images *behind* the mirrors of objects really situated in front. The **kaleidoscope** is another application of the same principle, only in this case the mirrors are usually inclined to one another at a smaller angle than 90°, and the object to be viewed is placed *between the two reflecting surfaces* (fig. 151). The effect of this is that an eye properly situated in front receives a number of impressions arising from the formation of a number of virtual images which are arranged symmetrically around the line where the surfaces of the two mirrors intersect; so that if the mirrors are inclined at an angle of 60°, there will be 5 images visible besides the real object, all 6 being arranged like a 6-pointed star radiating out from a common central point. If the angle of the mirrors be 45°, then 7 images will be seen as well as the real object, forming an 8-fold star.* A variety of optical toys producing geometrical patterns on this principle are sold; one of the oldest forms of **kaleidoscope** (from the Greek signifying "rendering visible of beautiful forms") consists of a pasteboard or metal tube with two mirrors arranged at an angle of 60°, so that the vertical section of the whole is like fig. 152. Fig. 153 represents a horizontal section across the line *mn*. At the far end is a round glass plate, *a*, and over this a cap, *cc*, with a glass end, the space between the two glass plates containing beads, coloured bits of glass, &c. This cap

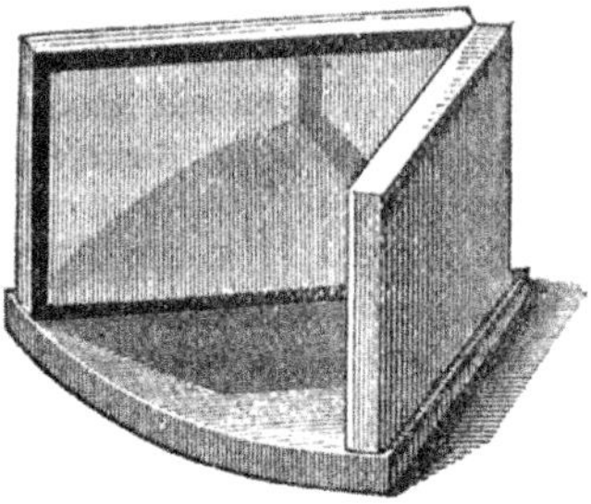

Fig. 151. Kaleidoscope.

Fig. 152. Kaleidoscope.

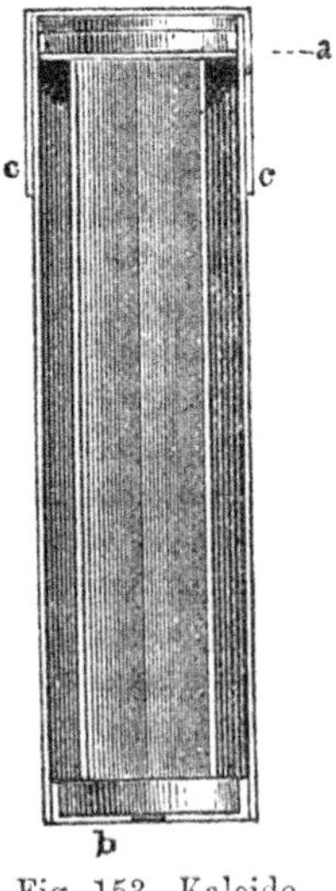

Fig. 153. Kaleidoscope.

* The rule guiding the number of images is that, if the angle between the mirrors is $\frac{1}{n}$ of 4 right angles, or $\frac{360°}{n}$, then an n-rayed star shape will be formed, one ray being the real object and the others virtual images. With an angle of 30° a 12-rayed star would consequently be formed, with 40° a 9-rayed star, and so on.

can be turned round, shaking up the beads, &c., as it revolves, so that they assume new positions with respect to one another. Every different position gives rise to an 8-fold (or a 6-fold) figure of extremely pleasing character when the bits of glass, &c., are brightly coloured and happen by chance to form an appropriate grouping, the figure being perceived on looking into the tube through the eyepiece *b*, at the end opposite to the cap. The best results are obtained when the tube is pointed slantingly downwards towards a candle flame, or a flat mirror laid on a table before a window, so that the bits of glass can be readily moved by rotating the cap, whilst a brightly illuminated field of view is observed.

Another form of the instrument contains three flat vertical mirrors inclined to one another at angles of 60°, so that the section of the three mirrors in position would be an equilateral triangle; a much brighter definition is thus obtained, as three reflecting surfaces are employed instead of two, and consequently more light rays actually meet the eye. In this case a 6-rayed figure results.

In another form of kaleidoscope, two mirrors are provided with an adjusting arrangement, so that the angle between them can be altered at will. Pieces of card with devices on them, or other objects, such as flowers, &c., are then placed between the mirrors and viewed by looking slightly downwards into the angle of the mirrors, when a 6-fold, 8-fold, 9-fold, 10-fold, 12-fold, &c., multiplication of the object will be visible star-fashion, according as the angle is 60°, 45°, 40°, 36°, 30°, &c. Arrangements on this principle are sometimes used for designing geometrically arranged patterns for wall papers, floor cloths, &c.

Expt. 345. The Endless Gallery.—When two mirrors are arranged exactly parallel to one another an observer situated between the two will see an apparently endless succession of images, formed by repeated reflections from the two mirrors alternately. If two large pierglasses or full length toilet mirrors be arranged perfectly vertical and parallel with one another, an observer standing between them with a lighted candle or lamp in his hand, held nearly level with his eye, will see an apparently interminable line of candles, extending far away into the distance until they become almost too faint to be visible. Any considerable degree of want of parallelism of the mirrors will greatly shorten the line of images formed. In Expt. 343, when a candle is looked at obliquely in a mirror consisting of a moderately thick plate of glass silvered behind, it will often happen that not only are two images seen (one by reflection from the front surface of the glass and the other from the silvered back surface), but in addition a succession of fainter images becomes visible, these being formed by a repetition of reflections from the two surfaces, just as in the endless gallery. If the mirror be so arranged that the two first reflections give images of about equal brightness, the next reflections will form pairs of images of successively decreasing brightness, because at each reflection (especially at the unsilvered surface) only a portion of the light is actually reflected, some being lost by absorption and some refracted outwards.

Expt. 346. How to see through a Brick Wall or a Board.—Procure 4 equal-sized pieces of flat silvered looking-glass, and arrange them as shown in fig. 154, at angles of 45°, with the corners of a square pasteboard or wooden tube shaped in 4 right angles, as shown. The front and back ends of the tube should be closed in, and a hole the size of a shilling cut in the closing in at *a* and *b*; opposite to them and behind the two upper mirrors should be cut two similar holes, *c* and *d*, so that an observer looking in at *a* apparently sees clear through all 4 holes in a line, so as to view objects beyond at *e*. Now place a brick, a hat, a piece of board or some similar opaque object, *f*, between *c* and *d*; the observer will still see whatever was in view just as well as before, thus apparently seeing straight through the opaque object. The image actually seen is really formed by a 4-fold reflection of such a nature that the rays ultimately reaching the eye practically come in the same direction as those originally emitted by the object viewed, the course of a given ray being along the line indicated.

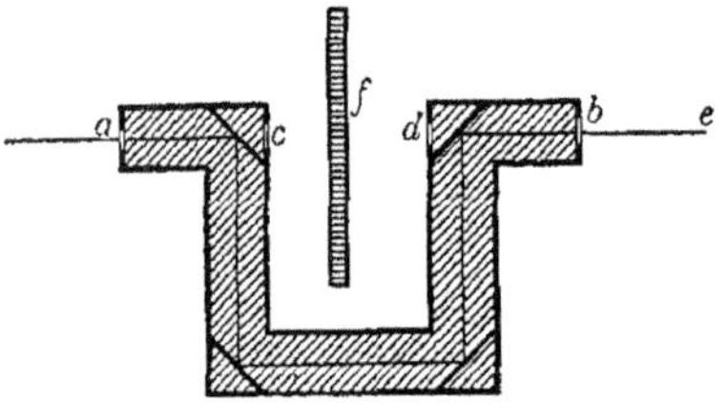

Fig. 154. To see through a Board.

Expt. 347. Alteration of Apparent Position by looking through a Medium different from air, with Parallel Faces.—It results from the circumstance that the path of a given ray is altered in one direction on passing from air *into* another medium and in the opposite direction in passing out again (Expt. 333), that on looking through a transparent medium with parallel faces (such as a pane of glass), the rays that meet the eye are usually parallel (or nearly so) with those that would have met the eye had nothing but air intervened, but actually produce the effect of causing the impression that the object seen is slightly displaced. If the light fall absolutely perpendicularly on the glass it will go onwards in exactly the same line through the glass, and subsequently on emerging again into the air; but if it fall at all obliquely, the path on emerging will only be parallel, and not identical with the original direction.

Arrange two pieces of straight metal tubing about 3 inches long and $\frac{1}{4}$ inch bore (or two pieces of quill glass tubing with brown paper pasted outside), holding them by stands and clamps in such a way that whilst the axes of the tubes are parallel to one another and in the same vertical plane they are not in the same straight line (fig. 155). Behind one of the tubes place a sheet of cardboard, *ab*, with a hole perforated in it just opposite the end of the tube at *a*, and behind this hole place a candle, *c*. If now you attempt to look through the other tube, placing the eye at *d*, you will not be able to see the candle; any rays of light passing through the hole in the pasteboard screen and then traversing the first tube will obviously be unable to pass through the second tube so as to reach the eye. Now place between the two tubes a thick piece of glass with parallel sides (a piece of plate glass, &c.) in the position indicated, *ef*. If the thickness and inclination of the glass plate be properly adjusted, as also the relative positions of the two tubes, it will now be possible to see through the whole combination, a ray of light from beyond the screen passing through the orifice in the screen and the front tube, and being then made to deviate to a parallel path by the refractive action of the glass plate, so as to pass through the second tube along the line *aghd*.

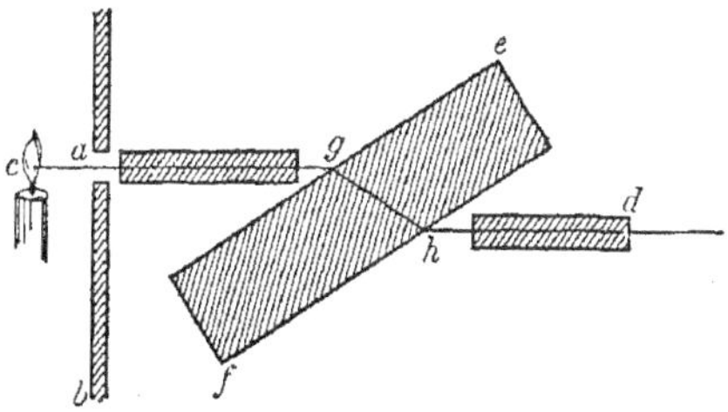

Fig. 155. Effect of transparent plate.

Expt. 348. Alteration of Apparent Position by looking through a Medium different from Air, with Faces not Parallel.—If, instead of a glass plate, a beam of light be made to traverse a **prism** of glass (*i.e.*, a mass of glass bounded by plane surfaces not parallel), the direction in which the light emerges will not only be displaced as compared with the original path, but, further, it will be no longer parallel with the former direction. Fig. 156 represents the path of a ray of light, *ab*, passing through a prism; at the first surface the light is refracted along *bc*, towards

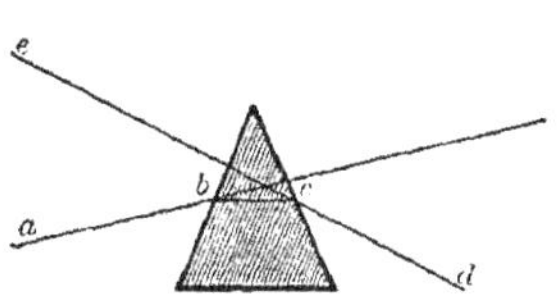

Fig. 156. Refraction through Prism.

the base of the prism, in accordance with the law of sines (Expt. 333); on emerging it is again refracted, but not into a line parallel with *ab*, the path taken being along *cd*; so that an observer whose eye catches the ray, perceives an image of the object from which the ray was emitted at *e*, the object being apparently deflected towards the edge of the prism, the extent of the deflection varying with the **refracting angle** of the prism, *i.e.*, the angle between the two inclined plane surfaces.

Expt. 349. Production of a Spectrum.—In the previous experiment, if the object viewed be illuminated with one kind of light only (*e.g.*, that emitted from the flame of spirits of wine and salt, Expt. 245), all that will be noticeable on viewing it through the prism will be displacement, as stated; but if sunlight or any other white light be employed, the image will be *fringed with colours.* This arises from the fact that differently coloured lights are differently refracted by one and the same medium, or more exactly, that *radiant energy generally consists of emanations of different degrees of refrangibility;* that is, bent out of the original path to different extents by one and the same refracting medium. Those rays that are least refracted are invisible to the human eye, but possess to a large extent heating qualities, and are accordingly often spoken of as **invisible heat rays** or **radiant heat** (Chapter XXIV.); of those rays that are visible to the eye, the least refrangible produce the sensation termed *red* colour, and those which are most refrangible the sensation termed *violet colour*, intermediate between which lie a number of other colours corresponding with intermediate degrees of refrangibility, the order being:—*

Red, . . .	Least refrangible.
Orange, . .	Intermediate.
Yellow, . .	
Green, . .	
Blue, . . .	
Indigo, . .	
Violet, . .	Most refrangible.

Beyond the violet rays come a series of others more refrangible still, but, like the heat rays, invisible to the human eye (not necessarily so to other animals); these rays possess, to the greatest extent of any, the power of setting up chemical action (Chapter XXV.), and of exciting fluorescence (*supra*), and are therefore generally designated the **actinic** or **ultra-violet** rays. Both the

* Taking the colours in the reverse direction, the order of succession may be conveniently remembered by means of the nonsense-word *vibgyor*, composed of the initial letters of the names of the tints.

visible rays and the less refrangible so-called "heat rays" can, however, produce some degree of chemical action; whilst the visible rays and the ultra-violet ones are by no means destitute of heating power; so that the division into heat rays, visible light, and actinic rays, although convenient, is not absolutely exact.

Let a room fronting the sun be closed by a shutter in which a small hole is perforated, so that a beam of sunlight can be admitted through the orifice. Hold a glass prism base upwards in the track of the beam of light (fig. 157), and at some feet (or even yards) distant hold vertically a sheet of white cardboard or other convenient white flat surface, so as to receive the refracted rays.

It will now be noticed that, instead of a white spot of light being produced, as would be the case were the white screen held in the beam of light *before* it meets the prism, a **band of colours** or **spectrum** will be developed in the order above enumerated.

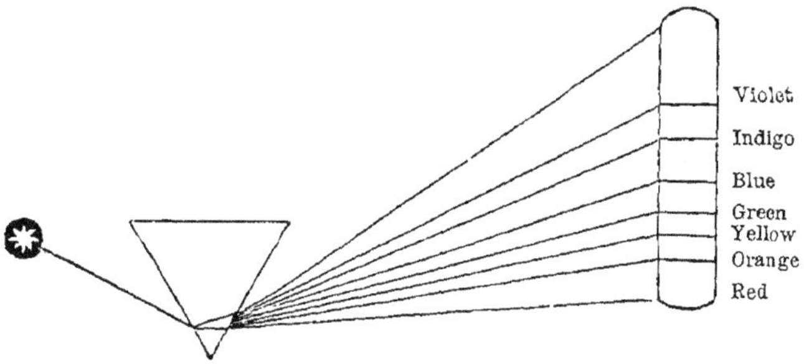

Fig. 157. Spectrum.

The lower end of the band will be red, being due to the least refrangible of the visible rays; whilst the upper end, obviously that due to the rays most deflected from the original path, will be violet. By holding the bulb of a thermometer in the different colours successively, and in the spaces just beyond the red and violet ends of the spectrum respectively, it can be shown that the greatest heating effect is produced by the invisible rays of slightly less refrangibility than the red; whilst by similarly holding a test-tube containing a solution of *sulphate of quinine* in the same different positions, the production of a pale blue light emitted from the liquid by fluorescence (Pa. 290) will be seen to take place to the greatest extent by the agency of the invisible ultra-violet rays. Similarly, *photographic paper* (*e.g.*, that prepared with silver chloride, Expt. 392) will be most acted upon by these same ultra-violet rays, and hardly at all, even after long exposure, by the less refrangible rays, especially the red, orange, and yellow visible rays, and the still less refrangible heat rays (*vide* also Chapter XXIV.).

Expt. 350. Combination of Colours.—The preceding experiment shows that sunlight produces on the eye a total effect due to the combined action of all the prismatic colours, and this combined effect is the sensation called **whiteness.** It should thence result that if all these separate colours could be recombined in proper proportions, white light should result. That this is so can be readily shown by breaking up a ray of sunlight by means of a prism, so as to give the prismatic colours as in the last experiment, and then viewing the spectrum through a precisely similar prism in an inverted position. By suitably adjusting the position of the screen and the two prisms, it can be shown that, whereas all the prismatic colours are in view when the screen is looked at directly, when viewed through the second prism they become all blended together and produce the sensation of whiteness. The same result may be produced by bringing all the rays to a focus by means of a concave mirror (Expt. 361) or convex lens (Expt. 365).

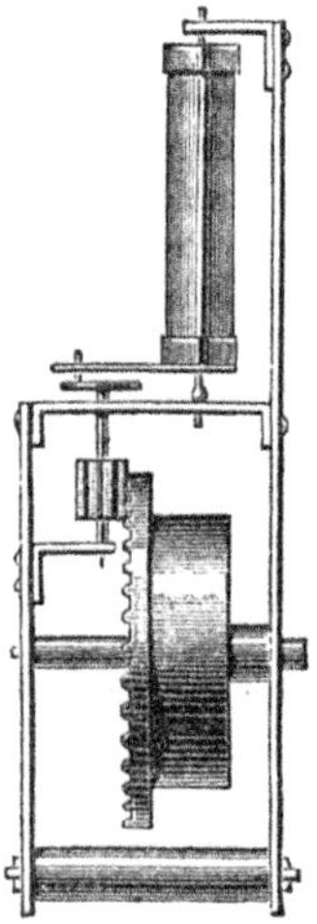

Fig. 158. Recomposition of White Light.

Another way of showing the same effect is to mount the prism on an axle, capable of rapidly revolving in such a way that each colour is thrown in succession on the same portion of a screen as the prism moves (fig. 158); in this way a band of white light becomes visible, all the tints being blended together. The combination of all the prismatic tints to white may also be shown by arranging a series of small mirrors instead of a screen to receive the spectrum, each mirror being so adjusted as to reflect the light falling upon it to the same spot on a screen properly placed; this spot will then appear white. But if any of the mirrors be removed or shaded by an opaque screen, a coloured spot will be produced, inasmuch as all the tints of the spectrum are not present, and consequently the effect on the eye is different from that of pure whiteness (*vide* Expt. 354).

Expt. 351. Newton's Colour Disc.—The effect of a prism in breaking up white light into colours was first studied by Sir Isaac Newton, to whom is also due a simple experiment illustrating the production of white light by the combination of colours. A disc is painted in sectors of the different prismatic colours as nearly as they can be matched, and is then set in rapid rotation (conveniently by means of a wheel and band); on looking at the disc all the colours will be seen blended together, producing the appearance of a greyish white disc, if the shades of colour are properly chosen and the sizes of the sectors properly apportioned.

In this experiment, as in that indicated by fig. 158, the result is mainly due to the fact that an impression conveyed to the eye lasts for an appreciable time before it fades; so that if a red sector is before the eye at any given moment, the sensation of redness lasts during the period whilst the other colours are successively presented to the eye by the revolution of the disc; and similarly with these, so that all are practically viewed simultaneously.

As in Expt. 350, if any one of the sectors be rendered invisible (*e.g.*, by pasting over it a piece of black cloth, &c.), the effect produced on rotating the disc will be that of a disc tinted with some kind of colour or other, the combination of sensations now produced by the light from the disc on reaching the observer's eye being no longer that producing the sensation of whiteness.

Expt. 352. Chromatropes and Colour Tops.—A variety of optical colour toys are in the market, depending on much the same principles as Newton's disc. Usually a wheel or top, or some similar rotating contrivance, is set spinning by means of a piece of string wound round the axle and then pulled away rapidly. Coloured discs, tinted plates shaped as sectors or otherwise, &c., are applied in fashions variable to some extent at will, so that as the apparatus revolves various combinations of colour sensations can be produced, according to the particular colours employed on the tinted surface rendered visible and the relative proportions of these surfaces as regards area. In all cases the ultimate colour effect is due to the combination of a number of separate colour effects, viewed one after the other so rapidly that the whole series is rendered visible before the effect of the first has subsided; just as a stick with a glowing ember at the end, when whirled through the air sufficiently rapidly, produces the sensation of a complete circle of fire.

Expt. 353. Thaumatropes and Zoetropes.—Several ingenious optical toys are also known, based on the above principle of persistence of visual impressions, although the phenomena of colour are not necessarily here involved. One of the oldest of these *thaumatropes* consists of a disc, capable of spinning about one of its diameters as an axis (fig. 159). On one side is painted part of a device or picture, &c., such as a bird or mouse, and on the other side the other half, *e.g.*, a cage or trap. On setting the disc spinning and looking at it, either directly or in a mirror, both sides are apparently seen simultaneously, and the complete device becomes visible, *e.g.*, the bird inside the cage, or the mouse inside the trap.

A modification of the same principle is employed in another class of toy illusions known as *zoetropes* and by various other fancy names. One form of this instrument consists of a pasteboard or metal hollow cylinder capable of revolution about its axis, which is placed vertically. In the sides of the cylinder are pierced a number of slits, so that an observer looking horizontally at the cylinder whilst it is revolving sees alternately the outer surface of the cylinder and the inner surface on the opposite side, the latter becoming visible through the slits as they successively pass in front of the eye. The outer surface should be painted black, so as to be practically invisible; whilst inside the cylinder are pasted or painted a series of figures representing, for instance, a horse in the various stages and attitudes of leaping over a hurdle, as many figures being used as there are slits. When properly adjusted, the

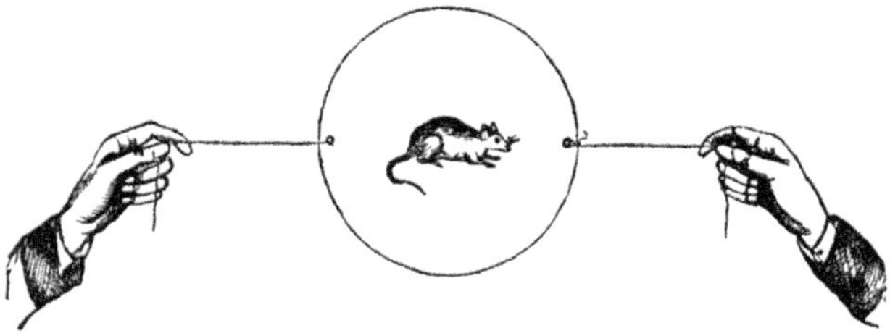

Fig. 159. Thaumatrope.

effect on the eye of an observer looking through the slits whilst the cylinder is in rapid rotation is that the succession of glimpses obtained of the horse in the various successive stages of the leap become blended one with the next, and so on, producing the same sensation as that of viewing a moving object; the horse appears to rise and leap over the hurdle, the action being the more natural and graceful the more accurately the various stages of the motion are depicted relatively to one another.

Fig. 160 represents another simple form of this kind of toy. A circular plate of pasteboard, &c., is perforated at the centre with a pinhole, so that by passing a pin through the hole and pressing the pin into a long stick of wood, which serves as a handle, the disc can be spun round in its own plane, the pin serving as axle. A series of slits is cut in the edge of the disc, and corresponding with each slit is painted one of the successive stages of the motion intended to be depicted (*e.g.*, two boys playing leap-frog). The toy is held by the handle, so that the upper part of the disc is about level with the eye, the painted side being away from the observer and facing a vertical mirror. On looking through the uppermost slits as the disc revolves (spun round by the other hand) a series of glimpses is obtained of the reflections in the mirror of the various stages of the motion, producing the ultimate effect of apparently viewing a moving object, the leaping boy seeming to rise and jump over the head of the other one, repeating the action for each revolution of the disc. Fig. 161 represents a metal holder for the revolving disc.

Fig. 160. Zoetrope.

Expt. 354. Complementary Colours.—For any given colour (*i.e.*, character of ray producing a certain kind of sensation) another colour can be found which is so related to the first that if the two colours be blended together the total effect on the eye is that of whiteness; such pairs of related colours are said to be **complementary**. Let the colours in a Newton's disc (Expt. 351), or a chromatrope, or colour top (Expt. 352), be so arranged as to produce on rotation as nearly as possible a sensation of whiteness (*i.e.*, no particular colour or shade being noticeable, but only a slight amount of greyness). Now, cover up with black any portion of the coloured disc, &c., so as to obliterate entirely that particular colour; on rotating the disc again a colour will be visible, *complementary to the one covered up.*

A peculiar illusion is readily producible by gazing fixedly for a minute or more at a device in some bright colour on a white (or black) ground, and then suddenly looking elsewhere against a uniform background, such as the sky, a cloth, &c. The device will be still apparent for a short time, but the tint seen will be *complementary to the original one;* thus, if large red letters on a white ground be viewed intently, on removing the eyes there will be seen the illusion of the same letters *in green on a black ground.* Similarly, orange letters on a black ground, in the first instance, will produce the subsequent impression of blue letters on a white ground; whilst white letters on an orange ground will give the impression of black letters on a blue ground. This device for attracting attention is utilised in various well known advertisements.

Expt. 355. Coloured Light Examined by the Prism.—If bright white sunlight or electric arc light be examined by means of the prism, the spectrum described in Expt. 349 is obtained; but if a source of some coloured light be substituted for the white beam the spectrum obtained is more or less different. Thus, from a beam of sunlight decomposed by the prism let all light be cut off by means of a screen, except some particular small pencil of coloured rays allowed to pass through a hole in the screen (fig. 157). On allowing this coloured pencil to pass through a second prism it will be deviated in direction in accordance with the law of refraction (Expt. 333), but will not be again split up into rays of different refrangibility, forming a second spectrum. In short, the selected pencil of rays is practically **monochromatic**; *i.e.*, it consists of rays of only one colour or one degree of refrangibility.

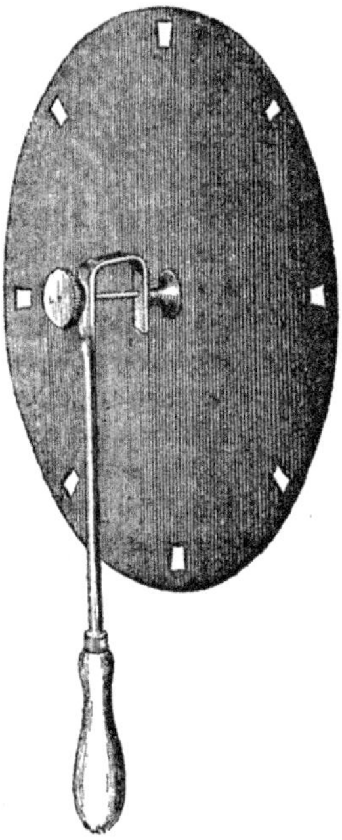

Fig. 161. Zoetrope holder.

If, however, some other natural or artificial source of coloured light be examined, different results will be noticed in different cases. The light emitted from a flame of *spirits of wine and common salt* (Expt. 245) is naturally monochromatic; so that if a beam of such light be made to pass through a prism, no coloured spectrum will be obtained, but only an alteration in the direction in which the light travels, due to refraction. Quite a different result will be obtained if a beam of white light be made to pass through a coloured transparent medium (tinted glass, &c.), and then examined by the prism. In this case a spectrum will be obtained, usually showing several colours, but by no means identical with the spectrum obtained from the original beam before passing through the coloured medium; certain tints will be *either entirely absent, or at least very much paler* than the others, this being due to the absorption of certain kinds of rays by the transparent medium to a greater extent than that taking place with other rays.

The examination of the **absorption spectrum** thus produced by coloured transparent media (*i.e.*, the determination of the kinds of light specially absorbed thereby) is employed in the analytical discrimination of different kinds of colouring matters which otherwise resemble one another. Thus the colouring matter of *blood*, that of genuine red grape wine, and those of various red coal tar dyestuffs, can be sharply distinguished from one another by noticing which particular rays are most readily absorbed; instruments for this purpose are termed *spectroscopes*, and are also employed, to discriminate the nature of bodies that become luminous when sufficiently heated, by examining the spectrum afforded by the light emitted.

Very few artificial sources of light produce exactly the same coloured light as that which would be developed by the sun were no intervening atmosphere present containing gases, watery vapour, &c., possessing the power of absorbing certain rays more than others. In foggy weather the sun looks red; a partial analogous extinction by absorption of the more refrangible rays takes place by the action of aqueous vapour in the air even in tropical climates and when the sun is viewed from a high elevation (such as a mountain top), so as to diminish as far as possible the quantity of moisture in the air traversed by its beams. In consequence the electric arc light always possesses a violet or bluish cast as compared with sunlight, owing to the presence of a larger relative proportion of highly refrangible visible rays, these not having been absorbed during passage through a long column of atmosphere; in other words, if the electric arc light be regarded as the standard of perfect whiteness, sunlight, as it practically reaches the surface of the earth, is more or less slightly yellowish in tint. Artificial sources of light, such as the lime light, the analogous Welsbach light (a kind of curtain of earthy substances heated to luminosity in an atmospheric gas burner something like a Bunsen burner), **incandescent** electric lamps (not arc lights), and more especially ordinary gas, oil, and petroleum lamps, candles, &c., are still yellower in shade, but for a different reason, viz., that the temperature is not so high as in the electric arc, and consequently rays of the highest refrangibility are not so freely emitted.*

* On gradually heating an infusible solid at first, it only emits invisible heat rays; by and by it emits visible rays of the lowest refrangibility and is then *red hot*; as it gets hotter it emits more refrangible rays, and the colour of the light approaches nearer and nearer to whiteness as it becomes *white hot*, but the less refrangible rays are always more or less in excess, as compared with the electric arc.

Expt. 356. Stereoscopes.—When an object at a moderate distance is looked at, the impression produced upon one eye is not identical with that produced on the other, as the two or three inches of horizontal difference in position of the two eyes corresponds with a different view of the relationships of surrounding objects to one another. Thus, a photograph taken at a particular spot would not be identical with one obtained by moving the camera a little to one side; objects that would be directly in line in the first case would not be so in the second, and *vice versâ.* When an individual in good health and possessing perfect vision looks at a view, &c., with both eyes, the two slightly different impressions produced on his two eyes become combined into one impression, giving the idea of **solidity** or **perspective.** In certain states of health, however, owing to derangement either of the optical constituents of the eyes, or of the nerves connected with them, this fusion into one of the two impressions does not always occur, so that **double vision** results. This can generally be brought about at will for a few seconds by gently pressing one of the eyeballs either towards the nose or away from it, when two images of an object looked at will usually be seen. Certain drugs, and more particularly alcohol (when given in greater quantities than certain limiting amounts varying with the constitution and habits of the individual), also produce the same result of "seeing double," along with others due to temporary derangements of the nervous mechanism through the presence of alcohol in the system.

If two pictures are obtained (conveniently by means of photography) one representing a view of a given object from one spot and the other another view of the same object from a slightly different spot, and these two pictures are presented, the one to the one eye and the other to the other, after a short time the two impressions will blend together and a single impression will result, exhibiting apparent solidity. Instruments for bringing about this result are termed **stereoscopes.** The best known form of stereoscope simply consists of a frame holding the two pictures in position in such a way that the one intended to be seen by the left eye cannot be seen by the right, and *vice versâ* (fig. 162); two lenses or portions of lenses are usually also supplied at the top of the frame, one for each eye, partly for the sake of magnification, partly to assist in "focussing" the two pictures into one. Fig. 163 represents the two

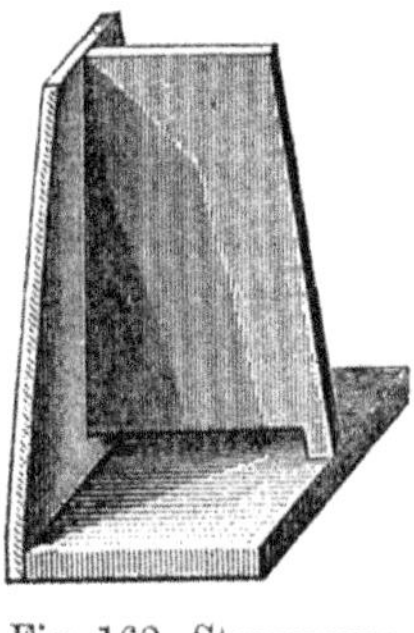

Fig. 162. Stereoscope.

views of a given pile of wooden blocks seen close at hand with the two eyes respectively; when these two different views are looked at through a stereoscope they blend together, and the eyes jointly convey to the brain the same sensations as would be produced by viewing the actual solid pile. Similarly, fig. 164 represents the corresponding views of an analogous out-door landscape. Fig. 165 represents one of the simplest forms of stereoscope used for producing this effect.

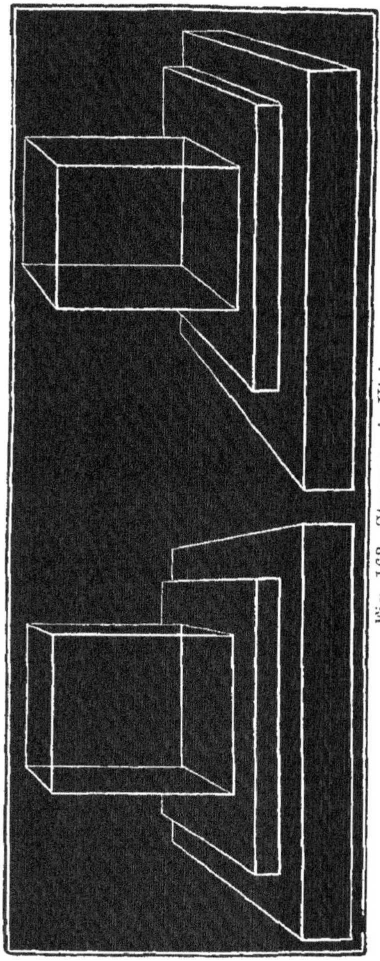

Fig. 163. Stereoscopic Vision.

Wheatstone's reflecting stereoscope is represented by fig. 166; here the two pictures are arranged parallel to one another, and are viewed by reflection from two mirrors placed at right angles to one another between the two pictures, so that the reflected rays from each pass forward to the right and left eyes respectively of the spectator. In this case the two pictures cannot be simply those that would be obtained by photographing from two spots a little apart; one must be reversed in position, as it would be seen in a mirror or in a transparency viewed from the other side. Fig. 167 illustrates this; the arrows A and B are obviously arranged in inverted directions, but on reflection from the two inclined mirrors, *abc*, each is seen at I by the right and left eyes situated at R and L respectively; the two superposed slightly different images thus seen simultaneously producing the sensation of perspective.

Expt. 357. The Pseudoscope.—When an observer regards his image in a looking glass, everything appears to be reversed, the

Fig. 164. Stereoscopic View of Landscape.

observer's right hand appearing to be the image's left, and so on. By holding a right-angled glass prism between the eye and any object in the position indicated by fig. 168, rays from an object, *ab*,

Fig. 165. Stereoscope.

Fig. 166. Wheatstone's Reflecting Stereoscope.

incident on one face of the prism are refracted to the back of the prism, where they are totally internally reflected to the other face, whence they are again refracted to the eye at *c*, which accordingly perceives an image at *de* reversed relatively to the object as any other reflected image would be (*i.e.*, as regards right and left, but not turned upside down). A pair of such prisms, one for each eye, set in a frame something like an opera-glass constitutes the *pseudoscope*, which reverses everything looked at. One effect of this is that, if a properly lighted

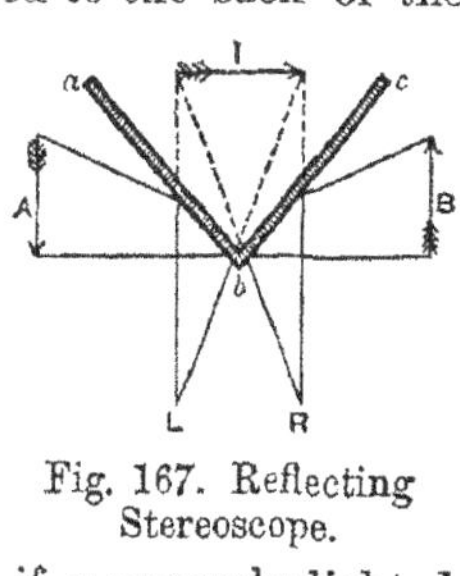

Fig. 167. Reflecting Stereoscope.

hollow be looked at, it generally appears to be in relief, and *vice versâ;* some people, however, are unable to perceive this effect, although perfectly able to distinguish the reversal of position as regards right and left.

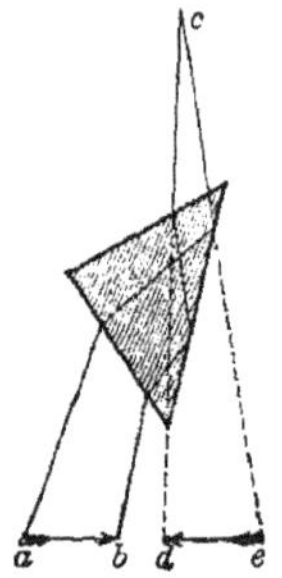

Fig. 168. Pseudoscope.

Many curious effects and illusions can be produced by taking advantage of the fact that the sense of vision under ordinary circumstances is based on the action of the eye as an optical instrument (Expt. 372), conjoined with the effect of habit and previous notions derived from experience. Thus most people have a pretty clear idea of what a length of 10 inches represents on a sheet of paper, such as a map placed in front of them. But if anyone put a "chimneypot" hat just 10 inches high on his head to be looked at, roughly speaking, level with the eye, and the observer be asked to mark off on the skirtingboard or wall of a room the height to which the brim of the hat will reach when standing on the floor on its crown, the judgments formed will in most cases differ considerably from each other and from the truth. Fig. 169 represents a somewhat different case, where, owing to the effect of custom and habit in allowing for perspective in looking at objects, the inclined lines, although really parallel, do not seem to be so. Similarly, in fig. 170 the lower halves of the letters S and figures 8 appear to most persons of the same size as the upper halves; but on turning the figure upside down it will be seen that the lower halves are really

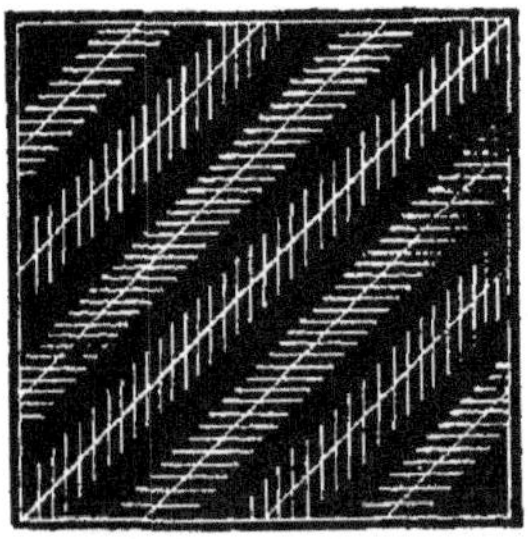

Fig. 169. Zollner's Lines.

S S S S S S S S   8 8 8 8 8 8 8 8

Fig. 170.

larger. Again, in fig. 171 the white spot on a black ground appears larger than the black spot on a white ground when viewed from a short distance, although both are really of the same size. If the centre of fig. 172 be looked at from a

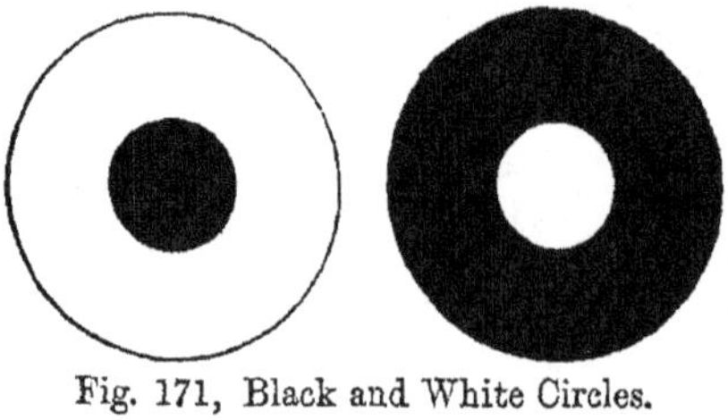

Fig. 171, Black and White Circles.

distance of only 2 or 3 inches, the curved lines bounding the alternate black and white areas will appear almost straight.

Fig. 172. Chequers from "Stirling's Outlines of Physiology."

These and others analogous peculiarities are to the eye very much what the imperfect perception of temperature is to the sense of touch (Expt. 328). A somewhat similar example of illusion through the sense of touch is afforded by the effect produced by crossing the first and second fingers, and placing a marble or other hard object between them (fig. 173), when *two* objects will be apparently felt.

Fig. 173. Aristotle's Experiment.

Expt. 358. Double Refraction.—Substances crystallised in certain shapes and of sufficient transparency, when interposed in certain positions between the eye and an object, allow *two* images to become visible; just as a pane of glass interposed obliquely between an object and the eye causes a certain amount of apparent displacement, on account of the different directions of the path pursued by the light whilst passing through the glass as compared with that whilst passing through the air (Expt. 347); so will a plate of certain kinds of crystallised matter, when cut in a particular way relatively to the planes bounding the crystal, **refract the light in two ways**, part passing along one path in the solid plate and part along another, so that two rays of light finally emerge, the directions of which are not coincident. The effect of this is that a **double image** is perceived through the plate. Thus fig. 174 indicates the effect of looking at the letters AB through a plate of "Iceland spar," a natural variety of crystallised carbonate of calcium; the letters appear in duplicate, owing to **double refraction** taking place.

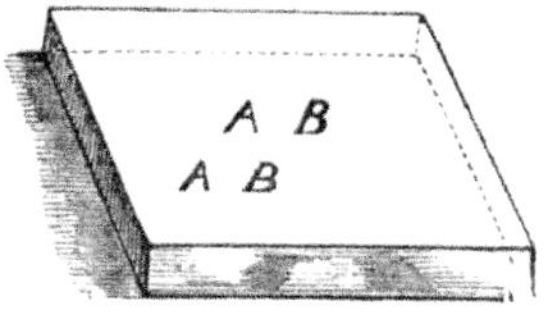

Fig. 174. Double Refraction.

With all doubly refracting crystals there is one (and sometimes more than one) position in which the crystal may be held, such that light passing through in a particular direction does *not* suffer double refraction; this position can easily be ascertained by trial in the case of Iceland spar. Of the two paths pursued by a ray of light when doubly refracted, one obeys the law of sines (Expt. 333), and is spoken of as the **ordinary ray**, whilst the other does not obey that law, and is termed the **extraordinary ray**. With *Iceland spar, ruby, sapphire, ferrocyanide of potassium*, and other substances, the former is the one more bent out of the original path than the other; with some other substances, such as *ice* and *quartz*, the opposite is the case. In consequence, if a dot of ink on a sheet of white paper be looked at through a piece of Iceland spar two images will be seen, of which one will appear slightly nearer the eye than the other; this is the **ordinary image** formed by the ordinary rays, which, being most refracted, will cause the greatest apparent raising of the image. If the dot be viewed in a direction at right angles to the paper and the crystal be turned round with the face still on the paper, the ordinary image will appear stationary, whilst the other will apparently revolve round it, the line joining the two being always in the direction of the shorter diagonal of the face of the crystal; if the edges are of equal length the face has the shape of a rhomb. The two rays also differ in certain other respects, the study of which is intimately connected with the phenomena termed **polarisation of light**, a branch into which the limits of the present work render it impossible to enter.

Expt. 359. Production of Colours by thin Films—Iridescence.—When a soap bubble is blown, as in Expts. 299 *et seq.*, prismatic colours are seen on its surface when the thinness is of sufficient extent; in the same kind of way a drop of oil, or tar, or other fluid lighter than water and insoluble therein, if placed on the surface of a large mass of water, will gradually spread over the surface, forming a very thin layer which will show prismatic colours. This effect is produced in consequence of the reflection of light from both the upper and under surfaces of the thin film, as in Expt. 343 with a glass plate, the two reflections causing an action termed **interference** to take place between the two rays when the film is of less than a certain thickness; in consequence of which rays of different refrangibility become more readily visible in one position than at another, producing the sensation of patches of different colours, somewhat like a spectrum received on a screen after a beam of white light has passed through a prism (Expt. 349). The "nacre" or peculiar coloured lustre of pearls and mother of pearl, the tints of opals and many other analogous substances, and the coloured films developed with melted lead or bismuth in Expts. 15 and 18 are all produced by actions of this kind; the production of rainbow colours in this fashion is generally termed **iridescence.** The formation of rainbow-tinted haloes during certain kinds of misty or foggy weather, and of coloured fringes round candles, &c., when looked at through glass covered with very fine globules of condensed dew, or dusted over with fine *lycopodium* powder, &c., are closely related phenomena, the full explanation of which must be sought for in larger works.

Expt. 360. The Rainbow.—Whenever small drops of water, such as rain or the spray of waterfalls, fountains, &c., are suspended for the time being in the air, and a beam of white light (natural, such as that from the sun and moon, or artificial, as the electric light, &c.) is allowed to fall thereon, the light enters each droplet and is refracted, the rays being partly reflected from the posterior surface of the drop and then again refracted outwards, ultimately forming an angle with their original path of less than a right angle. Under favourable conditions a ray of light may meet the eye of an observer in two ways by means of actions of this sort: firstly, when reflection takes place *once* in the drop;

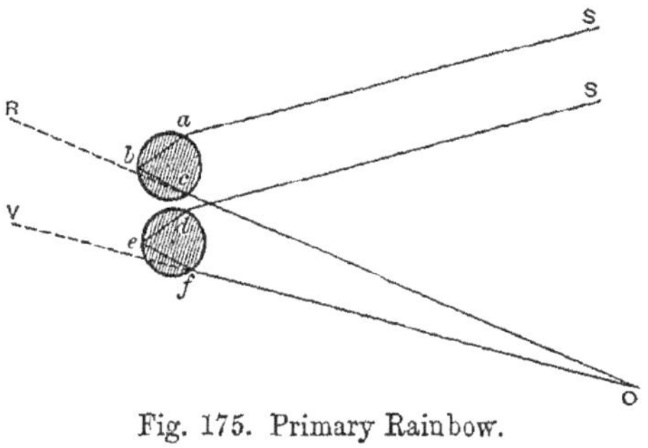

Fig. 175. Primary Rainbow.

and secondly, when it occurs *twice*. Fig. 175 represents the first case: a ray of sunlight S*a* meets the outside of a spherical drop of water at *a*, and is partly reflected and scattered; whilst part enters the drop, being refracted along *ab*. At *b* internal reflection takes place along *bc*; and at *c* part of the light emerges, being again refracted along *c*O to the eye of an observer at O, in such a fashion that the *red* part of the prismatic spectrum formed in consequence of refraction meets his eye. Another ray of light, S*d*, meets a second drop of water situated slightly lower down, and is similarly refracted and reflected along the course S*def*O in such fashion that the *violet* ray of the spectrum now formed is seen by the observer at O. Hence the observer perceives the *red colour of the* **primary bow** *outside and the violet inside*, the other colours being intermediate in due order.

Fig. 176 represents the second case. Here the ray S*a* is first refracted along *ab*, then twice successively internally reflected along *bc* and *cd*, and finally emerges at *d*, and is refracted to the eye of the observer at O, the ray reaching his eye being now the violet one; whilst a drop situated at a lower level acts similarly, the ray S*e* being refracted and reflected along the course S*efgh*O, so that the red ray of the spectrum now meets his eye. In this case, consequently, *the violet is outside and the red inside, i.e.*, the colours come in the reverse order in the **secondary bow** relatively to that in which they occur in the primary bow. Moreover, on account of the mode of formation of the two bows and the values of the index of refraction between water and air for the various coloured rays, it results that whilst in the case of the primary bow the angle between the lines S*a* and O*c* is about 42°, and that between S*d* and O*f* 40¼°, the corresponding angles for the secondary bow between S*a* and O*d* and S*e* and O*h* are respectively about 54° and 51°, or considerably greater; so that the secondary bow lies *above the primary one.*

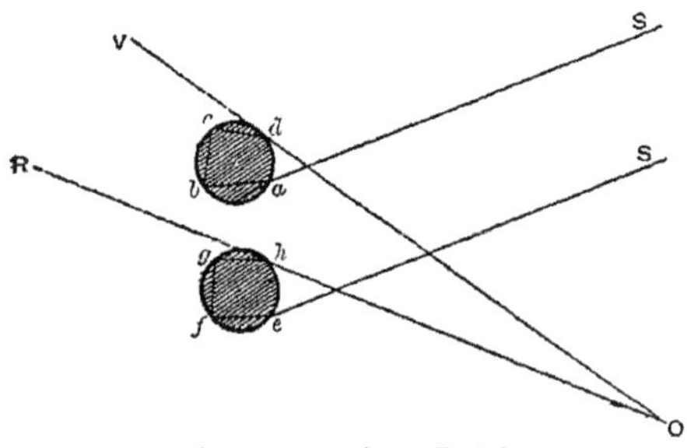

Fig. 176. Secondary Rainbow.

All drops situated on the surface of a cone, the apex of which is the observer's eye and the axis of which is parallel to the sun's rays, will act alike upon the light falling upon them, and produce the same effect on the observer's eye; so that when the angle between the axis of the cone and a generating line thereof is 40¼°, a portion of a violet circle will be visible; concentric with this and outside will lie the other colours, terminating with red, on the periphery of a wider cone where the angle is 42°, thus forming the primary bow; concentric with and outside this again, on the periphery of a still wider cone of 51°, will lie another red circle, external to which come the other colours, terminating with violet on the periphery of a cone of 54°, thus forming the secondary bow with inverted order of colours. The secondary bow is generally much fainter than the primary one, on account of the greater loss of light through reflection. Under favourable conditions a tertiary bow may be seen, concentric with the others and outside them; usually, however, the colours are so weakened by repeated reflections, &c., that the tertiary bow is imperceptible.

It results from the conical angles above mentioned that if the sun be elevated above the horizon more than 40° to 42°, the primary bow will fall wholly below the horizon, as will the secondary bow when the sun is more than 51° to 54° high in the heavens; so that, under these circumstances, no rainbow can be seen excepting when the spectator is so placed that the refracting and reflecting drops are *below* his level, as when standing on a mountain peak, or when looking downwards into the spray of a cataract. If, however, the sun be less elevated above the horizon, the arches will be correspondingly raised, so that when the sun is just setting or rising the summit of the primary bow will be elevated 40° to 42° high, and its "legs" will subtend an angle of 80° to 84° at the observer's point of view; whilst the secondary bow will be elevated 51° to 54°, and will subtend at its base an angle of 101° to 108°—corresponding variations being introduced according as the sun is at intermediate heights, and according to the position of the observer relatively to the refracting and reflecting droplets.

Water is by no means the only substance capable of forming "rainbows"; any transparent fluid reduced to fine spray by mechanical means or otherwise, and suitably situated in reference to a sufficiently brilliant source of light and to the eye of the observer, will produce similar results; the height and width of the arch seen will necessarily vary with the "index of refraction" of the particular fluid used as compared with air.

CHAPTER XXIII.

Reflection and Refraction at Curved Surfaces.

Expt. 361. Geometrical Focus of a Concave Mirror.—When a beam of light emanating from an object at a considerable distance (*e.g.*, the sun) falls upon a reflecting surface which is not flat (or *plane*), but curved regularly (*e.g.*, when the surface is part of a sphere), the rays which fall upon different parts of the surface will necessarily be reflected in different directions, even though originally parallel to one another. In such a case the law of reflection applies uniformly to every part of the surface, the *normal at the point of incidence* (Expt. 332) being in each case the line drawn perpendicular to the plane which is tangential to the surface at the point where the incident ray falls *.

Fig. 177 accordingly indicates the way in which two rays, SA, SB, would be reflected if incident upon opposite sides of the centre of a concave mirror when placed directly opposite to the sun. If C be the centre of the spherical surface, the normal at any given point on the surface will be a continuation of the radius passing through that point. Hence, for the points A and B the respective normals are the lines CAD, CBE. Hence, the lines AF, BF, representing the reflected rays (so situated that the angles SAC and CAF are equal, and also the angles SBC and CBF are equal), *will converge so as to cross one another at a point F.* The same

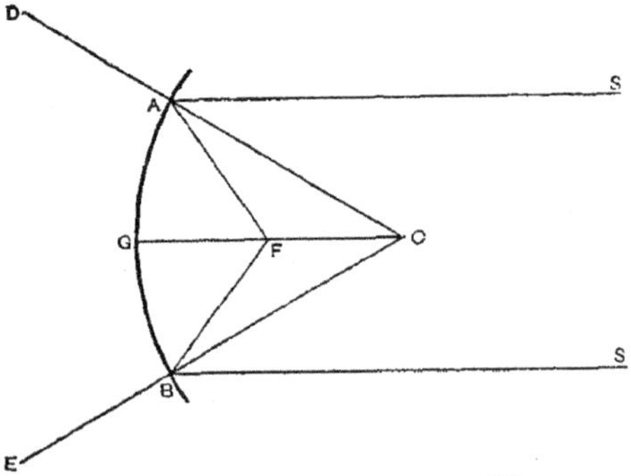

Fig. 177. Real Focus of Concave Mirror.

* A *plane tangential to a curved surface* may be illustrated by a flat slate touching a round cricket ball or orange; the plane of the slate's surface touches or is tangential to the curved surface of the ball at the point where the two surfaces meet. A line drawn perpendicular to the surface of the slate through this point would represent the normal at the point of incidence in reference to a ray of light incident upon the curved surface at that point.

remark applies wherever the points of incidence A and B are situated, so that all reflected rays meet at the point F; this point is accordingly called the **geometrical focus** or **principal focus** of the mirror, and is situated midway between C and G, the point on the surface of the mirror where the line CFG drawn through C and F cuts that surface.* The length CF = FG is called the **focal length** of the mirror.

Expt. 362. Virtual Focus of a Convex Mirror.—If the rays of light in the last experiment fall not on a **concave** mirror, but on a **convex** one, a different result will be brought about. Fig. 178 illustrates the track of the two rays SA, SB, reflected from the points A and B, in accordance with the law of reflection. Here

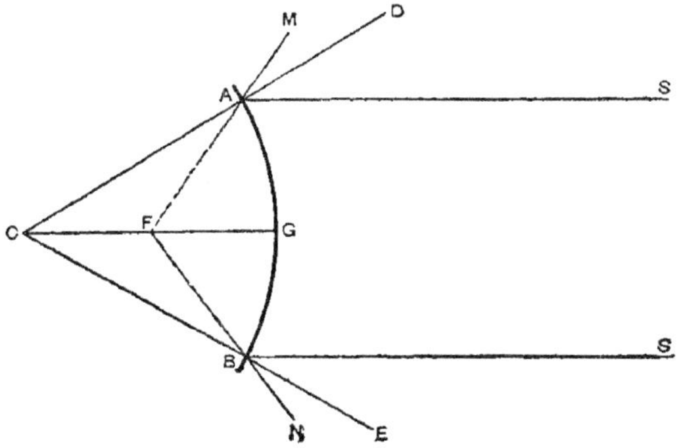

Fig. 178. Virtual Focus of Convex Mirror.

the reflected rays, AM and BN, make equal angles respectively, with the normals CAD and CBE, and *diverge from one another* as though they had emanated from a point, F, *behind* the mirror. This point is the **virtual geometrical focus**, and is situated just as far behind the mirror as the geometrical focus of a concave

* The term "focus" is derived from the circumstance that a concave mirror concentrates radiant heat at F as well as visible light (Expt. 378), so that inflammable bodies placed at F can be fired by a "burning mirror" (*focus*, Latin for fireplace). Strictly speaking the different reflected rays do not all pass through the same point, F, when the curvature of the mirror is that of a sphere, but only pass *very close to it;* this want of exact coincidence is termed **spherical aberration**, and introduces certain difficulties in the way of obtaining clear definition with optical instruments in which concave mirrors are employed. Certain other forms of curvature are free from this defect, especially "parabolic" mirrors (mirrors where the curved surface is that produced by the revolution about its axis of a parabolic curve); hence these are preferable to spherical mirrors for accurate optical instruments, particularly reflecting telescopes (Expt. 375).

mirror is in front, the radius of curvature being the same, so that CF = GF = ½ CG.

Expt. 363. Formation of Images in Curved Mirrors.—If an observer stand in front of a convex mirror, no matter what the distance between himself and the face of the mirror may be, he will always observe a reflected image (virtual image) of himself or of any other object situated in front of the mirror; and this image will always be *erect* (not upside down) and *diminished* in size (smaller than the actual object). Thus the arrow, AB, (fig. 179), in front of the convex mirror, when seen by reflection appears to be situated

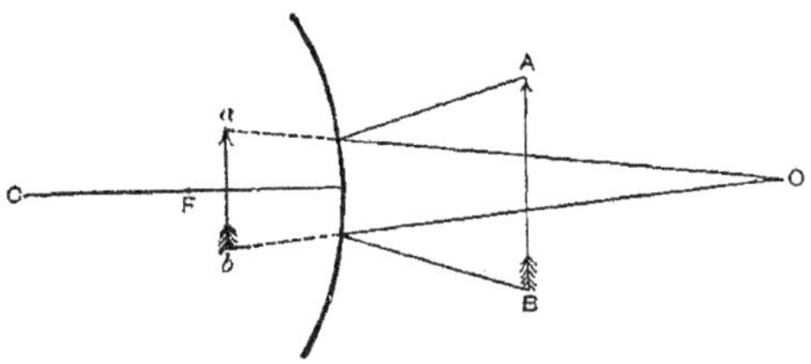

Fig. 179. Virtual Image formed by Convex Mirror.

at *ab*, *behind* the mirror, and nearer to it than its geometrical focus, F, and smaller in size, because it obviously subtends a lesser angle at the eye of the observer at O. But if a concave mirror be used instead of a convex one, the effect differs according to the situation of the object with respect to the mirror; if the object, AB, (fig. 180) (say a candle flame) be nearer to the mirror than the geometrical focus, F, an observer at O in front sees an **erect virtual image** at *ab*, not diminished, as in the case of a convex mirror, but *magnified*, because it now subtends a greater angle at the observer's eye than the actual object.

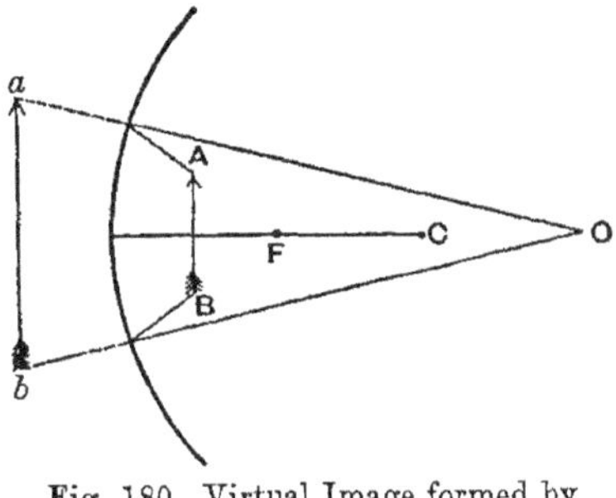

Fig. 180. Virtual Image formed by Concave Mirror.

If the flame be held as nearly as possible *at* the focus F, no distinct image of any kind is seen, but only a blaze and blurr of light; whilst, if the object AB be held between the focus F and the centre of curvature, C, a **real inverted and magnified image**, *ab*, is formed further from the mirror than the centre of curvature (fig. 181). Conversely, if held further from the mirror than the centre of curvature, a real inverted but diminished image is formed between the centre of curvature C and the focus F, AB and *ab* in fig. 181 now changing places. In either case the image can be received on a screen, magic lantern fashion; thus, if a piece of greased letter paper be moved about in front of the mirror together with a candle, numerous pairs of positions for the two can be found, such that a distinct image of the candle appears on the screen; the relative positions of candle and image for each pair are said to be "conjugate" with respect to one another; and the term "conjugate foci" is applied to represent the relative positions of any two given points similarly related to one another. The geometrical focus of a concave mirror has its conjugate focus at an infinitely great distance in front; that of a convex mirror at an infinite distance behind; whilst in all cases the relative positions of any pair of conjugate points is expressed by a simple algebraic formula depending on the focal length of the mirror.*

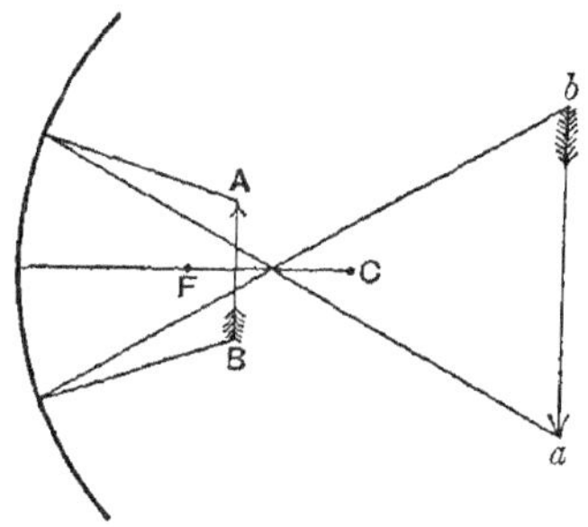

Fig. 181. Real Image formed by Concave Mirror.

It results from this, that a mirror with a surface of varying radius of curvature will give a varying degree of magnification or diminution; accordingly, by employing such mirrors, most amusingly distorted reflections can be obtained according to the curvatures of the surface. A vertical cylindrical mirror will not distort at all in a vertical line, but will reflect as a plane mirror would do arranged tangentially to the cylindrical surface; in reference to a horizontal line, on the other hand, it will behave precisely as a convex or concave mirror of the same radius of curvature.

* If x be the distance from the mirror of one point serving as object, y that of the other point or image, and f that of the focus from the mirror, then—

$$\frac{1}{x}+\frac{1}{y}=\frac{1}{f},$$

the values of y and f being reckoned as + when measured in the same direction as x from the surface of the mirror, but as − if measured in the opposite direction. In the case of a concave mirror and an object in front giving a real image (fig. 181), both y and f are of the same sign as x, or all three are +. But with a convex mirror (fig. 179) y and f are both opposite in sign to x, *i.e.*, both are − ; with a concave mirror forming a virtual image (fig. 180), y is −, whilst f is +.

Such a mirror, if convex to the observer, will accordingly give a reflection of the face of the natural length, but greatly diminished width; on the other hand, if placed horizontally, the width of the face will be natural, but the length greatly diminished. Analogous opposite results will be obtained with hollow (concave) cylindrical reflectors, provided the radius of curvature is sufficiently great to enable the observer's face to be at a less distance from the mirror surface than half that radius.

The curved sides of a polished smooth-faced silver teapot, cup, or goblet, &c., or a bright smooth block tin dish cover, &c., will often show curiously distorted reflections formed in this way. On the other hand, distorted drawings can be readily made which, when seen by reflection from a mirror of appropriate curvature, will no longer appear abnormal. Various amusing toys are constructed on these principles.

Expt. 364. Spectral Illusions by Concave Mirrors.—A large-sized properly shaped well-polished concave mirror may be employed to produce a spectral illusion, somewhat on the same principle as a magic lantern, but with still better effect. A highly illuminated object placed at one of two conjugate foci will form an image on a cloud of steam or other ground placed at the other focus so as to receive it conveniently. This image, being inverted with respect to the object, renders the production of stage phantoms and illusions by means of a concave mirror somewhat difficult, especially when the spectre is required to be in motion; hence the employment of such optical appliances is somewhat limited as compared with other methods. But it appears to be very probable that the properties of concave mirrors in this respect were known to the ancient soothsayers and utilised in their oracles, &c.; the more so as, by properly arranging the relative position of object and image, a magnified image can be obtained, so that the reflection of a man of ordinary stature would indicate a spectre of heroic proportions.

Expt. 365. Lenses of various kinds, and their Foci.—A lens means a portion of a transparent medium bounded by two intersecting surfaces of regular curvature, and so arranged that the light enters on one surface and issues at the other; one surface may be **plane** (*i.e.*, part of the surface of a sphere of indefinitely large radius). According as the lens is thickest or thinnest in the middle, it is spoken of as **convex** or **concave** respectively; although it may happen that one surface is actually convex and the other concave this definition is adhered to, the lens being said to be convex when the curvature of the convex surface is sharper (*i.e.*, the radius of curvature is less) than that of the concave surface;

and to be concave when the opposite is the case. Thus A, B, C (fig. 182) represent three forms of convex lens, and D, E, F three forms of concave lens, respectively termed **double convex** and **plano convex** (or **convexo plane**, according as the curved or plane surface meets the rays of light first), and **concavo convex** (or **convexo**

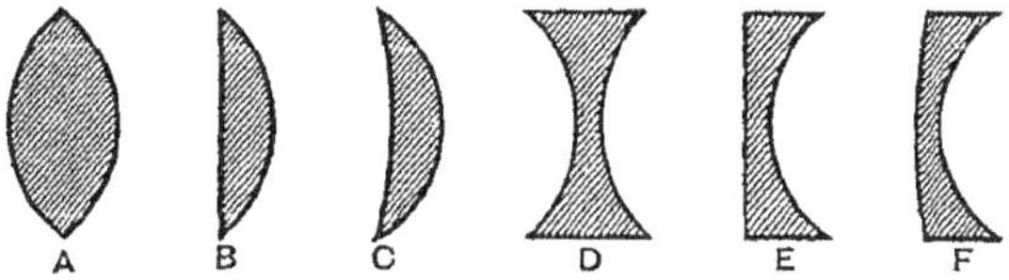

Fig. 182. Convex and Concave Lenses.

concave); and **double concave**, **plano concave** (or **concavo plane**), and **concavo convex** (or **convexo concave**). To distinguish the two forms of concavo convex lenses, the first is sometimes called a **meniscus**, and the terms concavo convex and convexo concave reserved for the other; sometimes the two kinds are distinguished as **converging** and **diverging** respectively.

When rays of light, SA, SB (fig. 183), from a source at so great a distance that the rays are practically parallel, meet any kind of convex lens, they are refracted so as to meet one another at a **focus**, F, just as with a concave mirror; whilst if the lens be concave the rays of light **diverge** after passing through, as if they had emanated from a **virtual focus**, as with a convex mirror (fig. 184). The refrangibility of differently coloured rays not being the same,

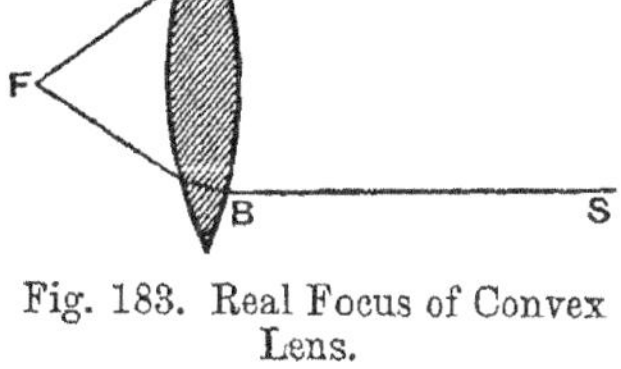

Fig. 183. Real Focus of Convex Lens.

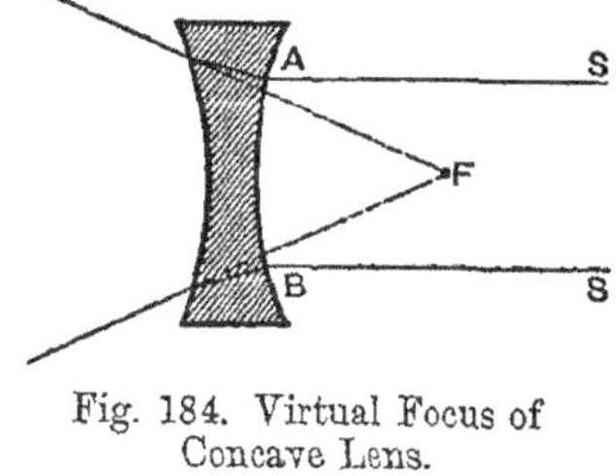

Fig. 184. Virtual Focus of Concave Lens.

it results that the more refrangible rays are brought to a focus by a convex lens a little nearer to the lens than the less refrangible ones, so that the position of the focus is not quite the same for all colours. This difference is termed **chromatic aberration**, and its existence produces a very inconvenient result with many kinds of optical instruments, viz., that images formed by means of lenses are apt to

be fringed with colours and so rendered indistinct, unless means are adopted for correcting the fault and rendering the lenses **achromatic.** For the same reason, the most refrangible actinic rays are brought to a focus by a convex lens at points nearer to the lens than the average of the visible rays, whilst the least refrangible heat rays similarly form a series of foci slightly further off. This difference renders it necessary to use perfectly achromatised lenses for photographic purposes, otherwise the picture formed by the action of the actinic rays would be more or less blurred and imperfect, even when the visible rays gave a clear image in the camera (Chapter XXV.). In the case of the heat rays, the difference is usually sufficiently glass" when exposed to the sun, so that some inflammable substance is placed at the spot where the image of the sun formed by the visible rays is fairly clear and distinct.*
small to render a convex lens a more or less powerful "burning

Expt. 366. Formation of Images by Lenses.—A concave lens behaves very similarly to a convex mirror in this respect, that the image is always **virtual, erect,** and **diminished.** Thus fig. 185 represents the virtual image, *ab*, seen by an observer at O when looking through a concave lens at an object, AB, a similar result taking place in all positions of the object as regards distance from the lens.

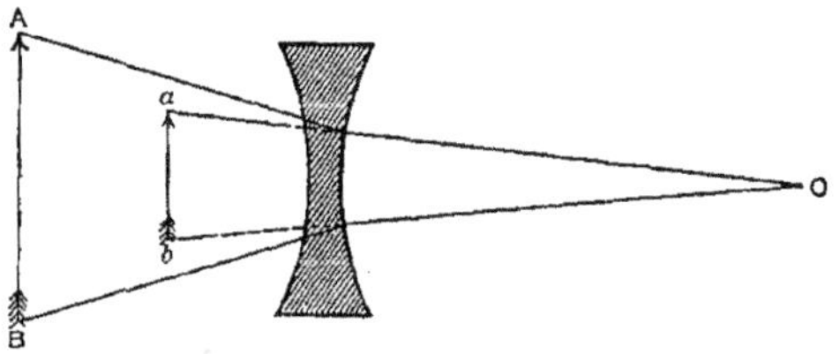

Fig. 185. Virtual Image formed by Concave Lens.

A convex lens, on the other hand, acts like a concave mirror, the nature of the image differing according to the position of the object with respect to the lens. When the object lies nearer to the lens than the **principal focus** (or point representing the average geometrical focus for all kinds of visible rays jointly) the image is **virtual, magnified, and erect,** this position corresponding to the use of a reading lens or simple magnifying glass (Expt. 367); when at the principal focus, no distinct image at all is formed (*i.e.*, the image is at an infinite distance); and when further away from the lens than the principal focus, a **real inverted** image is formed, which is **magnified** if the distance of the image from the lens on one side is greater than that of the object on the other, and

* As in the case of spherical mirrors, lenses the curvature of which are spherical are also subject to **spherical aberration,** which introduces another class of difficulties in constructing optical instruments.

diminished if less. The first case corresponds with the use of the convex lens in the magic lantern (Expt. 368), the second with that in the camera obscura (Expt. 370). When the object and image are equidistant from the lens each is at a distance equal to twice the **focal length** (the distance of the geometrical focus from the centre of the lens), on which circumstance is based a simple method of determining the focal length of a convex lens (Expt. 369). Just as with mirrors, a simple algebraic formula connects the mutual relationships to one another of the geometrical focus and the situations of any pair of given points forming conjugate foci.*

Expt. 367. Simple Microscopes or Magnifying Glasses.—A globular smooth transparent bead of glass, a drop of water, or any other convex **lenticular** (lens-shaped) transparent medium constitutes a more or less powerful magnifying glass, according as its **focal length** (the distance of the centre of the lens from the principal focus) is smaller or greater; thus a small drop of water hanging from a plate of glass forms a strongly magnifying convexo plane lens. To produce the effect of an ordinary magnifier, however, it is indispensable that the object should be placed at a less distance from the lens than its focal length, in which case the image is virtual and erect. Fig. 186 represents the formation of an image, *ab*, thus seen by an observer at O. The amount of linear magnification is expressed by the ratio of the angle *aOb*, subtended by the image at the eye, to AOB, that subtended by the object; this ratio may be made greater than 100 to 1 with a suitable lens; *i.e.*, the object becomes

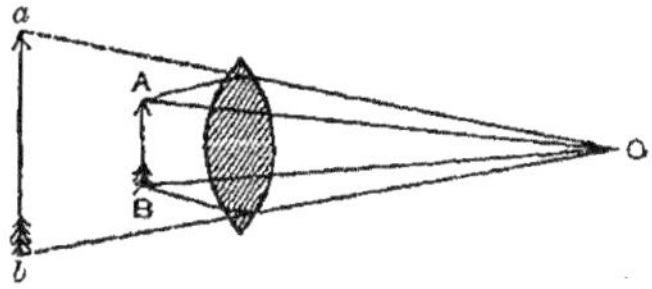

Fig. 186. Magnifying Glass. Virtual Image.

* This formula is the same as that for mirrors, viz.:—

$$\frac{1}{x}+\frac{1}{y}=\frac{1}{f},$$

x being the distance of the object from the lens (regarded as of negligible thickness), y that of the image, and f that of the geometrical focus. These latter two distances are considered as + or − according as they are measured from the mirror in the same direction as x, or in the opposite direction. In the case of a lens, as the rays must pass *through* it to form a real image, such an image can only be formed when x and y are opposite in sign; *i.e.*, when y is −; and for the same reason the value of f (when the sun or other bright object from which the light emanates is regarded as at an infinite distance) is − for a convex lens, and + for a concave one. Hence, in the case of a concave lens (fig. 185), x, y, and f are all three of the same sign, *i.e.*, y and f are +; with a convex lens used as a magnifying glass (fig. 186), y is + and f −; and when used as either a magic lantern or camera obscura, both y and f are − (fig. 187).

magnified upwards of 100 times in linear dimensions, or upwards of 10,000 times in area.

Expt. 368. The Magic Lantern.—The simplest form of magic lantern consists of an illuminated object, AB (fig. 187), placed at a distance from a convex lens only a little greater than its focal length, in which case a magnified image, *ab*, will be formed *on the other side of the lens*, further differing from that produced by the ordinary magnifying glass in being **inverted** and also **real**, so that it can be received on a screen, such as a strained wet sheet, &c. If the distance of the object from the lens be somewhat increased, that of the image when clearly defined will be proportionately decreased, as also will the degree of magnification; if the distance of the object be increased to twice the focal length, that of the image will diminish to the same value, and the object and image will be of the same size; whilst if the object be still further removed, the image will be formed at a distance from the lens greater than the focal length, but less than twice that amount, and will now be smaller than the object (*vide* Expt. 370).

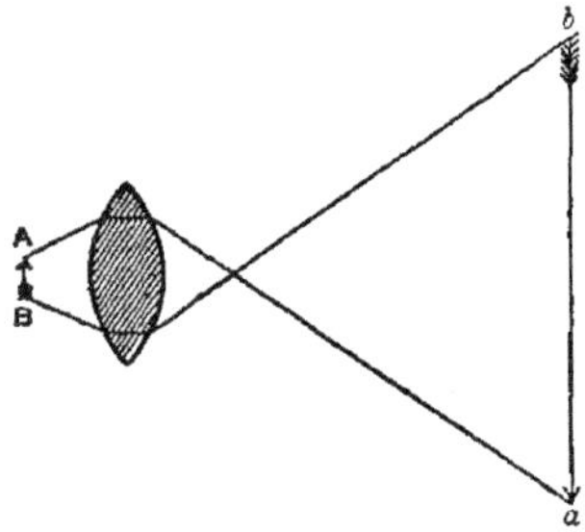

Fig. 187. Real Image formed by Convex Lens.

The better class of magic lanterns are illuminated by means of lime or electric lights, so that a transparent slide (*e.g.*, a photograph on a glass plate) can be very largely magnified without being rendered too faint owing to insufficient light. Specially arranged combinations of lenses are also employed to get rid of spherical and chromatic aberration (Expts. 361, 365), and to obtain a "flat field," *i.e.*, to ensure, as far as possible, that the outer portions of the image on the flat screen may be "focussed" and sharply defined equally with the central part. Cheaper lanterns with ordinary uncompensated lenses usually show defects of this kind to a greater or lesser extent; oil lamps are generally used as the source of light in such instruments, whence very highly enlarged pictures cannot be obtained, owing to lack of illuminating power.

Dissolving views are produced by having two lanterns side by side so arranged that one projects an image on the screen, whilst an opaque cap or shade cuts off the light of the second. This shade is so arranged that by slowly moving it the light from the first lantern is gradually cut off, and that from the second allowed to reach the screen, so that the one picture gradually fades as the light producing it is cut off, whilst the other appears at first faint and subsequently stronger, till at length the first picture is entirely obliterated, all light from the first lantern being cut off; whilst the second picture is visible to the fullest extent, the shade being now wholly removed from the front of the second lantern. Instead of using a shade, the same result can be obtained by turning the lights in the two lanterns up or down as required. For certain kinds of effects both lanterns may be used simultaneously, or even more than two lanterns, one being used to produce part of the effect, the second another part, and so on; *e.g.*, a phantom figure may be projected on the screen by the aid of one lantern, whilst the rest of the scene is shown by the second. By suddenly putting a shade in front of the first lantern, or by turning down the light therein, the phantom disappears, and so on.

In the *solar microscope* the source of light is the sun's rays reflected from a mirror, the general construction of the rest of the instrument being like that of the ordinary magic lantern.

Various instruments are also sold under different names, whereby the image of a highly illuminated **solid** (not transparent) object is projected on the screen, the formation of the image by lenses being formed in just the same way as with the ordinary lantern, but the lighting arrangements and other details being different, according to the circumstances of the case.

Expt. 369. To determine the Focal Length of a Convex Lens.—Arrange a lighted candle directly in front of the lens, so that the centres of the lens and flame may be in line, the light falling full upon one face of the lens, which should be held in position by a convenient clamp or other stand. Slightly grease a piece of note paper, and hold this behind the lens so as to receive the image of the flame, which will appear inverted and will be magnified or diminished according as the paper screen is further from or nearer to the lens than the candle. Now move the candle nearer to or further from the lens, along the direct line joining them, also moving the screen, until finally such positions are arrived at that whilst the inverted candle flame image is clear and distinct, the screen and candle are *at equal distances* from the lens. This happens when each is at a distance from the lens equal to twice the focal length, whence it follows that the focal length is *one-fourth* of the distance between the candle and screen.

Expt. 370. The Camera Obscura.—The *camera obscura* is optically the converse of the magic lantern, consisting in its simplest form of a lens so placed with reference to a receiving screen and surrounding objects that the distance of the latter from the lens is great as compared with that of the former, so that **a diminished and inverted real image** is formed. With the ordinary photographic "camera" no reflecting arrangements are employed to invert the image a second time or re-erect it, so that the image formed on a ground glass plate is upside down. When the camera is used for drawing purposes, or to obtain a picture of surrounding objects inside a dark room (whence the name *camera obscura* = dark chamber), an inclined mirror is employed to erect the image. In the case of the photographic camera, the essential points are that a perfectly achromatic combination of lenses must be used, such that the visible and chemical rays are focussed together at the same points (Expt. 365); and that the workmanship should be such that after the picture is focussed on the ground glass plate, and this is withdrawn and replaced by the plate carrying the materials to be acted on by light, the latter plate may be in exactly the same position as the former one, so as to produce a sharp photograph. The lenses must also be so constructed and

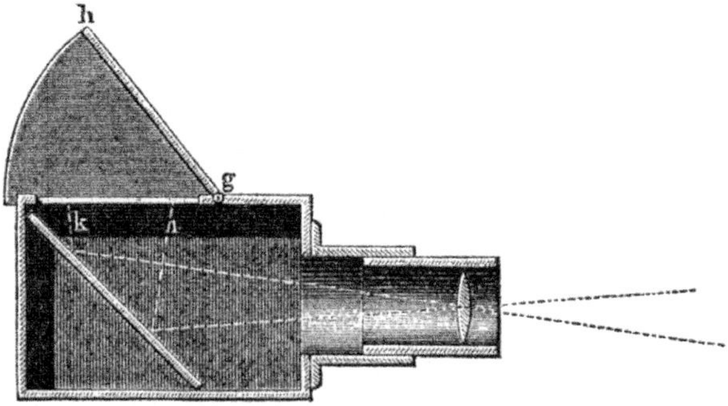

Fig. 188. Camera.

proportioned as to give a "flat field," all parts of the image formed on a vertical plane surface being as nearly as possible in focus, no matter whether the objects in focus are all at the same distance or not. A single uncorrected lens causes the image of a flat object in front to be curved.

Fig. 188 shows the use of the camera for drawing purposes; the rays of light, *k*, *i*, passing through the lens, fixed in a sliding tube at the end of a box painted dull black inside, are reflected upwards from an inclined mirror so as to form the image on a horizontal ground glass plate. A lid, *hg*, working on a hinge prevents the direct light from the object reaching the eye of the observer, a result still better accomplished by throwing a thick black cloth or sheet of black velvet over his head when in position, as usually done by the photographer when focussing his instrument. A sheet of thin semi-transparent paper being laid over the ground glass plate, the outlines of the image formed may be easily traced by a pencil.

For the dark chamber camera, a tent, summer house, or other convenient small building is used, at the top of which is arranged an inclined mirror, so as to throw the light *downwards* on to a horizontal table, painted dull white, or some form of screen capable of being moved slightly up and down, so as to focus the image according as near or distant objects are in view; a convex lens is placed either in front of or under the mirror, so as to form the image. A block of glass of triangular section, CDE, with a convex face on one side, the other two being plane, (fig. 189) may be used instead of a mirror and convex lens; the rays of light from a distant object, AB, are made to converge on meeting the curved front, and are then totally internally reflected by the inclined back surface, CD, so as finally to form an image at *ab*; the glass prism-lens is fixed in a frame capable of swinging round a vertical axis, so as to bring into view any part of the horizon.

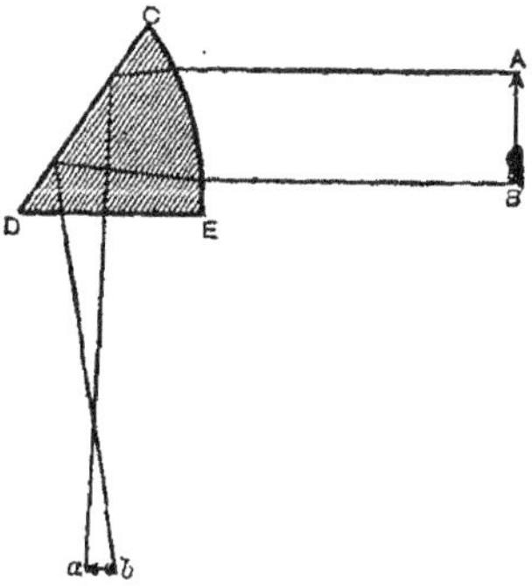

Fig. 189. Prism-lens for Camera.

Expt 371. The Pinhole Focus.—In the centre of one face of a rectangular wooden or pasteboard box make a small hole with a pin, and substitute a plate of ground glass or a sheet of greased paper for the opposite side of the box. Hold the box with the ground glass or paper screen in front of the eye, the pinhole being directed to a brightly illuminated object or landscape; an inverted image will then be seen on the screen resembling that formed in a photographic camera. The effect is heightened by throwing a thick black cloth over the head and box (not over the pinhole) so as to exclude all other light. With properly proportioned hole and box and good external illumination, a tolerably clear definition of the salient points of a landscape may be readily obtained; but images formed by a convex lens are generally much brighter and clearer. With skill, however, it is quite possible to obtain fairly well-defined photographs of scenery, &c., by means of a pinhole camera and highly sensitive plates.

Expt. 372. The Human Eye.—The human eye, like that of other animals, is essentially a kind of *camera obscura*, since the "crystalline lens" in the front part refracts rays to a focus at the back of the eyeball, where a diminished inverted image is formed. This may be shown without much difficulty by obtaining from the butcher a bullock's eye in a fresh condition, and very carefully cutting and paring away the opaque harder portion of the external coating. If this is done successfully, so that the remaining semi-transparent part of the coating can act as a screen, an inverted image of a candle flame, &c., can be readily seen to be formed on holding the eye with the front part facing the candle.

Short-sighted persons have naturally eyes of such a character that the lens forms images of far off objects *in front of* and not exactly upon the sensitive back part of the eye (or **retina**), by the production on which of an image the sensation of vision is produced; with nearer objects the image is formed further back, and more nearly upon the retina, so that such persons can see objects close to the eye much more distinctly than those at a greater distance. A concave lens interposed between the eye and a far off object causes the image to be formed further back than would otherwise be the case; hence short-sighted persons require spectacles where the glasses are thinnest in the middle, the degree of concavity being proportionate to the defect in the eye.

Long-sighted persons have the opposite defect; owing to flattening of the crystalline lens through advancing age, or other analogous causes, the image of a near object would be formed *further back* than the retina, and consequently the object is only seen indistinctly, whilst a far off object produces an image less far back, and is therefore seen more clearly. In this case a convex lens is requisite to cure the defect, the curvature, as before, being suitably proportioned to the degree of flattening of the crystalline lens.

It frequently happens that both eyes are not alike, and that the curvature requisite in the lens for the right eye is not the same as that necessary for the left eye, in order to produce equally distinct vision with both.

Combination of Lenses and Mirrors.—Several kinds of optical instruments are in common use, where the essential function of the arrangement is to produce an image largely magnified, as compared with the object viewed. According as the object is a long way off or close at hand, the instrument is generally known as a *telescope* (from the Greek for *vision afar off*) or a *microscope* (similarly from the Greek for *minute vision*). In order to obtain the best definition, simple lenses, &c., are usually replaced by combinations of lenses for the purpose of obtaining freedom from chromatic and spherical aberration, &c.; but the general principles involved are not affected by this circumstance.

Expt. 373. To Construct a Galilean Telescope.—Make a tube of pasteboard of convenient diameter by rolling brown paper, &c., on a cylindrical wooden rule, metal tube, &c., and pasting together the successive layers; preferably several such tubes may be made of such sizes that one will slide stiffly within the next when completely dry, so that by pulling out the series a considerable length of tube may be obtained, whilst by pushing each tube within the next the whole is reduced to a comparatively short length.* At one end of the widest tube fix a convex lens of from 6 to 12 inches focal length or even more. This is conveniently effected by selecting a tube, the diameter of which is a shade larger than that of the lens, and fixing therein a bit of the next-sized tube, only about $\frac{1}{4}$ inch in length, cut off for the purpose. This inner ring is cemented inside the wider tube, and furnishes a rest or support for the edges of the lens, which is kept in position by a piece of brass wire bent nearly into a circle of slightly larger diameter than the wider tube, so that its springiness fixes it tolerably firmly in position when it is slightly compressed, and so made to enter within the wider tube. In this way the lens is retained in position for use, but can be removed for cleaning when required by simply removing the brass spring ring and allowing the lens to drop out.

At one end of the smallest tube used a concave lens is to be similarly fixed, the focal length of this being about 1 inch or less; the tubes are then to be pulled out or pushed in until the two lenses are separated by a distance nearly equal to *the difference between their focal lengths;* so that if the convex lens have a focal length of 6 inches and that of the concave lens be 1 inch, the two lenses will be nearly 5 inches apart. If a distant object be now looked at through the lenses and tube, adjusting their distance apart accurately by pushing in or pulling out the tubes with a screwing motion until clear definition is obtained, an erect magnified virtual image will be perceived. Ordinary opera glasses consist of a pair of similar telescopes of this description, mounted side by side, forming a "binocular telescope." Owing to the comparatively considerable shortness of the instrument when the lenses are adjusted so that the focal lengths are respectively 6 or 8 inches and 2 to 3 inches, this form of telescope is extremely portable. Fig. 190 represents the course of the rays of light whilst passing through

* This device of preparing tubes (of metal or other materials), sliding one within the other, is extensively used for many purposes other than optical instruments; thus, sliding tube gasaliers, &c., are generally spoken of as "telescopic" on this account; similarly, a dining table that can be extended by pulling out the frame so as to admit of extra "leaves" being inserted is generally termed a "telescopic" table.

the two lenses; the distant object, AB, would form an inverted real image, *ab*, on the other side of the convex lens were it not that the rays tending to form this image are intercepted by the concave lens, and so refracted as to produce upon the eye of the observer the same effect as though they had been emitted by an object, $\alpha\beta$; *i.e.*, an erect virtual image, $\alpha\beta$, is formed. The linear magnification or magnifying power is the ratio of the angle subtended at the eye by the virtual image $\alpha\beta$ to that subtended by the actual object, AB, and is approximately equal to the quotient obtained by dividing the focal length of the convex lens by that of the concave one; *e.g.*, if these lengths be 8 and 2 inches respectively, the magnifying power will be about 4 diameters; if 12 and 1, about 12 diameters, and so on.

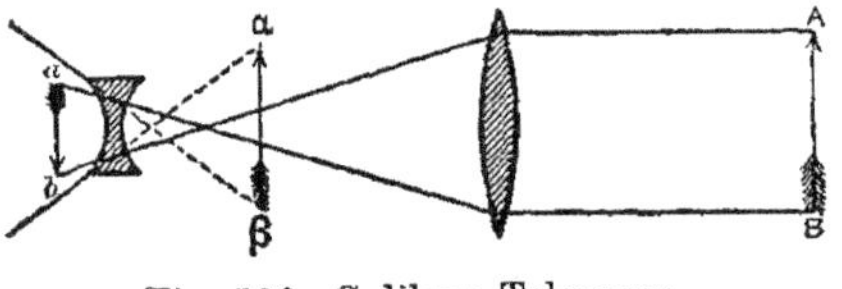

Fig. 190. Galilean Telescope.

Expt. 374. The Astronomical Telescope.—This telescope principally differs from the preceding form of Galileo's construction in that the lens nearest the eye (**eyepiece**) is convex and of short focal length, the two lenses being now placed at a distance apart nearly equal to *the sum of their focal lengths*, instead of the difference. As before, the magnifying power is nearly equal to the quotient of the focal length of the glass furthest from the eye (**object glass**), divided by that of the eyepiece; so that for the same magnification the astronomical telescope is longer than the Galilean.

Fig. 191 represents the formation of an image by this form of telescope; the object glass produces a real image, *ab*, situated between the two lenses and inverted, the object glass acting here

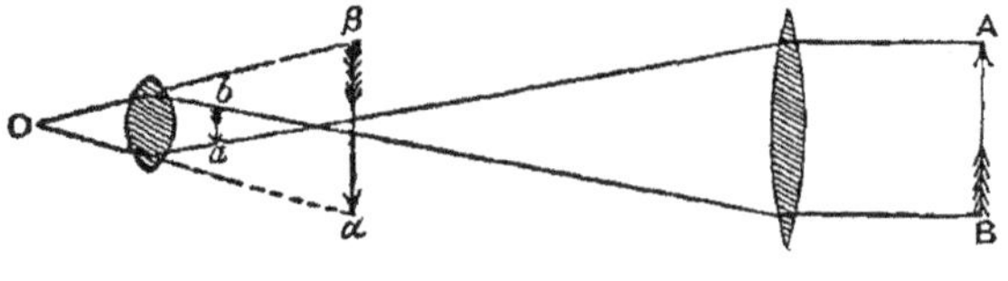

Fig. 191. Astronomical Telescope.

precisely as the lens of a *camera obscura* (Expt. 370). The eyepiece acts as a *simple magnifying lens* (Expt. 367), producing to the eye of an observer at O a virtual erect magnified image, $\alpha\beta$, of what serves to this lens as object, viz., the real image *ab*, formed by the object glass; as this is inverted with respect to the actual object viewed, it results that *the image ultimately perceived by the eye is upside down.*

For astronomical observations this inversion is of no particular consequence; for land use some extra lenses are introduced, the effect of which is to produce a second inversion of the image, and consequently to render it visible erect, much as the reflecting plane mirror in a camera obscura erects the inverted image formed by the lens (Expt. 370). Telescopes thus adjusted are termed *terrestrial telescopes.*

Expt. 375. Reflecting Telescopes.—By employing a concave mirror (preferably of parabolic curvature, Expt. 361, footnote) instead of a convex lens to receive light from a far off object, the rays of light are converged to a focus, and form a real image in front of the mirror. In what is termed the *front view reflecting telescope* the real image thus formed is magnified by lenses acting as eyepiece exactly as the analogous image formed by the convex lens of an astronomical telescope, the mirror being so large that the observer's head in front only blocks out a fraction of the total light (mirrors 6 feet and upwards in diameter being used). As with the astronomical telescope, the magnifying power is nearly the quotient of the focal length of the mirror divided by that of the eyepiece. In *Newton's reflecting telescope* fitted with a concave mirror not of such large dimensions, a small plane mirror is supported by a wire, &c., a little nearer to the concave mirror than its principal focus and inclined at 45° to its axis; the rays of light are then reflected to one side at right angles to the axis of the telescope, and are thus enabled to form an image in a position where it can be magnified by an eyepiece without requiring the observer's head to be in front of the mirror; very little light is obstructed by the plane mirror on account of its size, which is small in comparison with that of the concave mirror. Instead of a mirror a right-angled prism can be used acting like the prism-lens in Expt. 370.

In *Gregory's reflecting telescope* the concave mirror has a circular hole in its centre; the converging rays of light reflected from the outer annular portion are reflected back a second time from a smaller concave mirror, supported some distance in front of the larger one and directly facing it; the rays thus reflected back a second time, pass through the cavity in the larger mirror and form a real image, which is then magnified by an eyepiece as before. In *Cassegrain's telescope* a small convex mirror is used instead of a concave one to reflect back the rays, being placed in a different position in reference to the larger concave mirror.

In all kinds of telescopes the use of tubes to enclose the paths of refracted or reflected rays is chiefly to support the lenses, &c.,

and to prevent scattered light from surrounding objects from reaching the eye and so rendering it less sensitive to the impression produced by the rays passing through the instrument; they do not directly contribute to the formation of the image. A home-made astronomical telescope, with an eyepiece consisting of a double convex lens of about 1 inch focal length and half or $\frac{3}{4}$ inch diameter, together with an object-glass of $1\frac{1}{2}$ or 2 inches diameter and 30 inches focal length, will magnify some 30 or 40 diameters; a power sufficient to enable Jupiter's moons and other interesting celestial objects to be distinctly seen on a clear night, especially if the telescope is mounted on some kind of a stand to steady it, instead of holding it in the hands.

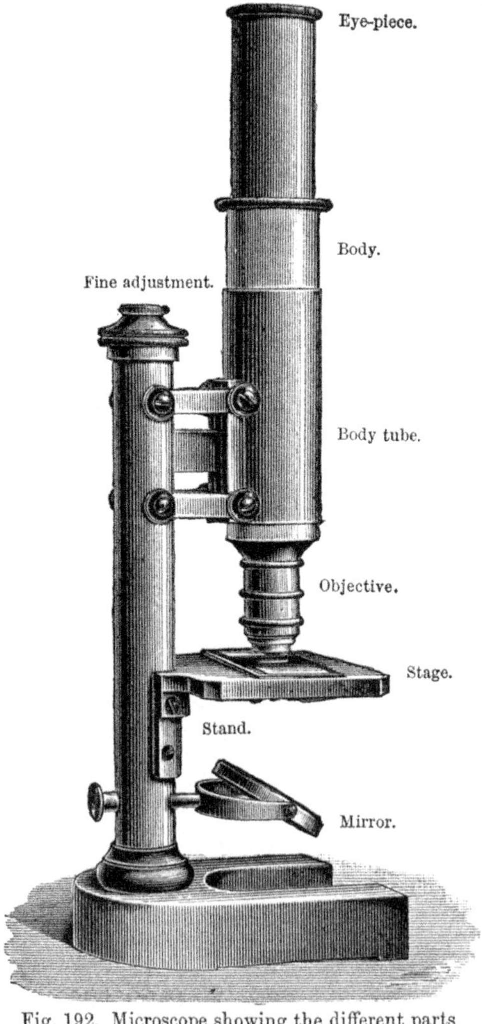

Fig. 192. Microscope showing the different parts of the Instrument.

Expt. 376. The Compound Microscope.—The astronomical telescope, as above stated (Expt. 374), is in principle a *camera obscura combined with a simple magnifier;* in the same kind of way a compound microscope is substantially a *magic lantern com-*

bined with a simple magnifier. The object to be viewed is brightly illuminated, if solid, by condensing upon it the light of a lamp by means of a subsidiary lens, or reflecting light on to it by a concave mirror, or both; if transparent, by similarly passing through it a condensed beam of light; the object glass being placed so near to the object that the distance between is a little greater than the focal length of the object glass, a real inverted magnified image is formed on the other side of the lens and further off precisely as in a magic lantern. This is then viewed through a magnifying eye piece after much the same fashion as the image formed by the object glass of an astronomical telescope. When the object glass is of short focal length ($\frac{1}{4}$-inch, $\frac{1}{8}$-inch, or less), the real image formed thereby at several inches distance on the other is largely magnified, and this magnification is multiplied again by the eyepiece. When "high powers" are used (object glasses of small focal length and highly magnifying eyepieces), the lenses and combinations of lenses used must be carefully proportioned and regulated to avoid chromatic aberration and similar sources of imperfect definition.

Fig. 192 represents a simple form of compound microscope, where the object to be viewed is placed on the stage, and illuminated from below by light reflected from a mirror.

Microscopes are sometimes fitted as a photographic camera at the eyepiece end, so that instead of employing the eye to observe directly, a small photographic glass slide, &c., is placed in position, and a highly magnified real image of the object viewed focussed on the slide, so as to obtain a permanent photograph of the object. Telescopes are also sometimes used in the same way, *eg.*, in obtaining photographs of *nebulæ*, &c.; in this case the telescope must be moved by a clockwork arrangement, so as to keep it constantly pointing in the same direction; otherwise the rotation of the earth would cause the position of the image of a given object to be always in motion, which would prevent a clearly defined photograph being obtained with any objects emitting but little light, and therefore requiring a considerable time of exposure (*vide* Chapter XXV.).

§ 9. Radiant Action: Invisible Light.

CHAPTER XXIV.

RADIANT HEAT.

One of the simplest possible observations is that when one stands in the sunshine in summer time so as to be illuminated by means of visible light a sensation of warmth is developed, which is considerably diminished by moving into the shade of some solid large object impervious to visible light, and consequently casting a "shadow;" in other words, a radiant action accompanying visible light, but producing a different sensation, is also present in sunlight. If an ordinary artificial fire burning in a grate be examined, it will be noticed that on interposing a pane of glass between the fire and the face, the sensation of warmth due to the radiation of heat will be diminished, although the fire is practically just as visible as before, showing that *glass is less transparent for that form of radiant energy termed heat than for visible light.* If, instead of a pane of glass, a deal board or metal teatray be used, still less heat will be felt, whilst all visible light will be cut off, both the wood and metal being quite opaque to either form of radiation.

If, whilst direct radiation is thus screened off, a mirror be so held by another person that the fire can be distinctly seen by reflection at not too great a distance, a sensation of warmth on the face will be noticeable, which will disappear on altering the position of the mirror, so that the blaze is no longer visible; therefore *radiant heat is reflected in the same way as visible light.* A delicate thermometer will indicate analogous results even more perfectly than the face.

Yet again, if a large convex lens (Expt. 365) be so held that a bright image of the sun is formed on a sheet of white paper, the paper will become strongly heated where the image is produced, and will be fired, the lens acting as a "burning glass" and converging the heat rays to a "focus" situated in much the same position as that of the visible rays.

In like manner all experiments possible with visible light may be performed with radiant heat with exactly analogous results when the methods of detection of heat employed are sufficiently delicate to enable such observations to be made. In short, that form of radiant energy known as heat only differs from the other form called visible light in certain respects of degree rather than of kind, and more especially in the particular quality known as **refrangibility** (Expt. 349).

Exactly the same remarks apply to that other form of radiant energy known as **actinism** (Chapter XXV.), or radiant chemical energy; the essential differences between the three being that an intensely heated source of radiant action (such as the electric arc or lime light amongst artificial sources, and the sun amongst natural ones) emits rays of different degrees of refrangibility; those of the highest refrangibility do not produce any effect on the ordinary human eye, and are consequently invisible; but they possess the power of setting up certain chemical changes to a much greater degree than rays of less refrangibility, and also when made to illuminate fluorescent bodies (Chapter XXV.) they enable these bodies to emit a visible form of light. Those of medium refrangibility are perceptible to the human eye, and produce the sensation of different kinds of colour, according to the degree of refrangibility, the tint termed *red* being due to less refrangible rays than those giving the colour *green*, which, again, is developed by rays less refrangible than those to which the sensations of *blue* and *violet* colours are due. Those rays the refrangibility of which is mostly too low to render them perceptible to the human eye are mainly such as affect the nerves of heat sensation and ordinary thermometric apparatus; so that it is quite possible to take a beam of bright sunlight or electric arc light, and by means of screens of suitable materials or other devices to absorb or remove one or other of the three kinds of rays, and obtain either a beam of invisible heat rays capable of producing large calorific effects, or a beam of visible light possessing but little heating power or chemical activity, or a beam of rays barely visible to the eye and almost destitute of heating power, but powerfully photographic, and capable of setting up a high degree of fluorescence. It must not, however, be supposed that the luminous rays are entirely devoid of heating action or photographic power, that the chemical rays are incapable of producing calorific effects, or that the invisible rays of low refrangibility are unable to effect chemical changes, for this is not at all the case; thus photographs of the invisible less refrangible part of the spectrum can be taken in virtue of the chemical changes capable of being set up by this class of rays, although it is true that they are much less active than more refrangible rays in this respect; and similarly, perceptible calorific effects are producible by means of the ultra-violet rays when sufficiently delicate means of observation are employed. Again, when rays of particular refrangibilities are absent from a spectrum, not only is there less luminous effect at that spot (producing the sensation of comparative darkness to the eye); but, further, the chemical and heating effects of that part of the spectrum are diminished relatively to the adjacent portions visible to the eye.

One remarkable point of similarity between visible and invisible light (whether of lowest or highest refrangibility) is that all kinds appear to be propagated with exactly the same velocity; whilst the same law as to decreasing power as the distance increases is also obeyed in all cases, viz., that the decrease goes on at a rate proportionate to the reciprocal of the square of the distance, so that the light is only one-fourth as intense at double the distance, one-ninth at treble, and so on.

Expt. 377. To obtain Rays of Low Refrangibility unmixed with Visible Light.—This effect is easily obtained by passing a beam of light from a powerful electric arc light, or from the sun (conveniently reflected by a *heliostat* or mirror kept moving by clockwork, so as always to throw the reflected beam along the same path, no matter how the sun varies in its position in the heavens), through a rectangular trough with glass sides (fig. 193), containing a solution of *iodine in carbon bisulphide.* If a sufficiently thick layer of this fluid be used, no visible light whatever passes through; but the invisible heat rays are only slightly absorbed; hence, if a convex lens be placed in the path of the beam of invisible light exactly the same result will be produced as with visible sunlight, viz., the rays are concentrated to a **focus** (Expt. 365), so that inflammable objects can be fired and solids heated red hot or even white hot by the accumulated heating effect, just as when the lens is used as an ordinary "burning glass." The best results are here obtained when the lens is made of clear *rock salt,* as glass and many other media, though transparent to visible light, largely absorb the heat rays of low refrangibility (Expt. 382); this absorption is especially marked with *alum* (Pa. 292).

Fig. 193. Trough with Glass Sides.

Another way of obtaining a beam of invisible heat rays is to pass the light from a powerful electric arc first through a rock salt lens to concentrate and parallelise it, and then through a rock salt prism or series of prisms; in this way the total beam is refracted, and the rays of different refrangibility made to traverse different paths mutually inclined to each other; by interposing dark screens in suitable positions, all the luminous rays may thus be cut off, and only invisible ones obtained. In this way any required portion of either the least refrangible or the most refrangible rays may be separated from the visible ones, so that any kind of invisible ray of particular refrangibility may be obtained as required.

For the most refrangible ultra-violet rays, absorption by a deep blue transparent medium, such as ammoniacal solution of copper, may also be employed; all visible light of lower refrangibility (red, orange, yellow, and green rays) being absorbed, and only the indigo, violet, and ultra-violet rays being allowed to pass.

Expt. 378. Reflection of Radiant Heat—Burning Mirror.—By the aid of a flat mirror and thermometer it may readily be shown, as above described, that radiant heat is reflected along the same path as the visible rays emanating from the same source. Accordingly a concave mirror will collect to a **focus** a beam of radiant heat, just as it will visible light. A *burning mirror* is simply a concave reflecting surface; if the curvature be spherical, and such a mirror be turned directly towards the sun, the heat rays will be reflected along with the visible ones, and will all become concentrated to a **focus** at or near a point midway between the surface of the mirror and its centre of curvature on the prolongation of the line joining the sun's centre and that centre of curvature. If the mirror be large enough, inflammable objects placed at this point (paper, gunpowder, &c.) will be fired, so that a cigar may be thus lit; even if it be small, the concentration of the heat rays is still readily perceptible on placing the bulb of a thermometer at this geometrical focus, or the head of a match, or a little bit of phosphorus.*

Expt. 379. Two Conjugate Mirrors.—Obtain two similar concave reflectors about 10 or 12 inches diameter, and with a radius of curvature of some 12 or 15 inches; the workmanship need not be very fine, circular sheets of copper roughly hammered into shape answering the purpose sufficiently well when the surface is tinned or silvered, or even rubbed over with quicksilver, and polished so as to reflect light pretty brightly. Arrange the two mirrors exactly

* Tradition states that Archimedes set fire to the Roman fleet at Syracuse by means of large concave mirrors. Buffon succeeded in firing a tarred wooden plank at 70 yards' distance by what was equivalent to a large concave mirror, viz., a number of comparatively small plane mirrors (128, each about 8 inches long and 5 broad), arranged close together and so inclined to the path of the sunlight that each one reflected the incident light to the same spot, 70 yards off.

facing one another (fig. 194), so that their centres of curvature and the centres of the mirrors may be in the same horizontal line, supporting the mirrors by any convenient stands; the distance apart may be 3 or 4 yards, or even more.

Midway between the centre of curvature and the centre of one mirror place a hot object, such as an iron ball, one or two inches in diameter, heated red hot, and on the further side of it place a circular screen a little larger in diameter, so as to cut off all radiation from the hot ball towards the further mirror. Notwithstanding, heat will reach this further mirror, being that radiated in the opposite direction by the hot ball, and then reflected in a nearly parallel beam by the first mirror; the heat thus reflected will be again condensed to a focus by the second mirror, so that if the bulb of a thermometer be placed at the geometrical focus of that mirror the mercury will quickly rise. If the mirrors are tolerably accurately shaped, the heat will be very perceptibly felt by the finger at that point, and a bit of phosphorus or a match may be fired there. This experiment, therefore, shows that the same laws govern the reflection from concave mirrors of invisible heat rays as of visible luminous rays.

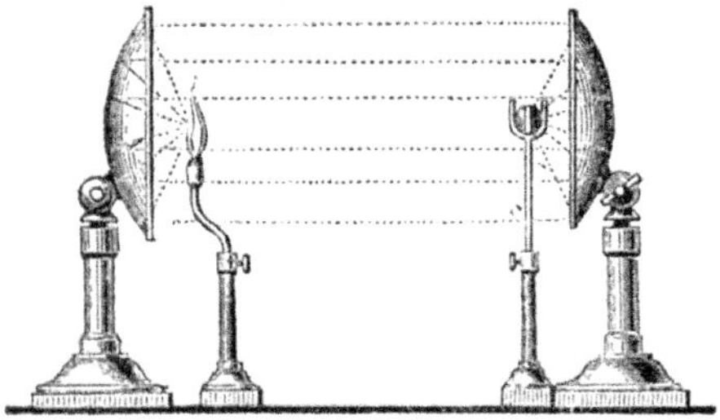

Fig. 194. Two Conjugate Mirrors (reflection of heat).

Expt. 380. Apparent Reflection of Cold.—Repeat the previous experiment in a moderately warm room (the mirrors not being too far apart), placing a lump of ice in the focus of the first mirror instead of a red-hot metal ball. A thermometer placed in the focus of the second mirror will now indicate a *fall* of temperature there, apparently indicating that *cold* is being radiated and reflected. This effect is due to the circumstance that every body is continually absorbing heat radiated from surrounding objects and losing heat by emission as radiant heat. When the rate of gain by absorption just equals that of loss by emission, the temperature remains constant; if the former exceed the latter, the temperature rises, and *vice versâ*. In Expt. 379 the thermometer bulb is placed in such circumstances that the total heat received by it (by radiation from the hot ball as well as other objects) considerably exceeds that parted with, wherefore an increase of temperature results;

but the opposite is the case when the ice is substituted for the hot metal, the thermometer bulb now radiating away more heat than it receives back in return.

On the other hand, if this experiment with ice be made in an excessively cold room during a hard frost, so that the temperature of the apartment is considerably below zero, the ice will behave as a comparatively *hot* object; a delicate thermometer placed in the focus of the second mirror (not too far away) will indicate a temperature below zero when nothing is placed in the focus of the first mirror, and all the arrangements are allowed to stand awhile to equalise the temperature; on placing a piece of ice in position at the focus of the first mirror, the thermometer will be seen to rise. In carrying out this experiment a screen must be so arranged as to prevent heat radiating from the warm person of the observer reaching the thermometer directly; the piece of ice should be placed in position by means of a long pair of tongs.

Expt. 381. Emission of Radiant Heat.—The emission of heat in the form of rays of low refrangibility from the surface of a hot body is greatly influenced by the nature of the surface. Obtain half a dozen circular canisters of polished tin plate, all of the same size, brighten the surface of one with oil and whiting, paint one white and another black, smoke one all over with lampblack by means of a large smoky flame, and cover the other two with tight-fitting jackets of calico and thin black cloth respectively. On filling the canisters with hot water from the same kettle, and allowing them to stand awhile to cool, it will be found, on placing a thermometer in each, that although each necessarily becomes cooled by convection to about the same extent (Chapter XX.), additional cooling by radiation goes on at very different rates in the different cases. The brightly polished tin radiates heat least readily and cools slowest, whilst the lampblacked canister cools very rapidly. The can painted black will cool quicker than the one painted white; similarly the one jacketed with black cloth will cool quicker than the one jacketed with white calico; in all cases the nature of the surface, and even its colour, influences the rate of radiation. The difference in radiant power thus possessed by different substances and surfaces is one great cause why dew deposits at night time more freely on some objects than on others. Those substances which are the best radiators naturally cool quickest, and therefore become cooler than others which do not radiate so freely. Consequently, air charged with moisture coming in contact with such bodies will be sooner chilled down below the dew point (Expt. 45), and will deposit moisture more copiously than when in contact with substances that have not become so much chilled on account of the lesser amount of radiation from their surfaces.

Expt. 382. Absorption of Radiant Heat.—In general, surfaces that radiate heat best absorb radiant energy in the form of rays of comparatively low refrangibility the best. Thus two canisters, such as those used in the preceding experiment, filled with the same quantity of cold water and placed side by side at equal distances from a good fire, will absorb heat and cause the water to become warmed at very different rates according to the nature of the outer coating; a polished bright smooth surface will absorb heat least freely, and the contained water will take a much longer time to rise through a given range of temperature, than will be the case with the lampblacked canister; and similarly in other instances.

It does not necessarily follow that a body which absorbs radiant energy of one degree of refrangibility more readily than another body will show the same difference with regard to radiant energy of another degree of refrangibility. Thus a copper vessel containing boiling water radiates forth rays of very low refrangibility, which are absorbed almost equally freely by a surface coated with powdered white lead, and by another covered with lampblack; but if a copper surface at a temperature of 400° be used as the source of radiant heat instead of boiling water (temp. = 100°, Expt. 28), the lampblack absorbs better than the white lead in the proportion of 9 to 8; whilst, if the radiation be from white hot platinum, the lampblack absorbs nearly double as much as white lead does.

Just as many bodies, when exposed to white light, scatter light of one refrangibility more than any other kind, and therefore appear to be **coloured** (Chapter XXII.), so bodies exposed to rays of lower refrangibility often scatter or **diffuse** certain of these heat rays more freely than others. Hence such bodies possess, as it were, a sort of **heat colour**; the other rays incident upon the body being more freely absorbed than the heat rays which are thus preferentially diffused.

Just also as transparent substances always absorb more or less of the visible form of radiant energy when this traverses their substance, the emergent light being **white** if all kinds are equally absorbed, and **coloured** if certain kinds are preferentially absorbed, so are the heat rays of different refrangibility analogously affected; thus different gases (*e.g.*, nitrogen, oxygen, sulphur dioxide, ammonia, &c.) absorb radiant heat with very different degrees of facility; ammonia, for example, absorbs about 1200, and sulphur dioxide about 700, times as much as either oxygen, nitrogen, or air when radiant heat is made to traverse equal lengths of column of gas in each case under the ordinary atmospheric pressure; whilst the vapours of many substances liquid at ordinary temperatures and pressures, exert even greater absorbing effects when examined either under very low pressures or disseminated through air (like aqueous vapour in the atmosphere). The aqueous vapour itself contained in ordinary air produces so great an absorptive action that the quantity present in the stratum of air lying between the levels of the top of Mont Blanc and Geneva (some 14,000 feet thick) is sufficient to absorb about $\frac{1}{5}$ of the radiant heat reaching the earth from the sun at the former elevation; and the difference in absorptive action, according as more or less aqueous vapour is present in the air of a room, is readily measurable by sufficiently delicate instruments.

Expt. 383. The Light Mill.—An instrument, known as the *radiometer* or *light mill*, has been constructed by Mr Crookes, consisting of a vertical axis with horizontal wires affixed thereto, at the extremities of which are vertical vanes of mica, platinum foil, or other substances, clean and bright on one side and smoked on the other. The whole is enclosed in a glass vacuum bulb, *i.e.*, a glass bulb from which the air is highly exhausted by means of a powerful mercurial pump. The difference in the nature of the front and back surfaces of each vane causes a difference in the effect of the radiations received by and emitted from the two surfaces respectively, the ultimate effect of which is that a mechanical force is produced, causing motion of the vane. As each vane is similarly situated as regards the sides blackened and not blackened, the effect of each one is to produce rotation of the whole axis and vanes in the same direction; hence, on setting the instrument in such a position that radiant energy meets it (*e.g.*, exposing it to diffused daylight or direct sunlight or electric arc light, &c.), a continuous revolution is produced, the whole revolving somewhat after the fashion of the sails of a windmill, whence the name "light mill."

Expt. 384. Passage of radiant heat through water.—Although aqueous vapour exerts a comparatively great absorptive action on radiant heat, still liquid water is sufficiently pervious to such rays to enable powerful burning glasses to be constructed by filling globular vessels with water and exposing them to the sun. A fishglobe or flask of distilled water can thus easily set fire to gunpowder or paper, &c., placed at the focus—a matter worth remembering as regards the possibility of setting fire to furniture, &c., since it has been proved that several serious fires have been thus originated. A "bull's eye" in the glass of a window has been known to act similarly, whilst occasionally a drop of dew or rain on a leaf on a summer's day acts as a burning glass and scorches the leaf.

Expt. 385. To make a Burning Glass of Ice.—Fill an evaporating basin, the shape of which is as near that of a hemisphere as possible, with cold water, and then surround it with a mixture of snow and salt (Expt. 21), renewed when requisite until the water is frozen to a solid mass. Detach this from the basin by turning it upside down and pouring a little warm water over it; as the heat penetrates through the porcelain or earthenware the ice in contact with the basin thaws; and by and by the mass of ice drops out in the form of a rough plano convex lens. Place this in the sun so that the beams fall directly upon either the flat or the convex side; the less irregularly shaped the rough ice lens is the more nearly will both the visible light and the heat rays be converged to a focus. If tolerably clear and accurate in form, the heat rays will be found to pass readily through the ice, and will cause the mercury to rise rapidly in a thermometer placed near the visible focus; with a lens some 8 or 10 inches in diameter and a moderately hot sun, paper, &c., may readily be fired.

A still better way of preparing the ice lens is to get a solid block of clear ice with one flat surface (either naturally flat, or rendered so by rubbing the block on a warm metal plate, &c.); with a hot poker carefully applied melt away some of the ice on the other side so as to make it roughly convex, and complete the shaping process by inverting a hemispherical basin over the lump, pouring hot water over it from time to time, moving the basin about and turning it round so as to thaw the upper surface of the ice block down to the proper shape.

Expt. 386. Sunshine Recorder.—A useful application of the principle of the burning glass is made in an instrument designed to measure the amount of sunshine visible during the day. A ball of glass or a hollow glass globe filled with water is set in the sun, supported by a light frame, with a piece of stout white paper so arranged that some part of the surface of the paper is in focus at all hours of the day, the particular part depending on the hour, *i.e.*, on the position of the sun in the heavens. When the sun shines its rays are focussed on the paper by the ball and a black spot is burnt there; if the sun continues shining, the focus alters its place as the sun moves, and the spot becomes lengthened into a black line. When the sun is hidden by clouds, &c., the burning action is stopped; so that at the end of the day the hours during which the sun has shone brightly, as compared with the period when it was under a cloud, can be readily measured by noticing the length of the part of the paper actually burnt as compared with the rest of the track that the focus would have followed and burnt had the sun shone all the time.

Somewhat analogous appliances are also used where the recording action is the production of chemical changes in photographic paper, &c., by the sunlight falling thereon. When the sunlight falls fully on properly prepared silvered photographic paper (Expt. 392) a large amount of darkening and blackening is produced; but when the sun is more or less obscured, the amount of darkening is proportionately lessened. In sunshine recorders of this kind it is usually not necessary to condense the light by means of a lens; a slit is made in a plate of metal, &c., and underneath this a strip of prepared paper is made to pass regularly by clockwork, being unwound from one roller and wound up on another, somewhat as the painting of a panorama exhibited to an audience. Thus suppose 12 inches of paper pass under the slit during 12 hours, and the sun shine at intervals during the time, the particular hours of the day when the sun was obscured will become visible by noticing whereabouts in the course of the 12 inches the light parts lie; whilst the relative intensity of the light is indicated by the greater or less amount of blackening produced at the darkened parts of the paper.

Automatic photographic registering appliances of this description are now largely used in meteorological and other observatories for the purpose of recording the height of the barometer, the fluctuations of the thermometer, the variations of the compass-needle, &c., &c.

CHAPTER XXV.

RADIANT CHEMICAL ACTION OR ACTINISM: PHOTOGRAPHY.

It has already been stated that whilst the most highly refrangible rays have the power of bringing about chemical changes to the greatest extent (Pa. 351), still the rays of lower refrangibility are not by any means destitute of this property. The amount of chemical change brought about by one kind of ray as compared with another is by no means the same for all kinds of substances. Just as a given coloured transparent medium absorbs visible radiations of certain kinds more freely than others, so that these are more readily transmitted, producing coloured emergent light; and just as a similar peculiarity is observed with certain substances in regard to rays of low refrangibility, producing the phenomena of **heat colour** (Expt. 382), so are various substances capable of undergoing chemical change under the influence of light more sensitive to rays of particular refrangibilities than to others, the energy of these particular rays being absorbed in doing the chemical work.

Of the numerous substances sensitive to this kind of radiant chemical action, certain silver compounds are, for a variety of reasons, the most convenient for the ordinary purposes of photographic reproduction of likenesses, views of landscapes, &c., &c. Accordingly, the most widely employed processes in photography are those involving the use of these materials. "Chromatised gelatin" and certain iron and platinum compounds, however, are also employed for special kinds of work to some considerable extent; whilst many other substances are known capable of use which have not hitherto come into extensive practical employment, on account either of their greater cost or of their requiring more troublesome manipulation, &c., as compared with silver compounds.

Expt. 387. To take a Photograph by the Collodion Process.— One of the most extensively used photographic processes consists in covering the flat surface of a sheet of glass or other supporting material with a film of "collodion," which is subsequently impregnated with silver compounds so as to obtain a flat surface sensitive to light. Collodion consists of a solution in a mixture of ether and alcohol of a variety of gun cotton (*pyroxylin*) obtained by treating cotton with sulphuric and nitric acids; the fluid thus obtained, when poured out on a flat surface and allowed to evaporate, leaves a film of collodion, somewhat resembling thin parchment when dry, or the white membrane lining an egg. A film thus obtained is often employed as a coating or protection for the skin when this has been severely injured by burning or scalding, an artificial protecting membrane being thus produced. In order to impregnate the film with silver compounds, *iodide of cadmium*, *bromide of ammonium*, or other corresponding salts (preferably those soluble in alcohol or ether) are also dissolved in the collodion, forming a compound fluid sold under the name of *iodised collodion*. A glass plate is made scrupulously clean and highly polished, care being taken not to touch the flat surface with the fingers, otherwise spots and smears will ultimately be produced; iodised collodion is then poured over the plate held horizontally, and allowed to flow all over the surface by dexterously inclining it, the surplus fluid being poured back into the bottle;

this operation requires a little practice to accomplish satisfactorily. In a few seconds the ether mostly evaporates, and a semi-transparent

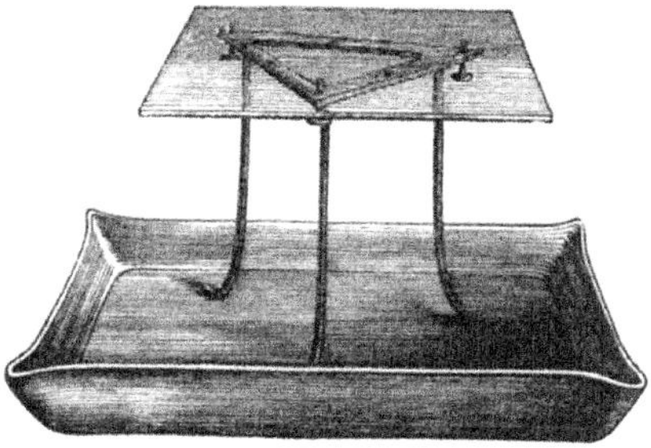

Fig. 195. Pouring Collodion on Plate.

Fig. 196. Glass Plate Holder.

film of soft collodion is obtained on the surface of the glass; better adherence of the film to the glass when dry is obtained if the glass be coated with albumin beforehand. The glass is now dipped into a watery solution of *nitrate of silver* (about 1 ounce of the solid salt to 10 or 12 of water). Fig. 195 represents one way in which the collodion may be poured over the glass plate so as to form a uniformly thick film; and fig. 196 represents the glass plate supported on a slab of glass or porcelain, with a ledge at the bottom so as to enable it to be dipped into the nitrate of silver bath (fig. 197). Double decomposition (Expts. 11, 12) now takes place between the iodide and bromide of cadmium contained in the collodion and the nitrate of silver, so that *nitrate of cadmium* is formed, which dissolves in the water, and *iodide* and *bromide of silver* are precipitated in small particles disseminated through the mass of the collodion, this salt being insoluble in water; the action being very similar to that whereby chloride of sodium and nitrate of silver give rise to soluble nitrate of sodium and insoluble chloride of silver in Expt. 37.

Fig. 197. Silver Bath.

The glass covered with collodion should be then **sensitised** in a dark room, *i.e.*, one illuminated only with yellow light, such as a sodium flame (Expts. 245, 355) or a candle, &c., placed in a lantern with yellow glass sides, so that actinic rays may be prevented from reaching the silver compound dispersed throughout the collodion. If the nitrate of silver solution be perfectly fresh it should be previously treated with a small quantity of iodide of potassium (about 2 grains for every ounce of solid silver nitrate), so as to form enough silver iodide to saturate the liquid with that salt, which, though nearly insoluble in pure water, is somewhat more soluble in solution of silver nitrate; if this precaution were not taken, the silver iodide formed by double decomposition would be more or less dissolved out of the film, which would thus be rendered much less sensitive. The plate should be dipped into the nitrate solution with a steady motion, resting in a glass or porcelain holder sold for the purpose, and after a minute lifted out, and again dipped in for another minute; finally, it is removed and placed in position in a dark slide * fitting into the camera (Expt. 370), supported on a tripod stand and previously properly focussed, so that a sharp

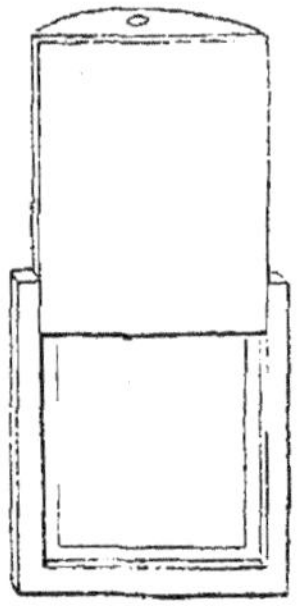

Fig. 198. Dark Slide.

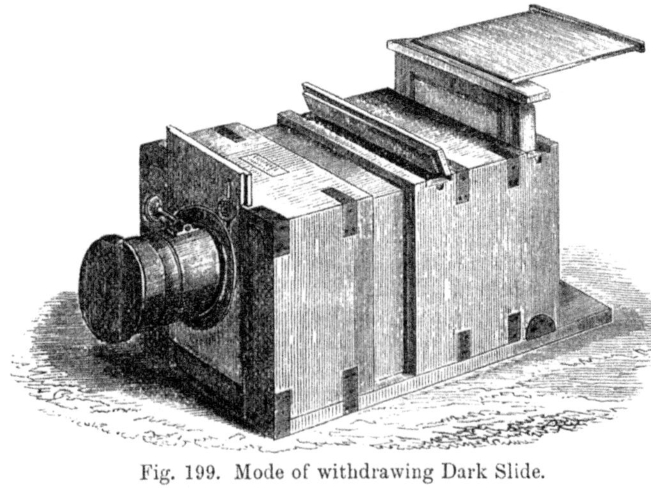

Fig. 199. Mode of withdrawing Dark Slide.

image of the landscape and sitter is produced on a ground glass

* The dark slide is simply a light wooden frame fitting into the camera, resembling a picture frame, excepting that the front part is of wood instead of glass, and can be removed by drawing it outwards (fig. 198), so that the sensitised plate is introduced into the camera inside the frame, with the sliding wooden panel closed to exclude light whilst placing the whole in position; the slide being then withdrawn (somewhat after the fashion of a dark lantern, fig. 199), the sensitised plate is ready to receive the impression on admitting light through the lens by uncovering the cap of the instrument.

plate occupying exactly the same position as that into which the collodion plate is subsequently introduced. The lens of the camera is then covered up with a cap to exclude light, and the slide withdrawn, a black cloth being thrown over the whole to exclude any trace of stray light that might get in through chinks and crevices. At a suitable moment the cap is withdrawn, and replaced after a sufficient period of exposure (from less than 1 second to 5, 10, or even more seconds in the case of a landscape).

The slide is then pushed in and the exposed plate removed for **development** (Expt. 390), which consists in pouring over the plate some reducing fluid which will cause the transformation of iodide of silver affected by light into metallic silver in quantity proportionate to the intensity of the action of the light during exposure.

After the picture has been developed sufficiently, it is **fixed** and rendered permanent by pouring over the plate (or immersing the plate in) a solution of some compound capable of dissolving away that part of the iodide of silver that has not been decomposed by the developer, without removing the particles of reduced silver. For this purpose a solution of *cyanide of potassium* (about 5 to 7 parts of salt to 100 of water) or one of *thiosulphate of sodium* (15 to 20 parts of the crystallised salt to 100 of water; this salt is frequently but unsystematically termed *hyposulphite of soda*), is generally employed, more especially the former. The final result, therefore, after washing and drying, is that the glass plate is coated with a film of collodion throughout which are interspersed particles of solid opaque metallic silver in greater or less proportion at different spots according as the light has acted there more or less.

Expt. 388. Dry Plate Processes.—In operating in this way many little details have to be carefully attended to, some of which can only be learnt thoroughly by practice; in any case, special handbooks on photography should be consulted before attempting any experiments with costly apparatus on a large scale, the limits of space in the present work preventing such details from being discussed in full. The collodion plates, prepared as above described, are found to be much more sensitive and to work far better if used fresh and still wet with nitrate of silver solution than if kept and allowed to dry; accordingly, such plates are always used whilst fresh and moist: but there are many modifications of the process (in some of which films of matter other than collodion are used, *e.g.*, *gelatin*) where *dry plates* are employed; *i.e.*, plates that have been prepared with sensitive material (such as an emulsion of collodion or gelatin and bromide of silver, &c.) and allowed to dry, so that they can be kept for a long time unchanged in perfect darkness. On account of their convenience in use for amateur and other photographers, dry plates are now extensively used, and are manufactured on a large scale at prices so low that, excepting for special purposes, it is hardly worth while to prepare them oneself.

A collodion plate prepared as in the last experiment is practically a film of organic material through which particles of silver iodide, or iodide mixed with bromide, are disseminated, these latter being precipitated in the body of the film *after* its application to the supporting plate. Just as iodide and bromide of cadmium, &c., can be added to collodion to "iodise" it, thus impregnating it with these salts; so, similarly, can a solution of nitrate of silver in alcohol be also added, with the result of precipitating in the body of the fluid particles of insoluble silver iodide and bromide, which thus form with the fluid a sort of "emulsion"; when this is poured out on a supporting plate a film containing particles of suspended silver compounds results, very similar to that obtained in the last experiment, chiefly differing in that, as the silver salt particles are already present in the film, no immersion in solution of silver nitrate is requisite. Dry plates of this kind are not very sensitive to light, and hence are but little used; but plates prepared in somewhat similar fashion, substituting a solution of gelatin in water for one of guncotton in mixed ether and alcohol, are largely employed, the sensitiveness to light being capable of being rendered extremely great by appropriately arranging the details of the manipulation, more especially by the careful application of heat to the gelatin emulsion containing silver bromide, either without admixture of iodide or containing only a small proportion thereof.

Expt. 389. Instantaneous Photographs.—The chemical action of light upon most substances requires to be exerted for some amount of time before measurable amounts of change are produced; but gelatin bromide plates prepared in particular ways are so extremely sensitive to light that an exposure for only a very small fraction of a second in the camera is sufficient to produce the required effect. Accordingly, by employing an instrument so constructed that light is admitted for only an extremely short interval of time (an orifice being rapidly uncovered and covered again by a rapidly moving slide, &c.), it is possible to take **instantaneous photographs**, *i.e.*, photographs where objects really in motion are depicted as though at rest, the time of exposure being so brief that no material change of position has occurred during that short period; thus a railway train in quick motion, a wave breaking on rocks and throwing up clouds of spray in a storm, the flight of a flock of birds in the air, &c., may be thus clearly and distinctly photographed; or what amounts to much the same thing, an operator may take a photograph of the surrounding objects whilst he himself, together with his apparatus, is in quick motion (on a coach, ship, railway train, &c.).

Expt. 390. Development of Latent Image—Negatives and Positives.—When photographs of the film class are obtained, it usually happens that on examining the "plate," &c., that has been exposed in the camera (Expt. 387) by the faint light of a sodium flame (Expt. 355, practically devoid of chemical rays), absolutely no visible trace of any action can be seen; but that a change in the silver compounds or other substances used has been effected is at once demonstrated by the application of the "developer," *i.e.*, a fluid containing substances in solution capable of producing further chemical action on those parts of the plate that have undergone the incipient invisible action, to an extent proportionate to the extent to which this invisible change has been effected; so that wherever the light has acted most during exposure (*i.e.*, at those parts of the picture which are lightest, *e.g.*, the shirt collar and front in the case of a portrait), the developer produces most further chemical action, and *vice versâ*. The picture or image on the exposed plate is in such cases said to be **latent**; and the action of the developer is to produce a formation of minute particles of metallic silver by means of a "reducing" action somewhat analogous to that occurring in Expt. 120, where silver chloride is converted into spongy metallic silver by means of caustic soda and milk sugar. As this action goes on most rapidly where most light has fallen, it results that in the "lights" of the picture most silver is thus reduced by the developer, and in the shadows least; hence, when the developed picture is held up to the light (especially after "fixing" or dissolving out the remaining unaltered bromide or iodide of silver), wherever the original light has been brightest, most silver is reduced, so that least light now shines through; *i.e.*, when the picture is viewed by transmitted light the lights and shadows of the object, &c., originally photographed are **reversed**; such a picture is spoken of as a **negative**. On the other hand, if the picture be looked at against a black background, such as a coating of black varnish or a piece of black cloth, &c., wherever least silver is deposited, the black background is most clearly seen, and wherever most is present the deposited metal appears white or light grey by contrast; in this case the lights and shadows are not reversed, and the picture thus viewed is termed a **positive**. The quickly produced cheap photographs executed by peripatetic photographers on holidays, on the sands at the seaside, &c., are mostly positives of this kind.

A landscape photograph directly projected on a screen by a magic lantern would accordingly give a negative picture. To obtain positive pictures, copies are taken from the first plate after fixing by placing a second sensitised plate immediately behind it, and exposing the two together to light. Wherever the silver deposit is thickest (*i.e.*, where the light originally fell to the greatest extent) less light penetrates through the negative to the other plate, and *vice versâ;* so that the picture finally obtained on the second plate is reversed in reference to the negative, and is consequently normal or positive in reference to the original landscape, &c., photographed. In the printing off of ordinary *cartes de visite* and photographic portraits generally, this double reversal takes place. A photograph is first taken on glass, which, after development and putting through subsequent processes to render permanent (Expt. 387), is a negative; positive prints are then taken from this negative on suitably prepared paper by the simple process of placing the paper behind the negative in a glass "printing frame" (something like a transparent slate—Fig. 201) and exposing to light until the chemical action has gone on to a sufficient extent, and then fixing the print to prevent further change (Expt. 392).

When the object is to produce a negative to be subsequently employed for printing positives a transparent supporting plate (glass, mica, &c.), is indispensable; but if a positive is intended to be directly produced the nature of the material supporting the film is of no particular consequence; thin sheets of metal, enamelled iron, smooth wood, &c., &c., may be employed successfully. What are frequently called "ferrotypes" are positives directly prepared upon thin sheets of iron coated with a suitable black varnish before applying a collodion film thereto; the term ferrotype, however, is more properly applied to those photographic processes where iron compounds are the bodies sensitive to light therein employed (Expt. 396).

The nature of the developing agent varies with that of the film and with the kind of photograph required; wet collodion plates are usually developed by pouring on to them a solution of *pyrogallic acid* or of *ferrous sulphate*, the reducing action being somewhat modified or retarded by the addition of strong acetic acid, &c.; gelatin bromide plates are usually dipped bodily into a solution of *double ferrous and potassium oxalate*, or one of *pyrogallic acid* rendered alkaline with *sodium carbonate; hydroquinone, hydroxylamine*, and various other organic reducing agents are also employed, mostly in alkaline solution. The "fixing" solution generally employed for wet collodion plates is one of *cyanide of potassium*, whilst *thiosulphate of sodium* is usually preferred for gelatin bromide plates.

A remarkable circumstance in connection with the latent image produced on a dry plate is that the incipient chemical action set up by exposure to light usually does not go on any further by long continued keeping in the dark, nor does it become materially obliterated or modified by so doing; so that whether the exposed plate be developed at once, or not till after several weeks or even months, makes little (if any) difference in the end result.

In one of the earliest photographic processes invented (now rarely if ever used), termed the *Daguerreotype*, after its inventor, M. Daguerre, the image is formed on a dry plate surface obtained by exposing a plate of silver (or copper coated with silver) to iodine vapour, whereby a film of iodide of silver is formed by direct combination (Chapter XIV.) of silver and iodine. After exposure in the camera little or no trace of a picture is visible; but a latent image exists, capable of being developed in a peculiar way. The exposed plate is placed in contact with hot mercury vapour, so that mercury may condense on its surface; wherever the light has acted most, the mercury condenses most readily, the altered silver iodide exerting a kind of attraction for the mercury, somewhat akin to that exerted by particles of metallic silver upon a mixture from which silver is being reduced (Expt. 391); whilst the unaltered silver iodide does not attract the mercury at all. Finally, the unaltered iodide of silver is dissolved away, and the picture fixed by means of sodium thiosulphate; so that when the picture is held in a proper position with respect to the light, the condensed quicksilver appears white or greyish white by contrast with the rest of the ground, thus giving a **positive** picture.

Expt. 391. Intensification and Reducing.—When an exposed plate is developed it sometimes happens that the lights and shadows are too strongly contrasted, owing to the reduction of silver in too great quantity, more especially when the action of the light has been somewhat too vigorous. A negative which is thus too dense, can sometimes be improved by subjecting it to the action of some substance that will dissolve away a portion of the deposited silver and thus thin down or "reduce" the image. This can be done by immersion in dilute nitric acid; preferably less energetic agents are employed, such as *ferric chloride, ferricyanide of potassium*, or other substance that will partly convert the deposited silver into chloride (or bromide, ferricyanide, &c.), which can then be dissolved out by the same reagents as are used for fixing (cyanide of potassium or thiosulphate of sodium).

On the other hand, when an image is too faint it requires **intensification.** In the case of wet collodion plates this is usually effected by pouring on to the plate a solution of silver nitrate mixed with ferrous sulphate or other reducing agent; this mixed fluid tends to deposit metallic silver, but the particles of silver already present on the developed plate exert a kind of special attractive action, in virtue of which the additional silver particles deposited from the intensifying solution are attracted towards those already there, and are not deposited indiscriminately all over the plate; the result of which is that the extra silver thus deposited does not form a uniform layer, but thickens up most those parts of the picture where most silver is already present, and so on in proportion, so that the gradations of light and shade are thus regularly intensified. This action is closely akin to that occurring during crystallisation of dissolved substances, especially from a slightly supersaturated solution (Expt. 63); if suitable nuclei, more particularly bits of solid crystals of the dissolved body, be introduced, rapid separation by crystallisation of the dissolved matter ensues, the substance thrown out of solution being deposited on the surface of the introduced solid nuclei, causing them to enlarge and grow bigger; whereas little or no such separation takes place in the absence of suitable nuclei; *e.g.* (Expt. 58), a strong aqueous solution of cane sugar readily deposits crystals of sugar-candy on strings or sticks immersed therein; but if kept in a clean smooth glass vessel will often form a supersaturated solution that will be long before it deposits any crystals at all. In somewhat similar fashion certain melted solids can be cooled down considerably below their normal solidifying points without setting solid, thus showing the phenomenon of **superfusion** (Expt. 30); but if a fragment of the substance in the solid state be brought in contact with the superfused mass, the whole almost instantly solidifies. In each case a sort of attractive action is exerted whereby the presence of the solid particle acting as nucleus facilitates, in one case the passage of the same kind of matter from the liquid to the solid state (superfusion); in another case the separation of the same kind of matter in the solid condition from a solution thereof in a solvent (supersaturation); and in the third case the separation of the same kind of matter from the condition of chemical combination with something else (intensification of photographic picture); the matter thus rendered solid being deposited upon and attracted to the nucleus in preference to other surrounding matters.

This method of intensifying images is not conveniently applicable to gelatin bromide plates, because staining of the gelatin is apt to occur; in this case other methods are generally employed, whereby the deposited silver is converted by simple chemical action into more opaque substances of different nature. Thus, by treating the plate with a solution of mercuric chloride the deposited silver is partially converted into silver chloride, whilst mercurous chloride (calomel) is also formed as a deposit; by treating the plate again with a solution of sodium sulphite, ammonia, or ferrous oxalate, a further change is produced whereby the silver chloride is either dissolved out or again reduced to metallic silver, whilst the calomel is converted into minute particles of metallic mercury or other mercurial compound; the end result in every case being that the total deposit is thicker, or at least more opaque, than at first, the more so where most silver originally existed, so that the lights and shadows are intensified relatively to one another just as with the wet collodion plate, but owing to different chemical and physical actions. Ferricyanide of potassium in conjunction with uranium or lead salts furnishes another set of intensifying agents acting in a somewhat similar fashion.

Expt. 392. To Prepare Silver Prints.—If a sheet of suitably prepared paper, a film of sensitised collodion, or any similar layer of material sensitive to photographic action is exposed to light with some opaque or semi-transparent object lying on its surface, wherever the light is not prevented from reaching the paper by the intervening object chemical action will take place, so that the paper is blackened or otherwise altered; where, however, the paper is covered with an entirely opaque portion of the object to be copied, the access of light is prevented, so that the paper underneath remains unchanged; whilst, similarly, a semi-transparent part of the object will allow more or less light to pass, giving a tint intermediate between that of the unaltered paper and that which has been exposed to the full power of the light.

Fig. 200. Skeleton Leaf (Silver Negative Print).

Fig. 200 represents the reproduction in this way of a skeleton leaf. A piece of ordinary note paper or of unglazed white paper is soaked in a solution of common salt (1 to 1½ ounce salt to 1 pint water) for a few minutes, and then taken out and hung up to dry, conveniently on a string stretched across a room after the fashion of a laundress's clothes line; or, instead of soaking the paper, it is floated on the surface of the salt solution; in either case care must be taken that no air bubbles adhere to the surface of the paper, as these will produce spots in the finished print.*

The dry salted paper can be kept for a long time unaltered; when required for use it is floated on a solution of nitrate of silver (1 part nitrate to 8 or 10 of distilled water), care being again taken to avoid adherence of bubbles; after remaining in contact with the silver solution for about five minutes it is removed and hung up to dry as before. This sensitising operation and the subsequent drying must be performed in a room into which no daylight enters, a candle or small gas flame surrounded by yellow semi-transparent paper, or preferably a yellow glazed lantern, being employed as source of light. When paper salted only on one side is used, that side must be the one in contact with the silver solution.

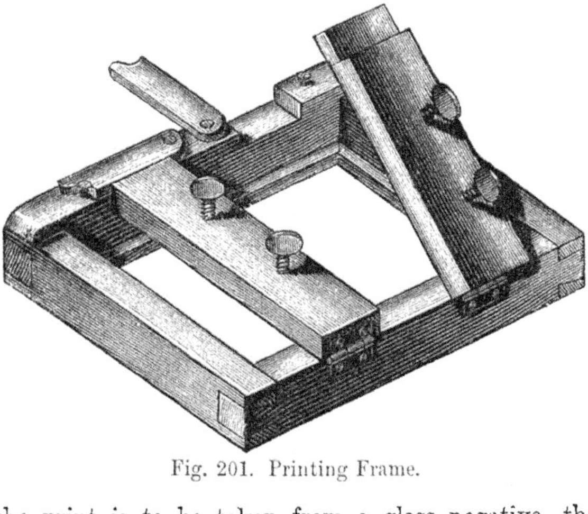

Fig. 201. Printing Frame.

If the print is to be taken from a glass negative, the latter (duly fixed, dried, and varnished with a thin film of transparent varnish to prevent rubbing of the surface) is placed in a wooden frame like a transparent slate, with a removable back kept in position by a spring clip, &c., the sensitised paper being next to it, with the sensitive side nearest the glass. Fig. 201 represents one form of printing frame. The whole is then exposed to day-

* "Albuminised paper," generally employed for *cartes de visite*, &c., is a superior kind of paper, coated on one side with albumen (prepared white of egg) by floating it on a strong solution of that substance to which some chloride of ammonium or other salt capable of acting on nitrate of silver has been added. Such paper consequently requires no further salting before use.

light, preferably at a window, until the paper is of a deep chocolate brown tint where directly exposed to the light. Usually the back of the frame is made in two parts (as represented) or hinged together, so that one half may be loosened and folded back and the print examined as to its progress without disturbing its position with reference to the negative when replaced, the glass and paper being held together firmly by the other half of the back. When the print is judged to be sufficiently deeply coloured, the paper is removed from the frame, and the print "fixed" by immersing it in a moderately strong solution of *thiosulphate of sodium* (hyposulphite of soda); the effect of this is to dissolve out unaltered silver chloride, &c., leaving behind particles of metallic silver in proportion as the action of the light passing through the negative has been more or less intense; so that the dark parts of the negative (corresponding with the light parts of the object photographed) furnish the light parts of the print, which is thus a positive.

If, however, the print to be prepared is not to be taken from a glass negative but from an actual object (*e.g.*, a skeleton leaf, a dried fern, a piece of lace, &c.), the printing frame is fitted with a piece of ordinary window glass; on this is laid the leaf, &c., and over this the sensitised paper, sensitive side next the object; the whole is then fixed in the frame by the wooden back and spring clip, &c., and exposed for a sufficient time, after which the print is fixed by steeping in thiosulphate of sodium solution, as before. In all cases most copious washing of the fixed print is requisite; steeping in a succession of waters, and for several hours in each, is desirable; or for two or three days in a dish into which a continuous small stream of fresh water runs from a tap. Silver prints are always more or less liable to fade on keeping for some years, even in a closed portfolio not exposed to light; this fading takes place far more rapidly when the prints have not been very thoroughly washed, so that it is probable that the fading is mainly due to slow chemical action taking place between the metallic silver and minute quantities of thiosulphate not completely washed out, even after long continued treatment with water, but retained in virtue of a surface attracting action analogous to that described in Expt. 253. The attracted thiosulphate can be got rid of by chemical means, *e.g.*, soaking the print in a highly diluted solution of *iodine*, and subsequently getting rid of the iodine by again soaking in a solution of *sodium sulphite*, or by treating with solution of *peroxide of hydrogen* or *hypochlorite of sodium*, &c.; but the chemicals thus used are rather apt to attack the metallic silver in the print itself and thus damage it; highly diluted *hydrochloric acid* solution is, perhaps, one of the least dangerous thiosulphate destroyers on this score.

Instead of treating paper with albumen, soluble chloride, and silver nitrate solution, it may be sensitised by coating with gelatin bromide emulsion like glass dry plates (Expt. 388); or this may be applied after albuminising to produce a glossy surface. Paper so prepared is sold ready for use by most dealers in photographic requisites; it is usually far more sensitive than ordinary albuminised paper sensitised by means of silver nitrate solution; the latter requires exposure to light (preferably direct sunlight) for a long time until the requisite depth of shade is produced; the former generally only requires a few seconds' exposure, after which the picture is developed in much the same way as a glass negative. Gelatin chloride emulsion (gelatin with finely divided silver chloride suspended therein) may be similarly employed.

Expt. 393. To print a Positive from a Paper Negative.—An impression of a fern, &c., prepared on suitable paper as a negative by the process just described may be transferred as a positive to another sheet of sensitised paper by repeating the process, using the first print as negative to print from. In most cases it is desirable to render the paper negative more transparent by dipping it in melted bees wax or paraffin wax, removing the surplus by pressing it between sheets of clean blotting paper with a hot flat iron. Fig. 202 represents a positive picture of a fern leaf thus prepared from the corresponding negative (fig. 203).

Expt. 394. Photo Enlargements, or Enlarged Photographs.—In order to prepare enlarged photographs, a collodion or dry plate negative is taken of convenient size in the camera (Expt. 387); after developing and fixing, &c., this is then used as a slide in a powerfully illuminated arrangement resembling a magic lantern (Expt. 368), the real magnified image formed being received upon a large sheet of sensitised paper. Precisely the same result is thus brought about as in the ordinary processes of printing positives from glass negatives (Expt. 392), excepting that the picture is magnified. Wherever the film of reduced silver is thickest in the negative, least light passes through, and therefore least chemical action is set up on the sensitised paper; so that the lights and shadows of the finished enlargement are reversed relatively to those of the negative, and a positive magnified picture is consequently obtained. This process is readily applicable in connection with "carbon" photographs prepared by the chromatised gelatin process (Expt. 398); when silver prints are thus prepared, paper treated with gelatin bromide emulsion (Expt.

388) is usually preferable on account of the shorter time of exposure requisite.

Expt. 395. Toning Silver Prints.—Particles of silver reduced from silver compounds by the action of light possess the property of attracting other silver particles to them from a mixture of solutions which would spontaneously cause the deposition of metallic silver in virtue of chemical changes set up between the dissolved substances; and this property is utilised in **intensifying** negatives, &c. (Expt. 391). In somewhat similar fashion, a paper silver print,

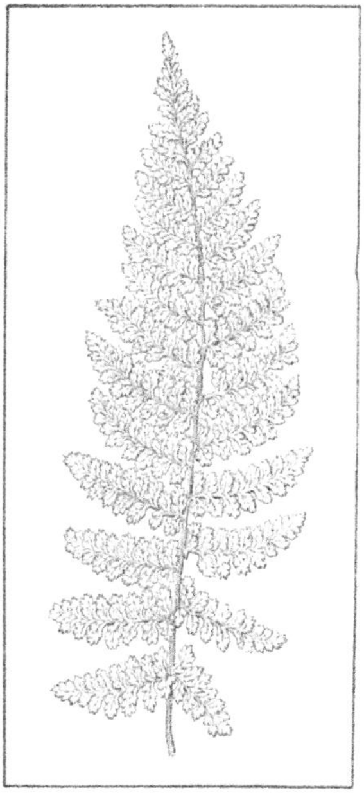

Fig. 202. Fern Leaf (Positive).

Fig. 203. Fern Leaf (Negative).

when placed in contact with certain fluids containing gold, platinum, and various other metals, possesses the property of attracting particles of gold, &c., to those parts where the silver deposit is thicker, to a greater extent than to those portions of the surface where it is thinner; so that gold, &c., is withdrawn from the fluid and deposited on the print, modifying the colour of the silver particles contained therein. Accordingly, silver prints are frequently "toned" before finally fixing them by immersing them in a "toning bath" consisting of a weak solution of chloride of gold mixed with acetate, carbonate, or borate of sodium, or other analogous fluids capable of allowing the required action to take place. Whether this action is simply a displacement of gold from gold chloride by silver forming metallic gold and silver chloride (as copper is displaced by iron from copper chloride, Expt. 9), or whether the organic matter of the paper, &c., is also involved in the action, seems somewhat uncertain; but the final result is that the untoned print gradually changes from a reddish hue to something more akin to purple black, owing to the coating over of the silver particles with gold in so fine a state of division as to appear purple or even blue, somewhat like the gold precipitated from solution by phosphorus (Expt. 136). Platinum chloride and various other salts of precious and certain rare metals are capable of giving somewhat analogous but different tones, a similar action probably taking place in each case, so that extremely fine particles of precious metal are deposited over the reddish silver particles.

Expt. 396. To prepare Iron Prints or Ferrotypes.—Although silver salts are the most convenient substances to use for the preparation of negatives by means of the camera, still various other metallic compounds may also be employed for the purpose of obtaining prints from a negative. Thus when compounds of iron of the "ferric" class are exposed to light, especially when in contact with organic matters of certain kinds, they become so acted upon as to form "ferrous" compounds. These latter, when brought into contact with *ferricyanide of potassium* (Expt. 83), produce a blue precipitate of "Turnbull's blue"; but the former do nothing of the kind. Hence if a piece of paper be soaked in a solution of a suitable ferric salt and be then exposed to light behind a negative for a sufficient time, the chemical change produced will be rendered evident by the production of Turnbull's blue to a greater or lesser extent on soaking the paper in solution of ferricyanide of potassium; so that the picture will appear in the deepest blue where the shadows occurred in the original object photographed, thus representing the least silver deposition in the negative, and consequently the most chemical action on the paper prepared with iron; and, conversely, the lights of the original object will be the palest blues in the print. If a copy of a drawing on tracing paper be thus made, the ground in the developed print will be blue and the lines white; or similarly, if a print of a piece of lace or a skeleton leaf, &c., be taken (Expt. 392), the ground will be dark blue and the lines faint blue or white. A convenient ferric salt for this purpose is *ferric ammonium citrate;* by soaking white paper in a solution of this salt to which a little *perfectly pure* crystallised *ferricyanide of potassium* has been added, and drying, a sensitive paper is obtained, which only requires washing with water to develop the print after exposure. Or the paper may be soaked in ferric ammonium citrate solution alone, and developed after exposure with a solution of ferricyanide of potassium. In this way *ferrotype* copies may be readily obtained of all objects capable of being printed from in silver, as in Expt. 392.

Expt. 397. Platinotypes.—The colour of the iron prints described in the last experiment renders them inapplicable for many kinds of photographic purposes, especially portraiture and views. On the other hand, the want of permanence frequently exhibited by silver prints renders it extremely desirable that some more stable material should be obtained. One means of overcoming this difficulty is by the use of chromatised gelatin (Expt. 398); another consists in treating paper with materials such as will cause the deposition of metallic platinum instead of silver in the print. One of the most successful of these *platinotype* processes consists of impregnating suitable paper with a mixed solution of *ferric oxalate* and *platinous chloride* and drying thoroughly; addition of a little potassium chlorate to the fluid improves the result. On exposing behind a negative of the ordinary kind the ferric oxalate becomes reduced to the ferrous state, as in the last experiment; when the exposed picture is developed by treating it with a hot solution of potassium oxalate, the ferrous oxalate formed reacts on the platinous chloride somewhat as ferrous sulphate on gold chloride in Expt. 135, setting free particles of metallic platinum. The iron compounds are partly removed from the paper by the solvent action of the hot developing fluid, partly by subsequent treatment of the print with dilute hydrochloric acid, so that finally the paper retains the print simply as platinum particles. Prints thus prepared are said to be as imperishable as the paper, and not in any way liable to the fading observed with silver prints. By suitable modification of the process, variations in colour and tone can be produced within certain limits. By printing a platinotype picture on to the surface of a glass or porcelain plate, and then applying heat in such a fashion as to burn away all organic matter from the film supporting the particles of reduced metal and *vitrify* or fuse the surface of the plate, a *photo enamel* picture is obtainable of imperishable nature, so far as fading is concerned, the kind of action which produces fading of an ordinary silver print being here impossible.

Expt. 398. To take Photographs by the Chromatised Gelatin Process.—Besides compounds of silver and iron, many other kinds of metallic derivatives are capable of being similarly affected by radiant energy in such fashion that chemical changes are brought about by exposure to light, either perceptible at once or capable of being rendered visible on "development" with appropriate reagents possessing the power of acting most rapidly on those parts where the action of the light has been greatest. In similar fashion many nonmetallic substances are known which are affected by light in various ways. In one of the earliest invented photographic processes the sensitive material was "bitumen of Judæa," which is readily soluble in naphtha before being exposed to light, but much less so afterwards, a change being produced somewhat analogous to that occurring in the white of an egg when boiled; previously to heating the white will readily mix with water, but after heating it becomes "coagulated," and insoluble in water. Gelatin, especially when mixed with certain metallic compounds (of which the *chromates* are some of the most convenient), is affected in a similar fashion, so that a picture can be readily obtained by exposing to light paper or glass, &c., coated with a thin film of gelatin dissolved in hot water to which a little potassium dichromate has been added, and a little lampblack or other insoluble pigment very finely divided; wherever the light acts the gelatin is rendered insoluble, but in the shadows it is hardly affected; so that on treating the exposed plate with warm water the gelatin is washed away in these parts, but not in those where the light has acted. Accordingly the picture appears as a negative, the residual film of insoluble gelatin containing lampblack being thickest where the light has acted most, when the action of the water is complete. As chromatised gelatin is far less sensitive to light than ordinary iodised collodion films, or gelatin bromide films, pictures are rarely taken directly in the camera by this process; negatives taken by silver processes of one kind or another are generally first prepared, and then prints (positives in reference to the original object) taken therefrom by the gelatin process. Chromatised gelatin films are apt to become spontaneously changed on keeping; so that most operators prefer to coat the paper employed with plain gelatin containing no chromate, but only pigment; and when required for photographic purposes, to render the films sensitive to light by dipping into a solution containing about 3 per cent. of potassium dichromate and a little ammonia, and drying in the dark. Paper thus coated with gelatin and carbon is prepared for sale by dealers in photographic requisites, and only requires sensitising with chromate solution to be ready for use.

It is obvious that where the light acts the gelatin is rendered insoluble, first on the outermost layers, and only later on in the underlying ones, because the former tend to absorb largely the active radiant energy, leaving but little to pass inwards; consequently there is always more or less danger of the warm water dissolving away, or at least softening the under less altered layers of gelatin below the outer most altered ones, and so undermining, as it were, the impression, and causing the film to be unduly removed where the light action has been but comparatively feeble. To obviate this it is usual to transfer the gelatin film from the original paper to a specially prepared "transfer paper" by moistening the two with cold water and then carefully pressing them together; on allowing the whole to remain in slightly warm water (about 37° C = 100° F.) the original paper (being more pervious to water than the transfer paper) allows the solvent action of the water to be exerted chiefly on the *under* surface of the gelatin film, and hence loosens its attachment to the original paper; so that by and bye the original paper can be carefully stripped off, leaving the gelatin film adhering to the transfer paper. When the development is complete and all soluble gelatin is washed away, the film of pigmented gelatin is further indurated or toughened by immersing in alum water (containing between 2 and 3 per cent. of alum) for a few minutes, and then washing with plain water and drying.

Instead of transfer paper, sheets of clear or opal glass, porcelain, ivory, metallic plates, &c., may be employed, so that ultimately the picture is formed upon them by transference of the gelatin film from the paper employed to support it in the first instance; all such transfers, although positive as regards light and shade, are *reversed as regards right and left*, just as an object seen by reflection in a mirror, or a transparent picture seen from the hinder side. To avoid this, a "double transfer" process is used; the first transfer is made to the surface of a sheet of glass coated firstly with wax (by pouring over it a solution of wax in benzine which evaporates like the ether from collodion) and then with ordinary (not iodised) collodion; the coated glass is placed in water to wash out the traces of alcohol and ether retained by the soft fresh collodion film, and the paper supporting the exposed gelatin film applied to it, so that the upper surface of the gelatin is next the collodion; on pressing the two closely together, and allowing them to remain in slightly warm water, the transfer is made to the collodion as above described. When the paper is stripped off and the development complete, the film is now transferred a second time from the collodion support to a sheet of prepared transfer paper, somewhat hotter water being employed; so that the wax film melts and allows the gelatin and collodion to leave the glass plate and adhere to the paper.

A good deal of dexterity and careful manipulation is requisite to produce a successful result in this way; but when properly effected with lampblack as pigment the "carbon prints" thus obtained are practically unalterable; if kept in a dry place, the hardened gelatin undergoes no change, and the picture is as indestructible as the paper on which it is formed, the influences tending to destroy ordinary silver prints (Expt. 371) having no influence on carbon prints.

Expt. 399. Printing Photographs. Woodburytype.—Gelatin pictures thus obtained possess some peculiar properties which lead to their being utilised for the reproduction of photographs by ordinary printing processes in more than one way. When a gelatin picture has been formed upon or transferred to a steel plate, the film when dry will be more or less in relief according to the thickness of the altered gelatin at each part of the surface; by pressing the picture against a plate of softer metal, or a sheet of tinfoil, in a powerful hydraulic press a similar picture is transferred to the metal, but in intaglio, or sunken in, instead of in relief. If the hollow mould thus obtained be treated with a warm solution of gelatin tinted with soluble colouring matter or insoluble finely divided pigment, the liquid being poured over it, and a truly flat scraper be then passed over the surface, all superfluous gelatin will be removed; where the hollows are least marked only an infinitesimal film will remain, but where they are deeper, a thicker and consequently a deeper tinted layer of coloured gelatin is left. In a short time the gelatin sets, and by then applying paper and passing through a press the gelatin adheres to the paper, and quits the metal mould, so that the paper becomes impressed with the picture thus printed off from the soft metal plate as a negative (in reference to the object from which the gelatin photograph was taken); the lights of the object corresponding with the highest relief (greatest insolubility of gelatin) in the first gelatin picture; with the deepest depressions in the soft transfer metal plate; and, consequently, with the thickest layer of coloured ink in the final print. If, therefore, the gelatin photograph first taken be a print from a silver negative, the final print will be a **positive** with reference to the object originally photographed; the light parts of the original yield the thickest silver film in the silver negative, and, consequently, the portions least in relief in the gelatin picture; these correspond with the least depressed portions of the metal transfer plate, and, consequently, with the thinnest layers of coloured ink, or lightest portions of the final print. This method of reproducing photographs by ordinary printing processes is called the *Woodbury-type*, having been invented by Mr Woodbury.

Expt. 400. Photolithography and Allied Processes.—In the ordinary process of lithography the design to be printed is sketched on a peculiar kind of stone with a greasy pencil, or with a pencil containing a composition of the nature of soap, which will form a free fatty acid on treatment with a mineral acid (Expt. 170); the stone is then treated with dilute nitric acid or other acid which will dissolve the stone in those parts where the surface is not protected by the greasy marks, just as in copper engraving (Expt. 272); after washing with water the marks are finally obtained in slight relief, and, being greasy, will take up printing ink or analogous compositions with ease, whereas the other portions, not being greasy but wet, will not take up the ink; accordingly, on passing an inking roller over the stone, ink is transferred to the marks, and by then pressing paper on the inked stone the ink is printed on to the paper, just as it is from the projecting portions of the types cut into the form of letters and figures in ordinary type printing. Now it happens that when a layer of chromatised gelatin is altered in places by light it becomes more easily wetted by printing ink and similar compositions and less readily wetted by water than at the unaltered portions; hence, by passing a printing ink roller over the exposed plate and treating the inked surface with water, the soluble parts of the gelatin are softened and can be wiped or washed away, carrying the ink with it, whilst the insoluble altered part remains with the ink adhering. The plate can thus be printed from, just as an ordinary lithographic stone. Various processes based on this property have been devised, all belonging to the class of *Collotypes*. One modification now largely employed, known as *photolithography*, consists in transferring the ink impression from the gelatin plate to a lithographic stone, by printing from the plate on to the stone with lithographic transfer ink; the stone is then used in the ordinary way. Another modification, termed *photozincography*, consists in coating a smooth bright plate of zinc with chromatised gelatin, exposing behind a negative, inking with lithographic ink, and dissolving away the unaltered gelatin by water so as to remove the ink from the unaffected parts, leaving it adhering to the altered gelatin. Powdered resin is then dusted over the surface, which is gently wiped; the resin sticks to the greasy parts, but not elsewhere; so that on heating the resin melts and forms a protecting varnish or "resist" (Expt. 272), preventing solution of the metal by acids when the varnish is applied. The plate is then treated with acid, and again resinised alternately several times until the metal is sufficiently eaten away from the unprotected parts to form a picture in relief capable of being printed from in the ordinary press.

A variety of other methods whereby photographs can be reproduced by ordinary printing processes have been introduced, the details of which cannot now be discussed. One simple method consists in transferring a gelatin film, &c., to a wood block, and using this instead of a pencil drawing on the wood, as a guide for the engraver to produce the requisite lines by cutting away the wood. Similarly, combinations of processes, involving treatment with acid, so as to dissolve the metal plate employed, and with ordinary engraving tools to finish and improve the work, are frequently employed, the picture being transferred to the plate or formed thereon by processes akin to those above described.

Expt. 401. Other kinds of Chemical Action produced by Actinic Rays.—Although the preparation of photographs in virtue of chemical changes set up or induced by radiant energy (and more especially by the most refrangible rays) are the best known and most frequently employed phenomena of this class, still many other different, though analogous, cases of chemical action thus induced are well known, *e.g.*, the *combination of free hydrogen and chlorine gases* determined by the action of light* (Expt. 215). Certain compound vapours are affected by light in such a fashion that if a clear transparent atmosphere of such a vapour be obtained (free from dust or fragments of floating solid matter, and from liquid particles suspended therein), and a powerful beam of light sent through it, chemical change and precipitation of *liquid* products, as finely divided fog or spray in the atmosphere, will be rapidly brought about, the effect of which will be that whereas at first the path of the light was practically invisible, there being no suspended particles to absorb and scatter the light (Pa. 291), in a short time the track of the beam of light will become brightly visible, like a ray of sunshine illuminating the motes in the air of a dusty room, owing to the development of liquid products of chemical change in the form of innumerable floating vescicles. One of the most important cases where chemical action is set up by actinic rays is afforded by the formation of the green colouring matter termed *chlorophyll* in the vegetable kingdom. This substance appears to be only capable of formation under the influence of sunlight, so that the hearts of tied up lettuces, the stalks of celery (trenched up with earth whilst growing), the inner leaves of cabbages, potato shoots sprouting in a dark cellar, and

* In similar fashion the gases carbon oxide (Expt. 219) and chlorine do not readily combine together in the dark; but if exposed to light they will combine tolerably readily, forming a compound gas sometimes termed *phosgene* gas on account of its thus being produced in virtue of the action of light. Similarly, chlorine can be easily substituted for hydrogen in many organic compounds containing that element if the action take place in sunlight, but only with great difficulty or not at all in the dark.

similar vegetable growths, are blanched and colourless, or nearly so, because the light does not reach the growing parts; whereas green coloured leaves and stalks, &c., are produced instead of white ones when the plant grows in such a form that the light freely reaches all parts of it. The evolution of free *oxygen gas* (Expt. 195) from growing vegetation by the decomposition of carbon dioxide appears generally to require the presence of chlorophyll, in order that the action may go on freely (mushrooms and naturally white vegetation forming exceptions in certain cases). Green plants that freely evolve oxygen in the daytime when exposed to sunlight cease to do so at night or in artificial darkness; on the other hand, by artificially illuminating such plants at night time (by means of powerful electric lights) continuous growth and evolution of oxygen may be kept up for an unlimited period, if the plant be sufficiently hardy to retain vigorous vitality under such altered conditions.

Expt. 402. To Illustrate Fluorescence.—The phenomenon of **fluorescence**, as already stated (Chapter XXII.), consists in the absorption of rays of such high refrangibility as to be invisible to the human eye, and the re-emitting of them with a lowered degree of refrangibility, so as to be visible, and of colour dependent on the particular degree of refrangibility of the emitted rays. Substances possessing this property in a high degree will often exhibit a peculiar coloured "sheen" or luminous appearance when viewed in certain positions in ordinary daylight; thus a bottle half full of solution of sulphate of quinine held level with the eye will generally exhibit a pale blue luminous appearance near the upper surface owing to the fluorescent emission of blue rays; similarly, many other substances glow in analogous fashion with other colours even in daylight. One of the most effective ways of showing this is to dissolve in a large bulk of water an appropriate fluorescent agent, and by suitable pumping machinery cause the water to form a fountain. The effect, however, is far more marked if visible rays of light other than those emanating from the fluorescent body examined are entirely excluded; this is most easily effected by passing a brilliant beam of arc light or sunlight through a prism (Expts. 349, 377) and cutting off by means of opaque screens all the less refrangible and visible part of the spectrum. On placing a piece of glass tinted with uranium, or a solution of quinine,* of the organic compound termed *fluorescein* (from its high

* It is curious that if quinine be dissolved in the watery solution of an acid not containing oxygen (such as hydrochloric acid), the liquid does not exhibit fluorescent action; whilst solutions in nitric or sulphuric acid, or other acids containing oxygen, fluoresce readily.

fluorescent qualities), or of the allied dyestuff *eosin*, in the track of the invisible ultra-violet rays thus separated from the visible rays, the particular coloured light emitted by each substance respectively is seen to advantage, a soft glow of light of peculiarly beautiful appearance becoming visible, somewhat resembling the luminosity of a mass of yellow phosphorus exposed to air in a dark room, but varying in tint.

The term "fluorescence" is derived from the circumstance that "fluor spar" (so named from its employment as a **flux** in certain metallurgical processes—*vide* also Expt. 271) was one of the first bodies in which this property was noticed; it is curious that this same substance is also capable of setting up, under suitable conditions, an apparently opposite phenomenon, sometimes spoken of as **phosphorescence by heat.**

Expt. 403. To render Fluor Spar Luminous by heating.—Heat a stout iron fireshovel over a good fire until it is nearly hot enough to glow dull red in the dark, but not quite; strew over it some coarsely powdered fluor spar, and then examine it in a perfectly dark room; the particles of spar will glow with a faint lambent light, somewhat resembling that emitted by phosphorus.

In this case the fluor spar appears to absorb the radiant heat rays of refrangibility too low to be visible and emit them again as rays of higher refrangibility so as to be visible, the action being somewhat akin to that taking place in Expt. 377, where invisible radiant heat is concentrated by a burning glass so as to heat a solid body placed at the focus sufficiently to render it red hot, *i.e.*, to cause it to emit visible light of higher refrangibility than the rays passing through the lens. In somewhat similar fashion certain fluorescent bodies can be made to fluoresce visibly when exposed to light of a less degree of refrangibility than that emitted whilst fluorescing. Thus if a beam of deep red light be isolated from arc light, &c., by means of a prism and screens, and allowed to fall on *naphthalene red* (a coal tar dyestuff) this latter fluoresces an orange yellow light; *chlorophyll* (the green colouring matter of vegetation) behaves similarly, the light emitted whilst fluorescing being less markedly orange than in the case of naphthalene red, but more so than the incident deep red light employed. The mineral known as *chlorophane* (a variety of fluor spar) is capable of fluorescing emerald green when illuminated by dark heat radiation of too low refrangibility to be visible at all.

Expt. 404. Phosphorescence.—The term **phosphorescence** is applied to two different kinds of phenomena which are not at all akin, excepting in the general result that a more or less feeble light is emitted by the phosphorescent body. When a piece of phosphorus is exposed to air in the dark it emits a pale light (Expts. 235, 236); this effect is due to oxidation simply, as may easily be shown by placing the phosphorus in an atmosphere incapable of acting chemically upon it, when the luminosity disappears. In the same kind of way decaying fish, certain kinds of rotten wood, and many other substances shine in the dark in virtue of chemical changes taking place between the oxygen of the air and the putrefying or decaying matter. The well-known phosphorescence of the sea on a summer's night, the emission of light by fireflies and glow-worms, and the natural phenomena of "will of the wisp" in marshy places and "corpse candles" in graveyards, are all due to analogous causes.* Besides this *chemical phosphorescence*, or production of visible light in consequence of chemical action, there exists a *physical phosphorescence*, consisting of the power of emitting light for a greater or lesser time after having been previously exposed to radiant energy. Some substances can exert this action very powerfully, but only for a short time; others when once powerfully excited by exposure to a strong light will glow for many hours in the dark. A highly phosphorescent substance of this class is the so-called "Bologna phosphorus," made by strongly heating in a crucible a mixture of powdered *heavy spar* (natural barium sulphate) and *gum tragacanth*, or other carbonaceous matter, made into a paste. *Canton's phosphorus* is similarly prepared by ramming into a clay crucible a mixture of 3 parts of finely-powdered *calcined oystershells* and 1 of *flowers of sulphur*, the crucible being well covered and heated to as high a temperature as possible in the centre of a good fire in a kitchen grate, &c. Of late years a somewhat similar preparation has come into use for a variety of purposes under the name of "Balmain's luminous paint." The phosphorescent matter is here mixed with oil, &c., and painted on to the object required to be rendered luminous (a buoy in the channel of a river, a clock face in a bedroom, &c., &c.); during the daytime, the direct sunlight or diffused daylight falling on the phosphorescent substance is, so to speak, absorbed by it, and re-emitted as phosphorescent light subsequently.

A pretty phosphorescent toy is sold, consisting of a glass slide,

* Somewhat similar appearances are occasionally due to peculiar forms of electric discharge whereby a glimmering light is produced instead of a lightning flash; currents of electrified air thus sometimes give rise to the appearance of flickering luminous masses of mist, &c., not unlike some forms of "Jack o' lantern." The formation of crystals from a supersaturated solution, and the sudden conversion of vitreous or glassy substances into crystalline masses, sometimes gives rise to the evolution of light, probably owing to temporary disturbances of electrical equilibrium during the action.

&c., on which is painted a butterfly or other device, using different phosphorescent powders for different parts of the body; a bit of magnesium wire is lighted near to the glass so as to illuminate it brightly for a few seconds; in the dark the device then shines forth in brilliant colours for a short time. In most cases (but not invariably) the actinic (ultra-violet) rays appear to be more active in exciting phosphorescence than the visible ones, just as in most instances they produce fluorescence more readily.

INDEX.

PRINTED BY NEILL AND COMPANY, EDINBURGH.

A

CATALOGUE

OF

STANDARD WORKS

PUBLISHED BY

CHARLES GRIFFIN & COMPANY.

LONDON:

12 EXETER STREET, STRAND.

INDEX.

CHARLES GRIFFIN & COMPANY'S
LIST OF PUBLICATIONS.

RELIGIOUS WORKS.

ANECDOTES (CYCLOPÆDIA OF RELIGIOUS AND MORAL). With an Introductory Essay by the Rev. GEORGE CHEEVER, D.D. *Thirty-sixth Thousand.* Crown 8vo. Cloth, 3/6.

*** These Anecdotes relate to no trifling subjects; and they have been selected, not for amusement, but for instruction. By those engaged in the tuition of the young, they will be found highly useful.

BIBLE HISTORY (A Manual of). By the Rev. J. WYCLIFFE GEDGE, Diocesan Inspector of Schools for Winchester. Small 8vo. Cloth, neat, 7d.

"This small but very comprehensive Manual is much more than a mere summary of Bible History."—*Church Sunday School Magazine.*

THE LARGE-TYPE BUNYAN.

BUNYAN'S PILGRIM'S PROGRESS. With Life and Notes, Experimental and Practical, by WILLIAM MASON. Printed in large type, and Illustrated with full-page Woodcuts. *Twelfth Thousand.* Crown 8vo. Bevelled boards, gilt, and gilt edges, 3/6.

CHRISTIAN YEAR (The): With Memoir of the Rev. JOHN KEBLE, by W. TEMPLE, Portrait, and Eight Engravings on Steel, after eminent Masters. *New Edition.* Small 8vo, toned paper. Cloth gilt, 5/.

Morocco elegant, ... 10/6.
Malachite, ... 12/6.

*** The above is the only issue of the "Christian Year" with Memoir and Portrait of the Author. In ordering, Griffin's Edition should be specified.

CRUTTWELL (REV. C. T., M.A.) A HISTORY OF EARLY CHRISTIAN LITERATURE. In large 8vo, handsome cloth. [*In preparation.*

*** This work is intended not only for Theological Students, but for General Readers, and will be welcomed by all acquainted with the Author's admirable "*History of Roman Literature,*" a work which has now reached its Fourth Edition.

DICK (Thos., LL.D.): CELESTIAL SCENERY; or, The Wonders of the Planetary System Displayed. This Work is intended for general readers, presenting to their view, in an attractive manner, sublime objects of contemplation. Illustrated. *New Edition.* Crown 8vo, toned paper. Handsomely bound, gilt edges, 5/.

DICK (Dr.): CHRISTIAN PHILOSOPHER (The); or, The Connection of Science and Philosophy with Religion. Revised and enlarged. Illustrated with 150 Engravings on Wood. *Twenty-eighth Edition.* Crown 8vo, toned paper. Handsomely bound, with gilt edges, 5/.

FOSTER (C.): THE STORY OF THE BIBLE, from Genesis to Revelation—including the Historical Connection between the Old and New Testaments. Told in Simple Language.

Now Ready. FOURTH AND GREATLY IMPROVED EDITION. In ROYAL 8VO. Cloth Elegant, 6s. Gilt and Gilt Edges, 7/6.

With Maps and over 250 Engravings

(Many of them Full-page, after the Drawings of Professor CARL SCHÖNHERR and others),

Illustrative of the Bible Narrative, and of Eastern Manners and Customs.

OPINIONS OF THE PRESS.

"A book which, once taken up, is not easily laid down. When the volume is opened, we are fairly caught. Not to speak of the well-executed wood engravings, which will each tell its story, we find a simple version of the main portions of the Bible, all that may most profitably be included in a work intended at once to instruct and charm the young—a version couched in the simplest, purest, most idiomatic English, and executed throughout with good taste, and in the most reverential spirit. *The work needs only to be known to make its way into families,* and it will (at any rate, it *ought* to) become a favourite Manual in Sunday Schools."—*Scotsman.*

"A HOUSEHOLD TREASURE."—*Western Morning News.*

"This attractive and handsome volume . . . written in a simple and transparent style. . . . Mr. Foster's explanations and comments are MODELS OF TEACHING."—*Freeman.*

"This large and handsome volume, abounding in Illustrations, is just what is wanted. . . . The STORY is very beautifully and reverently told."—*Glasgow News.*

"There could be few better Presentation Books than this handsome volume."—*Daily Review.*

"WILL ACCOMPLISH A GOOD WORK."—*Sunday School Chronicle.*

"In this beautiful volume no more of comment is indulged in than is necessary to the elucidation of the text. Everything approaching Sectarian narrowness is carefully eschewed."—*Methodist Magazine.*

"This simple and impressive Narrative . . . succeeds thoroughly in riveting the attention of children; . . . admirably adapted for reading in the Home Circle."—*Daily Chronicle.*

"The HISTORICAL SKETCH connecting the Old and New Testaments is a very good idea; it is a common fault to look on these as distinct histories, instead of as parts of *one grand whole.*"—*Christian.*

"Sunday School Teachers and Heads of Families will best know how to value this handsome volume."—*Northern Whig.*

*** The above is the original English Edition. In ordering, Griffin's Edition, by Charles Foster, should be distinctly specified.

STANDARD BIBLICAL WORKS

BY

THE REV. JOHN EADIE, D.D., LL.D.,

Late a Member of the New Testament Revision Company.

This SERIES has been prepared to afford sound and necessary aid to the Reader of Holy Scripture. The VOLUMES comprised in it form in themselves a COMPLETE LIBRARY OF REFERENCE. The number of Copies already issued greatly exceeds A QUARTER OF A MILLION.

I. EADIE (Rev. Prof.): BIBLICAL CYCLOPÆDIA (A); or, Dictionary of Eastern Antiquities, Geography, and Natural History, illustrative of the Old and New Testaments. With Maps, many Engravings, and Lithographed Fac-simile of the Moabite Stone. Large post 8vo, 700 pages. *Twenty-fifth Edition.*

Handsome cloth, ... 7/6.
Morocco antique, gilt edges, . . 16/.

"By far the best Bible Dictionary for general use."—*Clerical Journal.*

II. EADIE (Rev. Prof.): CRUDEN'S CONCORDANCE TO THE HOLY SCRIPTURES. With Portrait on Steel, and Introduction by the Rev. Dr. KING. Post 8vo. *Fifty-first Edition.* Handsome cloth, ... 3/6.

*** Dr. EADIE's has long and deservedly borne the reputation of being the COMPLETEST and BEST CONCORDANCE extant.

III. EADIE (Rev. Prof.): CLASSIFIED BIBLE (The). An Analytical Concordance. Illustrated by Maps. Large Post 8vo. *Sixth Edition.* Handsome cloth, . . . 8/6.

"We have only to add our unqualified commendation of a work of real excellence to every Biblical student."—*Christian Times.*

IV. EADIE (Rev. Prof.): ECCLESIASTICAL CYCLOPÆDIA (The). A Dictionary of Christian Antiquities, and of the History of the Christian Church. By the Rev. Professor EADIE, assisted by numerous Contributors. Large Post 8vo. *Sixth Edition.*

Handsome cloth, ... 8/6.

"The ECCLESIASTICAL CYCLOPÆDIA will prove acceptable both to the clergy and laity of Great Britain. A great body of useful information will be found in it."—*Athenæum*

V. EADIE (Rev. Prof.): DICTIONARY OF THE HOLY BIBLE (A); for the use of Young People. With Map and Illustrations. Small 8vo. *Thirty-eighth Thousand.*

Cloth, elegant, ... 2/6.

RELIGIONS OF THE WORLD (The): Being

Confessions of Faith contributed by Eminent Members of every Denomination of Christians, also of Mahometanism, Parseeism, Brahminism, Mormonism, &c., &c., with a Harmony of the Christian Confessions of Faith by a Member of the Evangelical Alliance. Crown 8vo. Cloth bevelled, 3/6.

SOUTHGATE (Henry): SUGGESTIVE

THOUGHTS ON RELIGIOUS SUBJECTS. (See page 36.)

SOUTHGATE (Mrs. H.): THE CHRISTIAN

LIFE: Thoughts in Prose and Verse from Five hundred of the Best Writers of all Ages. Selected and Arranged for Every Day in the Year. Small 8vo. With Red Lines and unique Initial Letters on each page. Cloth Elegant, 5/. *Second Edition.*

"A volume as handsome as it is intrinsically valuable."—*Scotsman.*
"The readings are excellent."—*Record.*
"A library in itself."—*Northern Whig.*

TAIT (Rev. James): MIND IN MATTER: A

Short Argument on Theism. *Second Edition.* Demy 8vo. Handsome Cloth, 8/6.

General Contents.—Evolution in Nature and Mind—Mr. Darwin and Mr. Herbert Spencer—Inspiration, Natural and Supernatural—Deductions.

"An able and original contribution to Theistic literature. . . . The style is pointed, concise, and telling to a degree."—*Glasgow Herald.*

"Mr. Tait advances many new and striking arguments . . . highly suggestive and fresh."—*Brit. Quarterly Review.*

THE MASSES: How Shall we Reach Them?

Some Hindrances in the way, set forth from the standpoint of the People, with Comments and Suggestions. By an Old Lay-helper. Cloth, 2s. 6d. *Second Edition.*

*** An attempt to set forth some deficiencies in our present methods of reaching the poor, in the language of the people themselves.

"So full of suggestiveness that we should reprint a tithe of the book if we were to transcribe all the extracts we should like to make."—*Church Bells.*

"'Hindrances in the way' exactly describes the subject-matter of the Book. Any one contemplating Missionary work in a large town would be helped by studying it."—*Guardian.*

"'The Masses' is a book to be well pondered over and acted upon."—*Church Work.*

"A very useful book, well worth reading."—*Church Times.*

"A most interesting book. . . . Contains a graphic description of work among the masses."—*English Churchman.*

WORDS AND WORKS OF OUR BLESSED

LORD: and their Lessons for Daily Life. Two Vols. in One. Foolscap 8vo. Cloth, gilt edges, 6/.

WORKS

By SIR WILLIAM AITKEN, M.D., Edin., F.R.S.,

PROFESSOR OF PATHOLOGY IN THE ARMY MEDICAL SCHOOL; EXAMINER IN MEDICINE FOR THE MILITARY MEDICAL SERVICES OF THE QUEEN; FELLOW OF THE SANITARY INSTITUTE OF GREAT BRITAIN; CORRESPONDING MEMBER OF THE ROYAL IMPERIAL SOCIETY OF PHYSICIANS OF VIENNA; AND OF THE SOCIETY OF MEDICINE AND NATURAL HISTORY OF DRESDEN.

SEVENTH EDITION.

The SCIENCE and PRACTICE of MEDICINE.

In Two Volumes, Royal 8vo., cloth. Illustrated by numerous Engravings on Wood, and a Map of the Geographical Distribution of Diseases. To a great extent Rewritten; Enlarged, Remodelled, and Carefully Revised throughout, 42/.

Opinions of the Press.

"The work is an admirable one, and adapted to the requirements of the Student, Professor, and Practitioner of Medicine. . . . The reader will find a large amount of information not to be met with in other books, epitomised for him in this. We know of no work that contains so much, or such full and varied information on all subjects connected with the Science and Practice of Medicine."—*Lancet.*

"Excellent from the beginning, and improved in each successive issue, Dr. Aitken's GREAT and STANDARD WORK has now, with vast and judicious labour, been brought abreast of every recent advance in scientific medicine and the healing art, and affords to the Student and Practitioner a store of knowledge and guidance of altogether inestimable value. . . . A classical work which does honour to British Medicine, and is a compendium of sound knowledge."—*Extract from Review in "Brian," by J. Crichton-Browne, M.D., F.R.*.

"The SEVENTH EDITION of this important Text-Book fully maintains its reputation. . . . Dr. Aitken is indefatigable in his efforts. . . . The section on DISEASES of the BRAIN and NERVOUS SYSTEM is completely remodelled, so as to include all the most recent researches, which in this department have been not less important than they are numerous"—*British Medical Journal.*

"The STANDARD TEXT-BOOK in the English Language. . . . There is, perhaps, no work more indispensable for the Practitioner and Student."—*Edin. Medical Journal.*

"The extraordinary merit of Dr. Aitken's work. . . . The author has unquestionably performed a service to the profession of the most valuable kind."—*Practitioner.*

OUTLINES OF THE SCIENCE AND PRACTICE OF MEDICINE.

A Text-Book for Students. *Second Edition.* Crown 8vo, 12/6.

"Students preparing for examinations will hail it as a perfect godsend for its conciseness."—*Athenæum.*

"Well-digested, clear, and well-written, the work of a man conversant with every detail of his subject, and a thorough master of the art of teaching."—*British Medical Journal.*

CAIRD (F. M., M.B., F.R.C.S.), and CATHCART (C. W., M.B., F.R.C.S.):

A SURGICAL HANDBOOK: For the use of Practitioners, Students, House-Surgeons, and Dressers. With Numerous Illustrations. *Third Edition.* Pocket-size, Leather, 8/6.

"This is a THOROUGHLY PRACTICAL MANUAL, well up to date, clear, accurate, and succinct. It is thoroughly trustworthy, handy, and well got up."—*Lancet.*

By PROFESSOR T. M'CALL ANDERSON, M.D.

Now ready, with two Coloured Lithographs, Steel Plate, and numerous Woodcuts. Royal 8vo, Handsome Cloth, 25s.

DISEASES OF THE SKIN

(A TREATISE ON),

WITH SPECIAL REFERENCE TO DIAGNOSIS AND TREATMENT, INCLUDING AN ANALYSIS OF 11,000 CONSECUTIVE CASES.

BY T. M'CALL ANDERSON, M.D.,

Professor of Clinical Medicine, University of Glasgow.

The want of a manual, embodying the most recent advances in the treatment of cutaneous affections, has made itself much felt of late years. PROFESSOR M'CALL ANDERSON'S Treatise, therefore, affording, as it does, a complete *résumé* of the best modern practice, will be doubly welcome. It is written—not from the standpoint of the University Professor—but from that of one who, during upwards of a quarter of a century, has been actively engaged both in private and in hospital practice, with unusual opportunities for studying this class of disease, hence the PRACTICAL and CLINICAL directions given are of great value.

Speaking of the practical aspects of Dr. ANDERSON'S work, the *British Medical Journal* says:—"Skin diseases are, as is well-known, obstinate and troublesome, and the knowledge that there are ADDITIONAL RESOURCES besides those in ordinary use will give confidence to many a puzzled medical man, and enable him to encourage a doubting patient. ALMOST ANY PAGE MIGHT BE USED TO ILLUSTRATE THE FULNESS OF THE WORK IN THIS RESPECT. . . . The chapter on Eczema, that universal and most troublesome ailment, describes in a comprehensive spirit and with the greatest accuracy of detail the various methods of treatment. Dr. Anderson writes with the authority of a man who has tried the remedies which he discusses, and the information and advice which he gives cannot fail to prove extremely valuable."

Opinions of the Press.

"Beyond doubt, the MOST IMPORTANT WORK on Skin Diseases that has appeared in England for many years. . . . Conspicuous for the AMOUNT AND EXCELLENCE of the CLINICAL AND PRACTICAL information which it contains."—*British Medical Journal.*

"Professor M'Call Anderson has produced a work likely to prove *very acceptable* to the busy practitioner. The sections on treatment are very full For example, ECZEMA has 110 pages given to it, and 73 of these pages are devoted to treatment."—*Lancet.*

WORKS by A. WYNTER BLYTH, M.R.C.S., F.C.S.,

Public Analyst for the County of Devon, and Medical Officer of Health for St. Marylebone.

I. FOODS: THEIR COMPOSITION AND ANALYSIS.

Price 16/. In Crown 8vo, cloth, with Elaborate Tables and Litho-Plates. *Third Edition.* Revised and partly rewritten.

General Contents.

History of Adulteration—Legislation, Past and Present—Apparatus useful to the Food Analyst—"Ash"—Sugar—Confectionery—Honey—Treacle—Jams and Preserved Fruits—Starches—Wheaten-Flour—Bread—Oats—Barley—Rye—Rice—Maize—Millet—Potato—Peas—Chinese Peas—Lentils—Beans—MILK—Cream—Butter—Cheese—Tea—Coffee—Cocoa and Chocolate—Alcohol—Brandy—Rum—Whisky—Gin—Arrack—Liqueurs—Beer—Wine—Vinegar—Lemon and Lime Juice—Mustard—Pepper—Sweet and Bitter Almond—Annatto—Olive Oil—Water. *Appendix:* Text of English and American Adulteration Acts.

"Will be used by every Analyst."—*Lancet.*

"STANDS UNRIVALLED for completeness of information. . . . A really 'practical' work for the guidance of practical men."—*Sanitary Record.*

"An admirable digest of the most recent state of knowledge. . . . Interesting even to lay-readers."—*Chemical News.*

*** The NEW EDITION contains many Notable Additions, especially on the subject of MILK and its relation to FEVER-EPIDEMICS, the PURITY of WATER-SUPPLY, the new MARGARINE ACT, &c., &c.

COMPANION VOLUME.

II. POISONS: THEIR EFFECTS AND DETECTION.

Price 16/.

General Contents.

Historical Introduction—Statistics—General Methods of Procedure—Life Tests—Special Apparatus—Classification: I.—ORGANIC POISONS: (*a.*) Sulphuric, Hydrochloric, and Nitric Acids, Potash, Soda, Ammonia, &c.; (*b.*) Petroleum, Benzene, Camphor, Alcohols, Chloroform, Carbolic Acid, Prussic Acid, Phosphorus, &c.; (*c.*) Hemlock, Nicotine, Opium, Strychnine, Aconite, Atropine, Digitalis, &c.; (*d.*) Poisons derived from Animal Substances; (*e.*) The Oxalic Acid Group. II.—INORGANIC POISONS: Arsenic, Antimony, Lead, Copper, Bismuth, Silver, Mercury, Zinc, Nickel Iron, Chromium, Alkaline Earths, &c. *Appendix:* A. Examination of Blood and Blood-Spots. B. *Hints for Emergencies:* Treatment—Antidotes.

"Should be in the hands of every medical practitioner."—*Lancet.*

"A sound and practical Manual of Toxicology, which cannot be too warmly recommended. One of its chief merits is that it discusses substances which have been overlooked."—*Chemical News.*

"One of the best, most thorough, and comprehensive works on the subject."—*Saturday Review.*

HYGIÈNE AND PUBLIC HEALTH (a Dictionary of):

embracing the following subjects:—

I.—SANITARY CHEMISTRY: the Composition and Dietetic Value of Foods, with the Detection of Adulterations.

II.—SANITARY ENGINEERING: Sewage, Drainage, Storage of Water, Ventilation, Warming, &c.

III.—SANITARY LEGISLATION: the whole of the PUBLIC HEALTH ACT, together with portions of other Sanitary Statutes, in a form admitting of easy and rapid Reference.

IV.—EPIDEMIC AND EPIZOOTIC DISEASES: their History and Propagation, with the Measures for Disinfection.

V.—HYGIÈNE—MILITARY, NAVAL, PRIVATE, PUBLIC, SCHOOL.

Royal 8vo, 672 pp., cloth, with Map and 140 Illustrations, 28/.

"A work that must have entailed a vast amount of labour and research. . . . Will become a STANDARD WORK IN PUBLIC HEALTH."—*Medical Times and Gazette.*

"Contains a great mass of information of easy reference."—*Sanitary Record.*

DAVIS (J. R. A., B.A., Lecturer on Biology at the University College of Wales, Aberystwyth):

AN INTRODUCTION TO BIOLOGY, for the use of Students. Comprising Vegetable and Animal Morphology and Physiology.

With Bibliography, Exam.-Questions, complete Glossary, and 158 Illustrations. Large Crown 8vo. 12s. 6d.

"The volume is literally packed with information."—*Glasgow Medical Journal.*

"As a general work of reference, Mr. Davis' Manual will be highly serviceable both to medical men and to amateur or professional scientists."—*British Medical Journal.*

THE FLOWERING-PLANT: First Principles of Botany. Specially adapted to the Requirements of Students preparing for Competitive Examinations. With numerous Illustrations. Crown 8vo, Cloth, 3s. 6d.

DUCKWORTH (Sir Dyce, M.D. (Edin.), F.R.C.P.):

GOUT (A Treatise on): for the Use of Practitioners and Students. With Chromo-Lithograph, Folding Plate, and Illustrations. Royal 8vo, 25s.

"At once thoroughly practical and highly philosophical. The practitioner will find an enormous amount of information in it."—*Practitioner.*

DUPRÉ (A., Ph.D., F.R.S.), and WILSON HAKE (Ph.D., F.I.C., F.C.S., of the Westminster Hospital Medical School):

INORGANIC CHEMISTRY (A Short Manual of): with Coloured Plate of Spectra. Crown 8vo, Cloth, 7s. 6d.

"The BEST MANUAL for Medical aspirants that has appeared on the subject of Inorganic Chemistry."—*Analyst.*

"A well-written, clear, and accurate Elementary Manual of Inorganic Chemistry. . . . We agree heartily in the system adopted by Drs. Dupré and Hake. WILL MAKE EXPERIMENTAL WORK TREBLY INTERESTING BECAUSE INTELLIGIBLE."—*Saturday Review.*

ELBORNE (W., F.L.S., F.C.S., late Assistant-Lecturer in Materia Medica and Pharmacy, Owens College, Manchester):

PHARMACY AND MATERIA MEDICA (A Laboratory Course of): Including the Principles and Practice of Dispensing. Adapted to the Study of the British Pharmacopœia and the requirements of the Private Student. With Litho-Plates, and numerous Illustrations. Large crown 8vo. Cloth, 8s. 6d.

EWART (J. COSSAR, M.D., F.R.S.E., Regius Professor of Natural History, University of Edinburgh):

THE PRESERVATION OF FISH, in reference to Food Supply Hints on). In Crown 8vo, Wrapper, 6d.

GARROD (A.E., M.A., M.D., Oxon., M.R.C.P., Assistant-Physician to the West London Hospital, &c.):

RHEUMATISM AND RHEUMATOID ARTHRITIS (A Treatise on): for the Use of Students and Practitioners. With Charts and Illustrations. Medium 8vo. Cloth, 21s.

GRIFFIN (John Joseph, F.C.S.):

CHEMICAL RECREATIONS: A Popular Manual of Experimental Chemistry. With 540 Engravings of Apparatus. *Tenth Edition.* Crown 4to. Cloth.

Part I.—Elementary Chemistry, 2/.
Part II.—The Chemistry of the Non-Metallic Elements, including a Comprehensive Course of Class Experiments, 10/6.
Or, complete in one volume, cloth, gilt top, . . 12/6.

HADDON (A. C., M.A., F.Z.S., Professor of Zoology at the Royal College of Science, Dublin):

AN INTRODUCTION TO THE STUDY OF EMBRYOLOGY, for the use of Students. With 190 Illustrations. In Royal 8vo, Cloth, 18s.

"An EXCELLENT RÉSUMÉ OF RECENT RESEARCH, well adapted for self-study. . . . Gives remarkably good accounts (including all recent work) of the development of the heart and other organs. . . . The book is handsomely got up."—*Lancet.*

"PROF. HADDON has the real scientific spirit for work of this kind. . . . The development of the various organs ably demonstrated. . . . Forms a handsome volume."—*Brit. Med. Journal.*

HUMPHRY (L., M.A., M.B., M.R.C.S., Assistant-Physician to, and Lecturer to Probationers at, Addenbrooke's Hospital, Cambridge):

NURSING (A Manual of): Medical and Surgical. With Numerous Illustrations. *Second Edition.* Crown 8vo, Cloth, 3s. 6d.

"We would advise all Nurses to possess a copy."—*The Hospital.*

JAKSCH (Prof. R. von, University of Graz):

CLINICAL DIAGNOSIS: A Text-book of the Chemical, Microscopical, and Bacteriological Evidence of Disease. Translated from the Second German Edition, by JAMES CAGNEY, M.A., M.D., St. Mary's Hospital. With additions by WM. STIRLING, M.D., Sc.D., Prof. of Physiology, Owens College, Manchester. With numerous Illustrations in Colour. Medium 8vo, 25s.

LANDIS (Henry G., A.M., M.D., Professor of Obstetrics in Starling Medical College):

THE MANAGEMENT OF LABOUR AND OF THE LYING-IN PERIOD. In 8vo, with Illustrations. Cloth, 7/6.

"Fully accomplishes the object kept in view by its author. . . . Will be found of GREAT VALUE by the young practitioner."—*Glasgow Medical Journal.*

LEWIS (W. Bevan, L.R.C.P., M.R.C.S., Medical Director of the West-Riding Asylum, Wakefield):

MENTAL DISEASES (A Text-book of): With Special Reference to the Pathological Aspects of Insanity. For the Use of Students and Practitioners. With 18 Litho-Plates, Charts, and Illustrations. Royal 8vo, 28/.

"A splendid addition to the literature of mental diseases. . . . Every page bristles with important facts. As a standard work on the pathology of mental diseases the volume should occupy a prominent place."—*Dublin Medical Journal.*

SEXTON (A. Humboldt, F.C.S., Professor of Metallurgy, Glasgow and West of Scotland Technical College):

1. QUANTITATIVE ANALYSIS (Outlines of). For the Use of Students. With Numerous Illustrations. *Third Edition.* Crown 8vo. Cloth, 3/.

"A good and useful book. . . . Really supplies a want."—*The Lancet.*
"This is an admirable little volume, and well fulfils its purpose."—*Schoolmaster.*

2. QUALITATIVE ANALYSIS (Outlines of). For the use of Students. With numerous Illustrations. *Second Edition.* Crown 8vo. Cloth, 3/6.

"The Work of a thorough practical Chemist . . . and one which may be unhesitatingly recommended."—*Brit. Medical Journal.*

STIRLING (William, M.D., D.Sc., Owens College, Manchester):

A TEXT-BOOK OF HUMAN PHYSIOLOGY (*See under Landois and Stirling, page* 10). *Third Edition.*

OUTLINES OF PRACTICAL PHYSIOLOGY: including Experimental and Chemical Physiology. With special reference to Clinical Medicine. A Laboratory Handbook for the use of Students. *Second Revised Edition.* With 234 Illustrations. In large Crown 8vo, 9/.

"An excellent treatise, which we can thoroughly recommend."—*Lancet.*
"May be confidently recommended as a guide to the Student of Physiology."—*Glasgow Medical Journal.*

OUTLINES OF PRACTICAL HISTOLOGY. With 344 Illustrations. Large Crown 8vo. Cloth, 12/6.

THORBURN (John, M.D., F.R.C.P., Late Professor of Obstetric Medicine, Owens College and Victoria University, Manchester; Obstetric Physician to the Manchester Royal Infirmary):

THE DISEASES OF WOMEN (A Practical Treatise on). Prepared with Special Reference to the Wants of the General Practitioner and Advanced Student. With Chromo-lithograph, and over 200 Illustrations. Royal 8vo, handsome cloth, 21/.

The ENTIRE WORK is IMPARTIAL and INSTRUCTIVE, and in every way worthy of its author."—*British Medical Journal.*

THORBURN (Wm., B.S., B.Sc., M.D., F.R.C.S., Assistant-Surgeon to the Manchester Royal Infirmary):

THE SURGERY OF THE SPINAL CORD (A Contribution to). With Diagrams, Tables, and Illustrations. Medium 8vo. Handsome cloth, 12s. 6d.

THORNTON (J. Knowsley, M.B., M.C., Surgeon to the Samaritan Free Hospital for Women, &c., &c.):

THE SURGERY OF THE KIDNEYS. Being the Harveian Lectures for 1889. With Illustrations and Tables. Demy 8vo. Cloth, 5s.

GENERAL SCIENTIFIC AND TECHNICAL WORKS.

BERINGER (C., F.I.C., F.C.S., late Chief Assayer to the Rio Tinto Company); and BERINGER (J. J., F.I.C, F.C.S., Public Analyst for, and Lecturer to the Mining Association of, Cornwall):

ASSAYING (A Text-Book of): for the Use of Students, Mine Managers, Assayers, &c. With numerous Tables and Illustrations. Crown 8vo. Cloth, 10/6.

"A REALLY MERITORIOUS WORK, that may be safely depended upon either for systematic instruction or for reference."—*Nature.*

BROWNE (Walter R., M.A., M. Inst. C.E., F.G.S., late Fellow of Trinity College, Cambridge):

THE STUDENT'S MECHANICS: An Introduction to the Study of Force and Motion. With Diagrams. Crown 8vo. Cloth, 4/6.

"Clear in style and practical in method, 'THE STUDENT'S MECHANICS,' is cordially to be recommended from all points of view. . . . Will be of great value to Students desirous to gain full knowledge." *Athenæum.*

"The merits of the work are especially conspicuous in its clearness and brevity . . . deserves the attention of all who have to teach or learn the elements of Mechanics. . . . An excellent conception."—*Westminster Review.*

—— FOUNDATIONS OF MECHANICS. Papers reprinted from the *Engineer.* In crown 8vo, 1/.

—— FUEL AND WATER: A Manual for Users of Steam and Water. By Prof. SCHWACKHÖFER and W. R. BROWNE, M.A. (See p. 23.)

BROUGH (B.H., F.G.S., Instructor of Mine Surveying, Royal School of Mines):

MINE SURVEYING (A Text-Book of): for the Use of Managers of Mines and Collieries, Students at the Royal School of Mines, City and Guilds of London Institute, &c. With Illustrations. *Second Edition.* Crown 8vo, cloth, 7/6.

"Supplies a long-felt want."—*Iron.*

"A valuable accessory to Surveyors in every department of commercial enterprise." —*Colliery Guardian.*

"The information is given in a concise manner."—*Engineering.*

CRIMP (W. Santo, A. M. Inst. C. E., F. G. S., Assistant-Engineer to the London County Council):

SEWAGE DISPOSAL WORKS. A Guide to the Construction of Works for the Prevention of the Pollution of Rivers and Estuaries. With 33 Lithographic Plates, Tables, and Illustrations in the Text. Medium 8vo, Cloth, 25/.

Medium 8vo, Handsome Cloth, 30s.

BRIDGE-CONSTRUCTION

(A PRACTICAL TREATISE ON):

Being a Text-Book on the Design and Construction of Bridges in Iron and Steel.

FOR THE USE OF STUDENTS, DRAUGHTSMEN, AND ENGINEERS.

BY

T. CLAXTON FIDLER, M. INST. C.E.

With Numerous Woodcuts and 17 Lithographic Plates.

THE object of this book is to describe the modern practice of Bridge-Construction, and to set forth in the simplest language the mechanical principles and experimental facts on which it is based. The design and arrangement of the work have been dictated by a desire to render it as useful as possible, not only to Engineers or Draughtsmen who may be engaged in the work of Bridge-Calculations and Bridge-Construction, but also to Students. With this object, the earlier chapters of the work are devoted to a simple demonstration of those mechanical principles which must of necessity form the beginning of any study of the subject, and which are more fully developed and applied in later portions of the book.

"Should prove not only an indispensable Hand book for the Practical Engineer, but also a stimulating Treatise to the Student of Mathematical Mechanics and Elasticity."—*Nature.*

"One of the VERY BEST RECENT works on the Strength of Materials and its application to Bridge-Construction. . . . Well repays a careful study."—*Engineering.*

"As an exposition of the LATEST advances of the Science, we are glad to welcome this well-written Treatise."—*Architect.*

"A Scientific Treatise of great merit, which cannot but prove useful."—*Westminster Review.*

"Mr. Fidler's book is one which every Student of Mechanics ought to possess, and which merits, as it will receive, the appreciative attention of all practical men."—*Scotsman.*

A full Prospectus of the above important work may be had on application to the Publishers.

GURDEN (Richard Lloyd, Authorised Surveyor for the Governments of New South Wales and Victoria):

TRAVERSE TABLES: computed to Four Places Decimals for every Minute of Angle up to 100 of Distance. For the use of Surveyors and Engineers. *Second Edition.* Folio, strongly half-bound, 21/.

*** *Published with Concurrence of the Surveyors-General for New South Wales and Victoria.*

"Those who have experience in exact SURVEY-WORK will best know how to appreciate the enormous amount of labour represented by this valuable book. The computations enable the user to ascertain the sines and cosines for a distance of twelve miles to within half an inch, and this BY REFERENCE TO BUT ONE TABLE, in place of the usual Fifteen minute computations required. This alone is evidence of the assistance which the Tables ensure to every user, and as every Surveyor in active practice has felt the want of such assistance, few knowing of their publication will remain without them."—*Engineer.*

"We cannot sufficiently admire the heroic patience of the author, who, in order to prevent error, calculated each result by two different modes, and, before the work was finally placed in the Printers' hands, repeated the operation for a third time, on revising the proofs."—*Engineering.*

JAMES (W. Powell, M.A.):

FROM SOURCE TO SEA: or, Gleanings about Rivers from many Fields. A Chapter in Physical Geography. Cloth elegant, 3/6.

"Excellent reading . . a book of popular science which deserves an extensive circulation."—*Saturday Review.*

JAMIESON (Andrew, C.E., F.R.S.E., Professor of Engineering, Glasgow and West of Scotland Technical College):

STEAM AND THE STEAM ENGINE (A Text-Book on): Specially arranged for the use of Science and Art, City and Guilds' of London Institute, and other Engineering Students. With 200 Illustrations and Four Folding-Plates. *Fifth Edition.* Crown 8vo. Cloth, 7/6.

"The BEST BOOK yet published for the use of Students."—*Engineer.*

"This is undoubtedly the MOST VALUABLE and MOST COMPLETE hand-book of reference on the subject that now exists."—*Marine Engineer.*

1. STEAM AND THE STEAM ENGINE (An Elementary Manual on), forming an introduction to the larger Work by the same Author. With numerous Illustrations and Examination Questions at the end of each Lecture. *Second Edition.* Crown 8vo. Cloth, 3/6.

2. MAGNETISM AND ELECTRICITY (An Elementary Manual on). With numerous Illustrations and Examination Questions. Crown 8vo. Part I.—Magnetism, 1s. Part II.—Voltaic Electricity, 1s. 6d. Part III.—Frictional Electricity, 1s. 6d. Or complete in Cloth, 3s. 6d.

3. APPLIED MECHANICS (An Elementary Manual on). With Diagrams and Examination Questions. Crown 8vo.

M'MILLAN (W. G., F.I.C., F.C.S.), Chemist and Metallurgist to the Cossipore Foundry and Shell-Factory, Calcutta.

ELECTRO-METALLURGY (A Treatise on): Embracing the Application of Electrolysis to the Plating, Depositing, Smelting, and Refining of various Metals, and to the Reproduction of Printing Surfaces and Art-Work, &c. With numerous Illustrations. Large Crown 8vo. Cloth, 10/6.

MUNRO (John, C. E.) and JAMIESON (Andrew, C.E., F.R.S.E.):

A POCKET-BOOK OF ELECTRICAL RULES AND TABLES, for the use of Electricians and Engineers. Pocket Size. Leather, 8/6. *Sixth Edition, revised and enlarged.* With numerous Diagrams.

*** The SIXTH EDITION has been thoroughly Revised and Enlarged by about 120 pages and 60 new Figures.

"WONDERFULLY PERFECT. . . . Worthy of the highest commendation we can give it."—*Electrician.*

"The STERLING VALUE of Messrs. MUNRO and JAMIESON'S POCKET-BOOK."—*Electrical Review.*

MUNRO (R. D.), STEAM BOILERS: Their Defects, Management, and Construction. A Manual for all concerned in the care of Steam Boilers, but written with a special view to the wants of Boiler-Attendants, Mill-Mechanics, and other Artisans. With Numerous Illustrations. Crown 8vo, Cloth 3/6.

"The volume is a valuable companion for workmen and engineers engaged about Steam Boilers, and ought to be carefully studied, and always at hand."—*Colliery Guardian.*

"The subjects referred to are handled in a trustworthy, clear, and practical manner. . . . The book is VERY USEFUL, especially to steam users, artisans, and young engineers."—*Engineer.*

PHILLIPS (J. Arthur, F.R.S., M. Inst. C.E., F.C.S., F.G.S., Ancien Elève de l'Ecole des Mines, Paris):

ELEMENTS OF METALLURGY: a Practical Treatise on the Art of Extracting Metals from their Ores. With over 200 Illustrations, many of which have been reduced from Working Drawings, and two Folding-Plates. Royal 8vo, 848 pages, cloth, 36/. *NEW EDITION by the Author and Mr. H. Bauerman, F.G.S.*

General Contents.

I.—A TREATISE on FUELS and REFRACTORY MATERIALS.
II.—A Description of the principal MINERALS, with their DISTRIBUTION.
III.—STATISTICS of the amount of each METAL annually produced throughout the World.
IV.—The METHODS of ASSAYING the different ORES, together with the PROCESSES of METALLURGICAL TREATMENT.

"'Elements of Metallurgy' possesses intrinsic merits of the highest degree. Such a work is precisely wanted by the great majority of students and practical workers, and its very compactness is in itself a first-rate recommendation. . . In our opinion, the BEST WORK EVER WRITTEN ON THE SUBJECT with a view to its practical treatment."—*Westminster Review.*

"The VALUE OF THIS WORK IS ALMOST INESTIMABLE. There can be no question that the amount of time and labour bestowed upon it is enormous. . . . There is certainly no Metallurgical Treatise in the language calculated to prove of such general utility."—*Mining Journal.*

MANY NOTABLE ADDITIONS
WILL BE FOUND IN THE SECTIONS DEVOTED TO
IRON, LEAD, COPPER, SILVER, AND GOLD,
Dealing with New Processes and Development.

Demy 8vo, Handsome cloth, 18s.

PHYSICAL GEOLOGY AND PALÆONTOLOGY,

ON THE BASIS OF PHILLIPS.

BY

HARRY GOVIER SEELEY, F.R.S.,

PROFESSOR OF GEOGRAPHY IN KING'S COLLEGE, LONDON.

With Frontispiece in Chromo-Lithography, and Illustrations.

"It is impossible to praise too highly the research which PROFESSOR SEELEY'S 'PHYSICAL GEOLOGY' evidences. IT IS FAR MORE THAN A TEXT-BOOK—it is a DIRECTORY to the Student in prosecuting his researches."—*Extract from the Presidential Address to the Geological Society*, 1885, *by Rev. Professor Bonney, D.Sc., LL.D., F.R.S.*

"PROFESSOR SEELEY maintains in his 'PHYSICAL GEOLOGY' the high reputation he already deservedly bears as a Teacher. . . . It is difficult, in the space at our command, to do fitting justice to so large a work. . . . The final chapters, which are replete with interest, deal with the Biological aspect of Palæontology. Here we find discussed the origin, the extinction, succession, migration, persistence, distribution, relation, and variation of species —with other considerations, such as the Identification of Strata by Fossils, Homotaxis, Local Faunas, Natural History Provinces, and the relation of Living to Extinct forms."—*Dr. Henry Woodward, F.R.S., in the "Geological Magazine."*

"A deeply interesting volume, dealing with Physical Geology as a whole, and also presenting us with an animated summary of the leading doctrines and facts of Palæontology, as looked at from a modern standpoint."—*Scotsman.*

"PROFESSOR SEELEY'S work includes one of the most satisfactory Treatises on Lithology in the English language. . . . So much that is not accessible in other works is presented in this volume, that no Student of Geology can afford to be without it."—*American Journal of Engineering.*

"Geology from the point of view of Evolution."—*Westminster Review.*

"PROFESSOR SEELEY'S PHYSICAL GEOLOGY is full of instructive matter, whilst the philosophical spirit which it displays will charm many a reader. From early days the author gave evidence of a powerful and eminently original genius. No one has shown more convincingly than the author that, in all ways, the past contains within itself the interpretation of the existing world."—*Annals of Natural History.*

Demy 8vo, Handsome cloth, 34s.

STRATIGRAPHICAL GEOLOGY AND PALÆONTOLOGY,

ON

THE BASIS OF PHILLIPS.

BY

ROBERT ETHERIDGE, F.R.S.,

OF THE NATURAL HIST. DEPARTMENT, BRITISH MUSEUM, LATE PALÆONTOLOGIST TO THE GEOLOGICAL SURVEY OF GREAT BRITAIN, PAST PRESIDENT OF THE GEOLOGICAL SOCIETY, ETC.

With Map, Numerous Tables, and Thirty-six Plates.

"In 1854 Prof. JOHN MORRIS published the Second Edition of his 'Catalogue of British Fossils,' then numbering 1,280 genera and 4 000 species. Since that date 3,000 genera and nearly 12,000 new species have been described, thus bringing up the muster-roll of extinct life in the British Islands alone to 3,680 genera and 16,000 known and described species.

"Numerous TABLES of ORGANIC REMAINS have been prepared and brought down to 1884, embracing the accumulated wealth of the labours of past and present investigators during the last thirty years. Eleven of these Tables contain every known British genus, zoologically or systematically placed, with the number of species in each, showing their broad distribution through time. The remaining 105 Tables are devoted to the analysis, relation, historical value, and distribution of specific life through each group of strata. These tabular deductions, as well as the Palæontological Analyses through the text, are, for the first time, fully prepared for English students."—*Extract from Author's Preface.*

*** PROSPECTUS *of the above important work—perhaps the* MOST ELABORATE *of its kind ever written, and one calculated to give a new strength to the study of Geology in Britain—may be had on application to the Publishers.*

It is not too much to say that the work will be found **to occupy a place entirely its own**, and will become **an indispensable guide** to every British Geologist.

"No such compendium of geological knowledge has ever been brought together before."—*Westminster Review.*

"If PROF. SEELEY'S volume was remarkable for its originality and the breadth of its views, MR. ETHERIDGE fully justifies the assertion made in his preface that his book differs in construction and detail from any known manual. . . . Must take HIGH RANK AMONG WORKS OF REFERENCE."—*Athenæum.*

SCIENTIFIC MANUALS

BY

W. J. MACQUORN RANKINE, C.E., LL.D., F.R.S.,

Late Regius Professor of Civil Engineering in the University of Glasgow.

THOROUGHLY REVISED BY W. J. MILLAR, C.E.,

Secretary to the Institute of Engineers and Shipbuilders in Scotland.

In Crown 8vo. Cloth.

I. RANKINE (Prof.): APPLIED MECHANICS: comprising the Principles of Statics and Cinematics, and Theory of Structures, Mechanism, and Machines. With numerous Diagrams. *Twelfth Edition*, 12/6.

"Cannot fail to be adopted as a text-book. . . . The whole of the information is so admirably arranged that there is every facility for reference."—*Mining Journal.*

II. RANKINE (Prof.): CIVIL ENGINEERING: comprising Engineering Surveys, Earthwork, Foundations, Masonry, Carpentry, Metal-work, Roads, Railways, Canals, Rivers, Water-works, Harbours, &c. With numerous Tables and Illustrations. *Seventeenth Edition*, 16/.

"Far surpasses in merit every existing work of the kind. As a manual for the hands of the professional Civil Engineer it is sufficient and unrivalled, and even when we say this, we fall short of that high appreciation of Dr. Rankine's labours which we should like to express."—*The Engineer.*

III. RANKINE (Prof.): MACHINERY AND MILLWORK: comprising the Geometry, Motions, Work, Strength, Construction, and Objects of Machines, &c. Illustrated with nearly 300 Woodcuts. *Sixth Edition*, 12/6.

"Professor Rankine's 'Manual of Machinery and Millwork' fully maintains the high reputation which he enjoys as a scientific author; higher praise it is difficult to award to any book. It cannot fail to be a lantern to the feet of every engineer."—*The Engineer.*

IV. RANKINE (Prof.): THE STEAM ENGINE and OTHER PRIME MOVERS. With Diagram of the Mechanical Properties of Steam, Folding-Plates, numerous Tables and Illustrations. *Twelfth Edition*, 12/6.

V. RANKINE (Prof.): USEFUL RULES and TABLES for Engineers and others. With Appendix: TABLES, TESTS, and FORMULÆ for the use of ELECTRICAL ENGINEERS; comprising Submarine Electrical Engineering, Electric Lighting, and Transmission of Power. By ANDREW JAMIESON, C.E., F.R.S.E. *Seventh Edition*, 10/6.

"Undoubtedly the most useful collection of engineering data hitherto produced."—*Mining Journal.*

"Every Electrician will consult it with profit."—*Engineering.*

VI. RANKINE (Prof.): A MECHANICAL TEXT-BOOK. by Prof. MACQUORN RANKINE and E. F. BAMBER, C.E. With numerous Illustrations. *Third Edition*, 9/.

"The work, as a whole, is very complete, and likely to prove invaluable for furnishing a useful and reliable outline of the subjects treated of."—*Mining Journal.*

*** THE MECHANICAL TEXT-BOOK forms a simple introduction to PROFESSOR RANKINE'S SERIES of MANUALS on ENGINEERING and MECHANICS.

PROF. RANKINE'S WORKS—(*Continued*).

VII. RANKINE (Prof.): MISCELLANEOUS SCIENTIFIC PAPERS. Royal 8vo. Cloth, 31/6.

Part I. Papers relating to Temperature, Elasticity, and Expansion of Vapours, Liquids, and Solids. Part II. Papers on Energy and its Transformations. Part III. Papers on Wave-Forms, Propulsion of Vessels, &c.

With Memoir by Professor TAIT, M.A. Edited by W. J. MILLAR, C.E. With fine Portrait on Steel, Plates, and Diagrams.

"No more enduring Memorial of Professor Rankine could be devised than the publication of these papers in an accessible form. . . . The Collection is most valuable on account of the nature of his discoveries, and the beauty and completeness of his analysis. . . . The Volume exceeds in importance any work in the same department published in our time."—*Architect.*

By SIR EDWARD REED.

Royal 8vo, Handsome Cloth, 25s.

THE STABILITY OF SHIPS.

BY

SIR EDWARD J. REED, K.C.B., F.R.S., M.P.,

KNIGHT OF THE IMPERIAL ORDERS OF ST. STANILAUS OF RUSSIA; FRANCIS JOSEPH OF AUSTRIA; MEDJIDIE OF TURKEY; AND RISING SUN OF JAPAN; VICE-PRESIDENT OF THE INSTITUTION OF NAVAL ARCHITECTS.

With numerous Illustrations and Tables.

THIS work has been written for the purpose of placing in the hands of Naval Constructors, Shipbuilders, Officers of the Royal and Mercantile Marines, and all Students of Naval Science, a complete Treatise upon the Stability of Ships, and is the only work in the English Language dealing exhaustively with the subject.

The plan upon which it has been designed is that of deriving the fundamental principles and definitions from the most elementary forms of floating bodies, so that they may be clearly understood without the aid of mathematics; advancing thence to all the higher and more mathematical developments of the subject.

The work also embodies a very full account of the historical rise and progress of the Stability question, setting forth the results of the labours of BOUGUER, BERNOULLI, DON JUAN D'ULLOA, EULER, CHAPMAN, and ROMME, together with those of our own Countrymen, ATWOOD, MOSELEY, and a number of others.

The modern developments of the subject, both home and foreign, are likewise treated with much fulness, and brought down to the very latest date, so as to include the labours not only of DARGNIES, REECH (whose famous *Mémoire*, hitherto a sealed book to the majority of English naval architects, has been reproduced in the present work), RISBEC, FERRANTY, DUPIN, GUYOU, and DAYMARD, in France, but also those of RANKINE, WOOLLEY, ELGAR, JOHN, WHITE, GRAY, DENNY, INGLIS, and BENJAMIN, in Great Britain.

In order to render the work complete for the purposes of the Shipbuilder, whether at home or abroad, the Methods of Calculation introduced by Mr. F. K. BARNES, Mr. GRAY, M. REECH, M. DAYMARD, and Mr. BENJAMIN, are all given separately, illustrated by Tables and worked-out examples. The book contains more than 200 Diagrams, and is illustrated by a large number of actual cases, derived from ships of all descriptions, but especially from ships of the Mercantile Marine.

The work will thus be found to constitute the most comprehensive and exhaustive Treatise hitherto presented to the Profession on the Science of the STABILITY OF SHIPS.

"Sir EDWARD REED'S 'STABILITY OF SHIPS' is INVALUABLE. In it the STUDENT, new to the subject, will find the path prepared for him, and all difficulties explained with the utmost care and accuracy; the SHIP-DRAUGHTSMAN will find all the methods of calculation at present in use fully explained and illustrated, and accompanied by the Tables and Forms employed; the SHIPOWNER will find the variations in the Stability of Ships due to differences in forms and dimensions fully discussed, and the devices by which the state of his ships under all conditions may be graphically represented and easily understood; the NAVAL ARCHITECT will find brought together and ready to his hand, a mass of information which he would otherwise have to seek in an almost endless variety of publications, and some of which he would possibly not be able to obtain at all elsewhere."—*Steamship.*

Medium 8vo, Handsome cloth, 25s.

HYDRAULIC POWER
AND
HYDRAULIC MACHINERY.

BY

HENRY ROBINSON, M. INST. C.E., F.G.S.,

FELLOW OF KING'S COLLEGE, LONDON; PROF. OF SURVEYING AND CIVIL ENGINEERING, KING'S COLLEGE, ETC., ETC.

With numerous Woodcuts, and 43 Litho. Plates.

GENERAL CONTENTS.

The Flow of Water under Pressure.
General Observations.
Waterwheels.
Turbines
Centrifugal Pumps.
Water pressure Pumps.
The Accumulator.
Hydraulic Pumping-Engine.
Three-Cylinder Engines and Capstans.
Motors with Variable Power.
Hydraulic Presses and Lifts.
Moveable Jigger Hoist.
Hydraulic Waggon Drop.
The Flow of Solids.
Shop Tools.
Cranes.
Hydraulic Power applied to Bridges.
Dock-Gate Machinery.
Hydraulic Coal-discharging Machines.
Hydraulic Machinery on board Ship.
Hydraulic Pile Driver.
Hydraulic Excavator.
Hydraulic Drill.
Hydraulic Brake.
Hydraulic Gun-Carriages.
Jets.
Hydraulic Ram.
Packing.
Power Co-operation.
Cost of Hydraulic Power.
Tapping Pressure Mains.
Meters.
Waste Water Meter.
Pressure Reducing Valves.
Pressure Regulator.

"A Book of great Professional Usefulness."—*Iron.*

A full Prospectus of the above important work—giving a description of the Plates—may be had on application to the Publishers.

ROBERTS-AUSTEN (W. C., F.R.S., Professor of Metallurgy, Normal School of Science and Royal School of Mines; Chemist and Assayer to the Royal Mint, &c).

METALLURGY (An Introduction to the Study of). With numerous Tables and Illustrations. Large Crown 8vo. Cloth.

SCHWACKHÖFER and BROWNE:

FUEL AND WATER: A Manual for Users of Steam and Water. By Prof. FRANZ SCHWACKHÖFER of Vienna, and WALTER R. BROWNE, M.A., C.E., late Fellow of Trinity College, Cambridge. Demy 8vo, with Numerous Illustrations, 9/.

"The Section on Heat is one of the best and most lucid ever written."—*Engineer.*

"Contains a vast amount of useful knowledge . . Cannot fail to be valuable to thousands compelled to use steam power."—*Railway Engineer.*

"Its practical utility is beyond question."—*Mining Journal.*

SEATON (A. E., Lecturer on Marine Engineering at the Royal Naval College, Greenwich, and Member of the Institute of Naval Architects):

A MANUAL OF MARINE ENGINEERING; Comprising the Designing, Construction, and Working of Marine Machinery. With numerous Illustrations. *Ninth Edition, revised and in part rewritten.* Demy 8vo. Cloth, 18/.

Opinions of the Press.

"The important subject of Marine Engineering is here treated with the thoroughness that it requires. No department has escaped attention. . . . Gives the results of much close study and practical work."—*Engineering.*

"By far the best MANUAL in existence. . . . Gives a complete account of the methods of solving, with the utmost possible economy, the problems before the Marine Engineer."—*Athenæum.*

"In the three-fold capacity of enabling a Student to learn how to design, construct, and work a modern Marine Steam-Engine, Mr. Seaton's Manual has no rival."—*Times.*

"The Student, Draughtsman, and Engineer will find this work the most valuable Handbook of Reference on the Marine Engine now in existence."—*Marine Engineer.*

SHELTON-BEY (W. Vincent, Foreman to the Imperial Ottoman Gun Factories, Constantinople):

THE MECHANIC'S GUIDE: A Hand-Book for Engineers and Artizans. With Copious Tables and Valuable Recipes for Practical Use. Illustrated. *Second Edition.* Crown 8vo. Cloth, 7/6.

"The MECHANIC'S GUIDE will answer its purpose as completely as a whole series of elaborate text-books."—*Mining Journal.*

TRAILL (Thomas W., F.E.R.N., M.Inst.C.E., Engineer-Surveyor-in-Chief to the Board of Trade):

BOILERS: THEIR CONSTRUCTION AND STRENGTH. A Handbook of Rules, Formulæ, and Tables for the Construction of Boilers; Safety-Valves; Material for Boilers; Tables of Areas, &c. Arranged for the Use of Steam-Users. *Second Edition*, Revised and Enlarged. Pocket-Size, Leather, 12s.; also for Office-Use, Cloth, 12s.

*** In the New Issue the subject matter has been considerably extended; Tables have been added for Pressures up to 200 lb. per square inch, and some of the Tables have been altered, besides which new ones and other matter have been introduced, which have been specially prepared and computed for the SECOND EDITION.

"Contains an enormous quantity of information—to be had nowhere else—arranged in a very convenient form. It is admirably printed, and reflects credit on the Publishers." *Engineer.*

"Will prove a welcome and valuable addition to the literature of the subject. . . . We can strongly recommend Mr. Traill's book as being the most complete, eminently practical, and most recent work on Boilers."—*Marine Engineer.*

"Will prove INVALUABLE to the Engineer and Practical Boiler-maker."—*Practical Engineer.*

WRIGHT (C. R. ALDER, D.Sc., F.R.S., Lecturer on Chemistry and Physics in St. Mary's Hospital Medical School).

THE THRESHOLD OF SCIENCE: Simple and Amusing Experiments illustrating some of the Chief Physical and Chemical Properties of Surrounding Objects and the Effect upon them of Light and Heat. With very numerous Illustrations. Large Crown 8vo, Handsome Cloth, 6/.; also Presentation Edition, gilt and gilt edges, 7/6.

*** In this work the object aimed at is to provide a kind of "Playbook," which in addition to affording the means of amusement, shall also to some extent tend in the direction of the course of mental education advocated by the British Association Committee; so that whilst the young philosopher finds pastime and entertainment in constructing simple apparatus and preparing elementary experiments, he may at the same time be led to observe correctly what happens, to draw inferences, and make deductions therefrom.

YEAR-BOOK OF THE SCIENTIFIC AND LEARNED SOCIETIES OF GREAT BRITAIN AND IRELAND.

Compiled from Official Sources, and giving, besides other necessary Official Information, Complete Lists of the PAPERS read during 1889 before all the leading Societies in every Department of Science throughout the Kingdom. *Seventh Annual Issue.* Cloth, 7/6.

"It goes almost without saying that a handbook of this subject would be in time one of THE MOST GENERALLY USEFUL WORKS for the library or the desk."—*Times.*

"AS A BOOK OF REFERENCE WE HAVE EVER FOUND IT TRUSTWORTHY. The value of the Lists of Papers can hardly be overrated."—*Lancet.*

*** *Copies of the previous Issues from 1884 may still be had.*

EDUCATIONAL WORKS.

*** *Specimen Copies of all the Educational Works published by Messrs. Charles Griffin and Company may be seen at the Libraries of the College of Preceptors, South Kensington Museum, and Crystal Palace; also at the depôts of the Chief Educational Societies.*

BRYCE (Archibald Hamilton, D.C.L., LL.D.,

Senior Classical Moderator in the University of Dublin):

THE WORKS OF VIRGIL. Text from HEYNE and WAGNER. English Notes, original, and selected from the leading German and English Commentators. Illustrations from the antique. Complete in One Volume. *Fourteenth Edition.* Fcap 8vo. Cloth, 6/.

Or, in Three Parts:

Part I. BUCOLICS and GEORGICS, . . 2/6.
Part II. THE ÆNEID, Books I.-VI., . 2/6.
Part III. THE ÆNEID, Books VII.-XII.,. 2/6.

"Contains the pith of what has been written by the best scholars on the subject. . . . The notes comprise everything that the student can want."—*Athenæum*

"The most complete, as well as elegant and correct edition of Virgil ever published in this country."—*Educational Times.*

"The best commentary on Virgil which a student can obtain."—*Scotsman.*

COBBETT (William): ENGLISH GRAMMAR,

in a Series of Letters, intended for the use of Schools and Young Persons in general. With an additional chapter on Pronunciation, by the Author's Son, JAMES PAUL COBBETT. *The only correct and authorised Edition.* Fcap 8vo. Cloth, 1/6.

"A new and cheapened edition of that most excellent of all English Grammars, William Cobbett's. It contains new copyright matter, as well as includes the equally amusing and instructive 'Six Lessons intended to prevent Statesmen from writing in an awkward manner.'"—*Atlas.*

COBBETT (William): FRENCH GRAMMAR.

Fifteenth Edition. Fcap 8vo. Cloth, 3/6.

"Cobbett's 'French Grammar' comes out with perennial freshness. There are few grammars equal to it for those who are learning, or desirous of learning, French without a teacher. The work is excellently arranged, and in the present edition we note certain careful and wise revisions of the text."—*School Board Chronicle.*

"Business men commencing the study of French will find this treatise one of the best aids. . . . It is largely used on the Continent."—*Midland Counties Herald.*

COBBIN'S MANGNALL: MANGNALL'S

HISTORICAL AND MISCELLANEOUS QUESTIONS, for the use of Young People. By RICHMAL MANGNALL. Greatly enlarged and corrected, and continued to the present time, by INGRAM COBBIN, M.A. *Fifty-fourth Thousand. New Illustrated Edition.* 12mo. Cloth, 4/.

COLERIDGE (Samuel Taylor): A DISSERTATION ON THE SCIENCE OF METHOD.
(*Encyclopædia Metropolitana.*) With a Synopsis. *Ninth Edition.* Cr. 8vo. Cloth, 2/.

CRAIK'S ENGLISH LITERATURE.

A COMPENDIOUS HISTORY OF ENGLISH LITERATURE AND OF THE ENGLISH LANGUAGE FROM THE NORMAN CONQUEST.
With numerous Specimens. By GEORGE LILLIE CRAIK, LL.D., late Professor of History and English Literature, Queen's College, Belfast. *New Edition.* In two vols. Royal 8vo. Handsomely bound in cloth, 25/.

GENERAL CONTENTS.

INTRODUCTORY.

I.—THE NORMAN PERIOD—The Conquest.

II.—SECOND ENGLISH—Commonly called Semi-Saxon.

III.—THIRD ENGLISH—Mixed, or Compound English.

IV.—MIDDLE AND LATTER PART OF THE SEVENTEENTH CENTURY.

V.—THE CENTURY BETWEEN THE ENGLISH REVOLUTION AND THE FRENCH REVOLUTION.

VI.—THE LATTER PART OF THE EIGHTEENTH CENTURY.

VII.—THE NINETEENTH CENTURY (*a*) THE LAST AGE OF THE GEORGES.
(*b*) THE VICTORIAN AGE.

With numerous Excerpts and Specimens of Style.

"Anyone who will take the trouble to ascertain the fact, will find how completely even our great poets and other writers of the last generation have already faded from the view of the present, with the most numerous class of the educated and reading public. Scarcely anything is generally read except the publications of the day. YET NOTHING IS MORE CERTAIN THAN THAT NO TRUE CULTIVATION CAN BE SO ACQUIRED This is the extreme case of that entire ignorance of history which has been affirmed, not with more point than truth, to leave a person always a child. . . . The present work combines the HISTORY OF THE LITERATURE with the HISTORY OF THE LANGUAGE. The scheme of the course and revolutions of the language which is followed here is extremely simple, and resting not upon arbitrary, but upon natural or real distinctions, gives us the only view of the subject that can claim to be regarded as of a scientific character."—*Extract from the Author's Preface.*

"Professor Craik has succeeded in making a book more than usually agreeable."—*The Times.*

CRAIK (Prof.): A MANUAL OF ENGLISH LITERATURE,
for the use of Colleges, Schools, and Civil Service Examinations. Selected from the larger work, by Dr. CRAIK. *Tenth Edition.* With an Additional Section on Recent Literature, by HENRY CRAIK, M.A., Author of "A Life of Swift." Crown 8vo. Cloth, 7/6.

"A Manual of English Literature from so experienced and well-read a scholar as Professor Craik needs no other recommendation than the mention of its existence."—*Spectator.*

"This augmented effort will, we doubt not, be received with decided approbation by those who are entitled to judge, and studied with much profit by those who want to learn. . . . If our young readers will give healthy perusal to Dr. Craik's work, they will greatly benefit by the wide and sound views he has placed before them."—*Athenæum.*

"The preparation of the NEW ISSUE has been entrusted to Mr. HENRY CRAIK, Secretary to the Scotch Education Department, and well known in literary circles as the author of the latest and best Life of Swift. . . . A Series of TEST QUESTIONS is added, which must prove of great service to Students studying alone."—*Glasgow Herald.*

WORKS BY REV. C. T. CRUTTWELL, M.A.,

Late Fellow of Merton College, Oxford.

I.—A HISTORY OF ROMAN LITERATURE:

From the Earliest Period to the Times of the Antonines. *Fourth Edition.* Crown 8vo. Cloth, 8/6.

"Mr. CRUTTWELL has done a real service to all Students of the Latin Language and Literature. . . . Full of good scholarship and good criticism."—*Athenæum.*

"A most serviceable—indeed, indispensable—guide for the Student. . . . The 'general reader' will be both charmed and instructed."—*Saturday Review.*

"The Author undertakes to make Latin Literature interesting, and he has succeeded. There is not a dull page in the volume."—*Academy.*

"The great merit of the work is its fulness and accuracy."—*Guardian.*

"This elaborate and careful work, in every respect of high merit. Nothing at all equal to it has hitherto been published in England."—*British Quarterly Review.*

Companion Volume. Second Edition.

II.—SPECIMENS OF ROMAN LITERATURE:

From the Earliest Period to the Times of the Antonines. Passages from the Works of Latin Authors, Prose Writers, and Poets:

Part I.—ROMAN THOUGHT: Religion, Philosophy and Science, Art and Letters, 6/.

Part II.—ROMAN STYLE: Descriptive, Rhetorical, and Humorous Passages, 5/.

Or in One Volume complete, 10/6.

Edited by C. T. CRUTTWELL, M.A., Merton College, Oxford; and PEAKE BANTON, M.A., some time Scholar of Jesus College, Oxford.

"'Specimens of Roman Literature' marks a new era in the study of Latin."—*English Churchman.*

"A work which is not only useful but necessary. . . . The plan gives it a standing-ground of its own. . . . The sound judgment exercised in plan and selection calls for hearty commendation."—*Saturday Review.*

"It is hard to conceive a completer or handier repertory of specimens of Latin thought and style."—*Contemporary Review.*

*** KEY to PART II., PERIOD II. (being a complete TRANSLATION of the 85 Passages composing the Section), by THOS. JOHNSTON, M.A., may now be had (by Tutors and Schoolmasters only) on application to the Publishers. Price 2/6.

III.—A HISTORY OF EARLY CHRISTIAN LITERATURE.

For the use of Students and General Readers. 8vo, Handsome Cloth. [*In Preparation.*

CURRIE (Joseph, formerly Head Classical Master of Glasgow Academy):

THE WORKS OF HORACE: Text from ORELLIUS. English Notes, original, and selected from the best Commentators. Illustrations from the antique. Complete in One Volume. Fcap 8vo. Cloth, 5/.

Or in Two Parts:

Part I.—CARMINA, ... 3/.

Part II.—SATIRES AND EPISTLES, . . . 3/.

"The notes are excellent and exhaustive."—*Quarterly Journal of Education.*

—— EXTRACTS FROM CÆSAR'S COMMENTARIES;

containing his description of Gaul, Britain, and Germany. With Notes, Vocabulary, &c. Adapted for Young Scholars. *Fourth Edition.* 18mo. Cloth, 1/6.

DOERING (E.) AND GRÆME (E.):

HELLAS: AN INTRODUCTION TO GREEK ANTIQUITIES, comprising the Geography, Religion, and Myths, History, Art, and Culture of Old Greece. On the basis of the German work by E. Doering, with additions by Elliott Græme. In large 8vo, with Map and Illustrations.

PART I.—The Land and the People: the Religion and Myths of Old Greece.

*** In the English version of Mr. Doering's work, the simple and interesting style of the original—written for young Students—has been retained; but, throughout, such additions and emendations have been made, as render the work suitable for more advanced Students, and for all who desire to obtain, within moderate compass, more than a superficial acquaintance with the great People whose genius and culture have so largely influenced our own. The results of the latest researches by Dr. SCHLIEMANN, MM. FOUQUÉ, CARAPANOS, and others, are incorporated.

D'ORSEY (Rev. Alex. J. D., B.D., Corpus Christi Coll., Cambridge, Lecturer at King's College, London):

SPELLING BY DICTATION: Progressive Exercises in English Orthography, for Schools and Civil Service Examinations. *Sixteenth Thousand.* 18mo. Cloth, 1/.

FLEMING (William, D.D., late Professor of Moral Philosophy in the University of Glasgow):

THE VOCABULARY OF PHILOSOPHY: PSYCHOLOGICAL, ETHICAL, AND METAPHYSICAL. With Quotations and References for the Use of Students. Revised and Edited by HENRY CALDERWOOD, LL.D., Professor of Moral Philosophy in the University of Edinburgh. *Fourth Edition, enlarged.* Crown 8vo. Cloth, 10/6.

"The additions by the Editor bear in their clear, concise, vigorous expression, the stamp of his powerful intellect, and thorough command of our language. More than ever, the work is now likely to have a prolonged and useful existence, and to facilitate the researches of those entering upon philosophic studies."—*Weekly Review.*

JEVONS (Frank B., M.A., University of Durham, sometime Scholar of Wadham College, Oxford):

A HISTORY OF GREEK LITERATURE, from the Earliest Times to the Death of Demosthenes. *Second Edition.* Crown 8vo. Cloth, 8/6.

"It is beyond all question the BEST HISTORY of Greek literature that has hitherto been published."—*Spectator.*

"An admirable text-book."—*Westminster Review.*

"Mr. Jevons' work supplies a real want."—*Contemporary Review.*

"Mr. Jevons' work is distinguished by the Author's THOROUGH ACQUAINTANCE with THE OLD WRITERS, and his DISCRIMINATING USE of the MODERN LITERATURE bearing upon the subject. . . . His great merit lies in his EXCELLENT EXPOSITION of the POLITICAL AND SOCIAL CAUSES concerned in the development of the Literature of Greece." —*Berlin Philologische Wochenschrift.*

"As a Text-Book, Mr. Jevons' work from its excellence deserves to SERVE AS A MODEL."—*Deutsche Litteraturzeitung.*

——— AND DR. O. SCHRADER:

THE PREHISTORIC ANTIQUITIES OF THE ARYAN PEOPLES (See p. 30).

McBURNEY (Isaiah, LL.D.,): EXTRACTS FROM OVID'S METAMORPHOSES. With Notes, Vocabulary, &c. Adapted for Young Scholars. *Third Edition.* 18mo. Cloth, 1/6.

MENTAL SCIENCE: S. T. COLERIDGE'S celebrated Essay on METHOD; Archbishop WHATELY'S Treatises on LOGIC and RHETORIC. *Tenth Edition.* Crown 8vo. Cloth, 5/.

MILLER (W. Galbraith, M.A., LL.B., Lecturer on Public Law, including Jurisprudence and International Law, in the University of Glasgow):

THE PHILOSOPHY OF LAW, LECTURES ON. Designed mainly as an Introduction to the Study of International Law. In 8vo. Handsome Cloth, 12/. *Now Ready.*

"Mr. MILLER'S 'PHILOSOPHY OF LAW' bears upon it the stamp of a wide culture and of an easy acquaintanceship with what is best in modern continental speculation. . . . Interesting and valuable, because suggestive."—*Journal of Jurisprudence.*

WORKS BY WILLIAM RAMSAY, M.A.,

Trinity College, Cambridge, late Professor of Humanity in the University of Glasgow.

A MANUAL OF ROMAN ANTIQUITIES. For the use of Advanced Students. With Map, 130 Engravings, and very copious Index. *Fourteenth Edition.* Crown 8vo. Cloth, 8/6.

"Comprises all the results of modern improved scholarship within a moderate compass."—*Athenæum.*

—— AN ELEMENTARY MANUAL OF ROMAN ANTIQUITIES. Adapted for Junior Classes. With numerous Illustrations. *Eighth Edition.* Crown 8vo. Cloth, 4/.

—— A MANUAL OF LATIN PROSODY, Illustrated by Copious Examples and Critical Remarks. For the use of Advanced Students. *Seventh Edition.* Crown 8vo. Cloth, 5/.

"There is no other work on the subject worthy to compete with it."—*Athenæum.*

—— AN ELEMENTARY MANUAL OF LATIN PROSODY. Adapted for Junior Classes. Crown 8vo. Cloth, 2s.

THE SCHOOL BOARD READERS:

A SERIES OF STANDARD READING-BOOKS.

EDITED BY A FORMER H.M. INSPECTOR OF SCHOOLS.

Adopted by many School Boards throughout the Country.

Elementary Reader, Part I.,	1d.	Standard III.,	9d.
,, ,, ,, II.,	2d.	,, IV.,	1s. 0d.
Standard I.,	4d.	,, V.,	1s. 6d.
,, II.,	6d.	,, VI.,	2s. 0d.

Key to the Questions in Arithmetic in 2 Parts, each 6d.

"THE BOOKS GENERALLY ARE VERY MUCH WHAT WE SHOULD DESIRE."—*Times.*
"The Series is DECIDEDLY ONE OF THE BEST that have yet appeared."—*Athenæum.*

THE SCHOOL BOARD MANUALS

ON THE SPECIFIC SUBJECTS OF THE REVISED CODE,

BY A FORMER H.M. INSPECTOR OF SCHOOLS,

Editor of the "School Board Readers."

64 pages, stiff wrapper. *6d.*; neat cloth, *7d.* each.

I.—ALGEBRA.
II.—ENGLISH HISTORY.
III.—GEOGRAPHY.
IV.—PHYSICAL GEOGRAPHY.
V.—ANIMAL PHYSIOLOGY. (Well Illustrated with good Engravings.)
VI.—BIBLE HISTORY. (Entirely free from any Denominational bias.)

SCHRADER (Dr. O.) and JEVONS (F. B., M.A.): THE PREHISTORIC ANTIQUITIES OF THE ARYAN PEOPLES: Translated from the SECOND GERMAN EDITION by F. B. Jevons, M.A., author of "*A History of Greek Literature.*" Demy 8vo. Handsome cloth, gilt top, 21/.

*** The Publishers have pleasure in announcing that to the English Translation of Dr. Schrader's well-known "*Sprachvergleichung und Urgeschichte*"—a work which commends itself alike to the Scholar by its thoroughness and the moderation of its tone, and to the General Reader by its clear and interesting style—an INTRODUCTION is furnished by the AUTHOR.

SENIOR (Nassau William, M.A., late Professor of Political Economy in the University of Oxford): A TREATISE ON POLITICAL ECONOMY: the Science which treats of the Nature, the Production, and the Distribution of Wealth. *Sixth Edition.* Crown 8vo. Cloth. (*Encyclopædia Metropolitana*), 4/.

THOMSON (James): THE SEASONS. With an Introduction and Notes by ROBERT BELL, Editor of the "Annotated Series of British Poets." *Third Edition.* Fcap 8vo. Cloth, 1/6.
"An admirable introduction to the study of our English classics."

WHATELY (Archbishop): LOGIC—A Treatise on. With Synopsis and Index. (*Encyclopædia Metropolitana*), 3/.

—— RHETORIC—A Treatise on. With Synopsis and Index. (*Encyclopædia Metropolitana*), 3/6.

WORKS IN GENERAL LITERATURE.

BELL (Robert, Editor of the "Annotated Series of British Poets"):

GOLDEN LEAVES FROM THE WORKS OF THE POETS AND PAINTERS. Illustrated by Sixty-four superb Engravings on Steel, after Paintings by DAVID ROBERTS, STANFIELD, LESLIE, STOTHARD, HAYDON, CATTERMOLE, NASMYTH, Sir THOMAS LAWRENCE, and many others, and engraved in the first style of Art by FINDEN, GREATBACH, LIGHTFOOT, &c. *Second Edition.* 4to. Cloth gilt, 21/.

"'Golden Leaves' is by far the most important book of the season. The Illustrations are really works of art, and the volume does credit to the arts of England."—*Saturday Review.*

"The Poems are selected with taste and judgment."—*Times*

"The engravings are from drawings by Stothard, Newton, Danby, Leslie, and Turner, and it is needless to say how charming are many of the above here given."—*Athenæum.*

THE WORKS OF WILLIAM COBBETT.

THE ONLY AUTHORISED EDITIONS.

COBBETT (William): ADVICE TO YOUNG Men and (incidentally) to Young Women, in the Middle and Higher Ranks of Life. In a Series of Letters addressed to a Youth, a Bachelor, a Lover, a Husband, a Father, a Citizen, and a Subject. *New Edition. With admirable Portrait on Steel.* Fcap 8vo. Cloth, 2/6.

"Cobbett's great qualities were immense vigour, resource, energy, and courage, joined to a force of understanding, a degree of logical power, and above all a force of expression, which have rarely been equalled. . . . He was the most English of Englishmen."—*Saturday Review.*

"With all his faults, Cobbett's style is a continual refreshment to the lover of 'English undefiled.'"—*Pall Mall Gazette.*

—— COTTAGE ECONOMY: Containing information relative to the Brewing of Beer, Making of Bread, Keeping of Cows, Pigs, Bees, Poultry, &c.; and relative to other matters deemed useful in conducting the affairs of a Poor Man's Family. *Eighteenth Edition,* revised by the Author's Son. Fcap 8vo. Cloth, 2/6.

WILLIAM COBBETT'S WORKS—*(Continued).*

COBBETT (Wm.): EDUCATIONAL WORKS.

(See page 25.)

—— A LEGACY TO LABOURERS: An

Argument showing the Right of the Poor to Relief from the Land. With a Preface by the Author's Son, JOHN M. COBBETT, late M.P. for Oldham. *New Edition.* Fcap 8vo. Cloth, 1/6.

"The book cannot be too much studied just now."—*Nonconformist.*

"Cobbett was, perhaps, the ablest Political writer England ever produced, and his influence as a Liberal thinker is felt to this day. . . . It is a real treat to read his strong racy language."—*Public Opinion.*

—— A LEGACY TO PARSONS: Or, have the

Clergy of the Established Church an Equitable Right to Tithes and Church Property? *New Edition.* Fcap 8vo. Cloth, 1/6.

"The most powerful work of the greatest master of political controversy this country has ever produced."—*Pall Mall Gazette.*

DALGAIRNS (Mrs.): THE PRACTICE OF

COOKERY, adapted to the business of Every-day Life. By Mrs. DALGAIRNS. *he best book for Scotch dishes.* About Fifty new Recipes have been added to the present Edition, but only such as the Author has had adequate means of ascertaining to be valuable. *Seventeenth Edition.* Fcap 8vo. Cloth. *(In preparation.)*

GILMER'S INTEREST TABLES; Tables for

Calculation of Interest, on any sum, for any number of days, at ½, 1, 1½, 2, 2½, 3, 3½, 4, 4½, 5 and 6 per Cent. By ROBERT GILMER. Corrected and enlarged. *Eleventh Edition.* 12mo. Cloth, 5/.

GRÆME (Elliott): BEETHOVEN: a Memoir.

With Portrait, Essay, and Remarks on the Pianoforte Sonatas, with Hints to Students, by DR. FERDINAND HILLER, of Cologne. *Third Edition.* Crown 8vo. Cloth gilt, elegant, 5/.

"This elegant and interesting Memoir. . . . The newest, prettiest, and most readable sketch of the immortal Master of Music."—*Musical Standard.*

"A gracious and pleasant Memorial of the Centenary."—*Spectator.*

"This delightful little book—concise, sympathetic, judicious."—*Manchester Examiner.*

"We can, without reservation, recommend it as the most trustworthy and the pleasantest Memoir of Beethoven published in England."—*Observer.*

"A most readable volume, which ought to find a place in the library of every admirer of the great Tone-Poet."—*Edinburgh Daily Review.*

—— A NOVEL WITH TWO HEROES.

Second Edition. In 2 vols. Post 8vo. Cloth, 21/.

"A decided literary success."—*Athenæum.*

"Clever and amusing . . . above the average even of good novels . . . free from sensationalism, but full of interest . . . touches the deeper chords of life . . delineation of character remarkably good."—*Spectator.*

"Superior in all respects to the common run of novels."—*Daily News.*

"A story of deep interest. . . . The dramatic scenes are powerful almost to painfulness in their intensity."—*Scotsman.*

THE EMERALD SERIES OF STANDARD AUTHORS.

Illustrated by Engravings on Steel, after STOTHARD, LESLIE, DAVID ROBERTS, STANFIELD, Sir THOMAS LAWRENCE, CATTERMOLE, &c., Fcap 8vo. Cloth, gilt.

Particular attention is requested to this very beautiful series. The delicacy of the engravings, the excellence of the typography, and the quaint antique head and tail pieces, render them the most beautiful volumes ever issued from the press of this country, and now, unquestionably, the cheapest of their class.

BURNS' (Robert) SONGS AND BALLADS.

With an Introduction on the Character and Genius of Burns. By THOMAS CARLYLE. Carefully printed in antique type, and illustrated with Portrait and beautiful Engravings on Steel. *Second Thousand.* Cloth, gilt edges, 3/.

BYRON (Lord): CHILDE HAROLD'S PILGRIMAGE.

With Memoir by Professor SPALDING. Illustrated with Portrait and Engravings on Steel, by GREATBACH, MILLER, LIGHTFOOT, &c., from Paintings by CATTERMOLE, Sir T. LAWRENCE, H. HOWARD, and STOTHARD. Beautifully printed on toned paper. *Third Thousand.* Cloth, gilt edges, 3/.

CAMPBELL (Thomas): THE PLEASURES OF HOPE.

With Introductory Memoir by the Rev. CHARLES ROGERS, LL.D., and several Poems never before published. Illustrated with Portrait and Steel Engravings. *Second Thousand.* Cloth, gilt edges, 3/.

CHATTERTON'S (Thomas) POETICAL WORKS.

With an Original Memoir by FREDERICK MARTIN, and Portrait. Beautifully illustrated on Steel, and elegantly printed. *Fourth Thousand.* Cloth, gilt edges, 3/.

GOLDSMITH'S (Oliver) POETICAL WORKS.

With Memoir by Professor SPALDING. Exquisitely illustrated with Steel Engravings. *New Edition.* Printed on superior toned paper. *Seventh Thousand.* Cloth, gilt edges, 3/.

GRAY'S (Thomas) POETICAL WORKS.

With Life by the Rev. JOHN MITFORD, and Essay by the EARL of CARLISLE. With Portrait and numerous Engravings on Steel and Wood. Elegantly printed on toned paper. *Eton Edition, with the Latin Poems. Sixth Thousand.* Cloth, gilt edges, 5/.

HERBERT'S (George) POETICAL WORKS.

With Memoir by J. NICHOL, B.A., Oxon, Prof. of English Literature in the University of Glasgow. Edited by CHARLES COWDEN CLARKE. Antique headings to each page. *Second Thousand.* Cloth, gilt edges, 3/.

KEBLE (Rev. John): THE CHRISTIAN YEAR.

With Memoir by W. TEMPLE, Portrait, and Eight beautiful Engravings on Steel. *Second Thousand.*

Cloth, gilt edges, ...	5/.
Morocco, elegant, ...	10/6.
Malachite, ...	12/6.

THE EMERALD SERIES—(*Continued*).

POE'S (Edgar Allan) COMPLETE POETICAL

WORKS. Edited, with Memoir, by JAMES HANNAY. Full-page Illustrations after WEHNERT, WEIR, &c. Toned paper. *Thirteenth Thousand.*

Cloth, gilt edges, ... 3/.
Malachite, ... 10/6.

Other volumes in preparation.

MACKEY'S FREEMASONRY:

A LEXICON OF FREEMASONRY. Containing a definition of its Communicable Terms, Notices of its History, Traditions, and Antiquities, and an Account of all the Rites and Mysteries of the Ancient World. By ALBERT G. MACKEY, M.D., Secretary-General of the Supreme Council of the U.S., &c. *Eighth Edition*, thoroughly revised with APPENDIX by Michael C. Peck, Prov. Grand Secretary for N. and E. Yorkshire. Handsomely bound in cloth, 6/.

"Of MACKEY'S LEXICON it would be impossible to speak in too high terms; suffice it to say, that, in our opinion, it ought to be in the hands of every Mason who would thoroughly understand and master our noble Science . . . No Masonic Lodge or Library should be without a copy of this most useful work."—*Masonic News.*

HENRY MAYHEW'S CELEBRATED WORK ON THE STREET-FOLK OF LONDON.

LONDON LABOUR AND THE LONDON

POOR: A Cyclopædia of the Condition and Earnings of *those that will work and those that cannot work.* By HENRY MAYHEW. With many full-page Illustrations from Photographs. In three vols. Demy 8vo. Cloth. Each vol. 4/6.

"Every page of the work is full of valuable information, laid down in so interesting a manner that the reader can never tire."—*Illustrated London News.*

"Mr. Henry Mayhew's famous record of the habits, earnings, and sufferings of the London poor."—*Lloyd's Weekly London Newspaper.*

"This remarkable book, in which Mr Mayhew gave the better classes their first real insight into the habits, modes of livelihood, and current of thought of the London poor."—*The Patriot.*

The Extra Volume.

LONDON LABOUR AND THE LONDON

POOR: *Those that will not work.* Comprising the Non-workers, by HENRY MAYHEW; Prostitutes, by BRACEBRIDGE HEMYNG; Thieves, by JOHN BINNY; Beggars, by ANDREW HALLIDAY. With an Introductory Essay on the Agencies at Present in Operation in the Metropolis for the Suppression of Crime and Vice, by the Rev. WILLIAM TUCKNISS, B.A., Chaplain to the Society for the Rescue of Young Women and Children. With Illustrations of Scenes and Localities. In one large vol. Royal 8vo. Cloth, 10/6.

"The work is full of interesting matter for the casual reader, while the philanthropist and the philosopher will find details of the greatest import."—*City Press.*

MR. MAYHEW'S LONDON LABOUR—(*Continued*).

Companion volume to the preceding.

THE CRIMINAL PRISONS OF LONDON,

and Scenes of Prison Life. BY HENRY MAYHEW and JOHN BINNY. Illustrated by nearly two hundred Engravings on Wood, principally from Photographs. In one large vol. Imperial 8vo. Cloth, 10/6.

This volume concludes Mr. Henry Mayhew's account of his researches into the crime and poverty of London. The amount of labour of one kind or other, which the whole series of his publications represents, is something almost incalculable.

*** This celebrated Record of Investigations into the condition of the Poor of the Metropolis, undertaken from philanthropic motives by Mr. HENRY MAYHEW, first gave the wealthier classes of England some idea of the state of Heathenism, Degradation, and Misery in which multitudes of their poorer brethren languished. His revelations created, at the time of their appearance, universal horror and excitement—that a nation, professedly *Christian*, should have in its midst a vast population, so sunk in ignorance, vice, and very hatred of Religion, was deemed incredible, until further examination established the truth of the statements advanced. The result is well known. The London of Mr. MAYHEW will, happily, soon exist only in his pages. To those who would appreciate the efforts already made among the ranks which recruit our "dangerous" classes, and who would learn what yet remains to be done, the work will afford enlightenment, not unmingled with surprise.

POE'S (Edgar Allan) COMPLETE POETICAL

WORKS. Edited, with Memoir, by JAMES HANNAY. Full-page Illustrations after WEHNERT, WEIR, and others. In paper wrapper. Illustrated, 1/6.

SOUTHGATE (Mrs. Henry): THE CHRIS-

TIAN LIFE: Thoughts in Prose and Verse from the Best Writers of all Ages. Selected and Arranged for Every Day in the Year. *Second Edition.* Cloth Elegant, 5/.

THOMSON (Spencer, M.D., L.R.C.S., Edinburgh,

and J. C. STEELE, M.D., of Guy's Hospital):

DOMESTIC MEDICINE AND HOUSEHOLD SURGERY (A Dictionary of). Thoroughly Revised and in part Re-Written by the Editors. With a Chapter on the Management of the Sick-room, and many Hints for the Diet and Comfort of Invalids. With many new Engravings. *Twenty-sixth Edition.* Royal 8vo. Cloth, 10/6. (See page 3 of Wrapper.)

MR. SOUTHGATE'S WORKS.

"No one who is in the habit of writing and speaking much on a variety of subjects can afford to dispense with Mr. SOUTHGATE'S WORKS."—*Glasgow News.*

FIRST SERIES—THIRTY-FIFTH EDITION. SECOND SERIES—NINTH EDITION.

MANY THOUGHTS OF MANY MINDS:

Selections and Quotations from the best Authors. Compiled and Analytically Arranged by

HENRY SOUTHGATE.

In Square 8vo, elegantly printed on Toned Paper.

Presentation Edition, Cloth and Gold, ...	Each Vol.	12/6.
Library Edition, Roxburghe, ...	,,	14/.
Ditto, Morocco Antique, ...	,,	21/.

Each Series complete in itself, and sold separately.

"The produce of years of research."—*Examiner.*

"A MAGNIFICENT GIFT-BOOK, appropriate to all times and seasons."—*Freemasons Magazine.*

"Not so much a book as a library."—*Patriot.*

"Preachers and Public Speakers will find that the work has special uses for them."—*Edinburgh Daily Review.*

BY THE SAME AUTHOR.

Now Ready, THIRD EDITION.

SUGGESTIVE THOUGHTS ON RELIGIOUS SUBJECTS:

A Dictionary of Quotations and Selected Passages from nearly 1,000 of the best Writers, Ancient and Modern.

Compiled and Analytically Arranged by HENRY SOUTHGATE. In Square 8vo, elegantly printed on toned paper.

Presentation Edition, Cloth Elegant, ...	10/6.
Library Edition, Roxburghe, ...	12/.
Ditto, Morocco Antique, ...	20/.

"The topics treated of are as wide as our Christianity itself: the writers quoted from, of every Section of the one Catholic Church of JESUS CHRIST."—*Author's Preface.*

"This is another of Mr. Southgate's most valuable volumes. . . . The mission which the Author is so successfully prosecuting in literature is not only highly beneficial, but necessary in this age. . . . If men are to make any acquaintance at all with the great minds of the world, they can only do so with the means which our Author supplies."—*Homilist.*

"A casket of gems."—*English Churchman.*

"Mr. Southgate's work has been compiled with a great deal of judgment, and it will, I trust, be extensively useful."—*Rev. Canon Liddon, D.D., D.C.L.*

"Many a busy Christian teacher will be thankful to Mr. Southgate for having unearthed so many rich gems of thought; while many outside the ministerial circle will obtain stimulus, encouragement, consolation, and counsel, within the pages of this handsome volume."—*Nonconformist.*

"Mr. SOUTHGATE is an indefatigable labourer in a field which he has made peculiarly his own. . . . The labour expended on 'Suggestive Thoughts' must have been immense, and the result is as nearly perfect as human fallibility can make it. . . . Apart from the selections it contains, the book is of value as an index to theological writings. As a model of judicious, logical, and suggestive treatment of a subject, we may refer our readers to the manner in which the subject 'JESUS CHRIST' is arranged and illustrated in 'Suggestive Thoughts.'"—*Glasgow News.*

Printed in Dunstable, United Kingdom

91220090R00258